FATAL INVENTION

Sponsored by

www.mafp.org

ALSO BY DOROTHY ROBERTS

Killing the Black Body: Race, Reproduction, and the Meaning of Liberty

Shattered Bonds: The Color of Child Welfare

FATAL INVENTION

How Science, Politics, and Big Business Re-create Race in the Twenty-first Century

Dorothy Roberts

THE NEW PRESS

NEW YORK
LONDON

Originally published in the United States by The New Press, New York, 2011
This paperback edition published by The New Press, New York, 2012
Distributed by Two Rivers Distribution

ISBN 978-1-59558-834-0 (pbk.)
ISBN 978-1-59558-691-9 (e-book)

LIBRARY OF CONGRESS CATALOGING-IN-PUBLICATION DATA

Roberts, Dorothy.
Fatal invention : how science, politics, and big business re-create race in the twenty-first century / Dorothy Roberts.
p. cm.
Includes bibliographical references and index.
ISBN 978-1-59558-495-3 (hc)
1. Race—Social aspects. 2. Race—Political aspects. 3. Race—Economic aspects. 4. Physical anthropology. 5. Human population genetics. 6. Genomics.
I. Title.
GN269.R64 2011
305.8—dc22

2011012830

The New Press publishes books that promote and enrich public discussion and understanding of the issues vital to our democracy and to a more equitable world. These books are made possible by the enthusiasm of our readers; the support of a committed group of donors, large and small; the collaboration of our many partners in the independent media and the not-for-profit sector; booksellers, who often hand-sell New Press books; librarians; and above all by our authors.

www.thenewpress.com

Composition by Westchester Book Composition

Printed in the United States of America

10 9

For my parents, Iris and Robert Roberts,
who taught me that there is only one human race.

Contents

Preface

The principal human races presumably emerged as the populations of each continent responded to different evolutionary pressures.

—Nicholas Wade, "A New Look at Race and Natural Selection," *New York Times*, April 2, 2009.

The Food and Drug Administration (FDA) approved BiDil, a drug for the treatment of heart failure in self-identified black patients, representing a step toward the promise of personalized medicine.

—FDA News Release, June 23, 2005.

To chip away at an overwhelming budget deficit, Miami's public hospital system stopped paying for kidney dialysis for the indigent this week, officials said, leaving some patients to rely on emergency rooms for their life-sustaining treatments.

—Kevin Sack, "Hospital Cuts Dialysis Care for the Poor in Miami," *New York Times*, January 8, 2010.

In the agency's confidential files was a jail video showing Mr. Bah face down in the medical unit, hands cuffed behind his back, just before medical personnel sent him to a disciplinary cell. The tape shows him crying out repeatedly in his native Fulani, "Help, they are killing me!"

—Nina Bernstein, "Officials Obscured Truth of Migrant Deaths in Jail," *New York Times*, January 10, 2010.

News stories about race, genetics, life, and death in the first decade of the twenty-first century reflect an ominous trend overtaking the social and political life of this nation. We are witnessing the emergence of a new form of racial politics in America, in which the state's power to control the life and death of populations relies on classifying them by race. Defining the political system of race in biological terms has been a constant feature of U.S. society for centuries, but the precise mechanisms for re-creating race have changed to reflect current sociopolitical realities. This book examines the role of genomic science and biotechnologies in today's reinvention of our enduring racial order.

The emerging biopolitics of race has three main components. First, some scientists are resuscitating biological theories of race by using cutting-edge genomic research to modernize old racial typologies that were based on observations of physical differences. Science is redefining race as a biological category written in our genes. Second, the biotechnology and pharmaceutical industries are converting the new racial science into products that are developed and marketed according to race and that incorporate assumptions of racial difference at the genetic level. Finally, government policies that are officially color-blind are stripping poor minority communities of basic services, social programs, and economic resources in favor of corporate interests while simultaneously imposing on these communities harsh forms of punitive regulation. These dehumanizing policies of surveillance and control are made invisible to most Americans by the emerging genetic understanding of race that focuses attention on molecular differences while obscuring the impact of racism in our society.

Only a decade ago, the biological concept of race seemed finally to have met its end. The Human Genome Project, which mapped the entire human genetic code, proved that race could not be identified in our genes. On June 26, 2000, when President Bill Clinton unveiled the draft genomic sequence, he famously declared that "human beings, regardless of race, are 99.9 percent the same." Contrary to popular misconception, we are not naturally divided into genetically identifiable racial groups. Biologically, there is one human race. Race applied to human beings is a *political* division: it is a system of governing people that classifies them into a social hierarchy based on invented biological demarcations.

But reports of the demise of race as a biological category were premature.

Instead of hammering the last nail in the coffin of an obsolete system, the science that emerged from sequencing the human genome was shaped by a resurgence of interest in race-based genetic variation. Some scientists are claiming that clusters of genetic similarity detected with novel genomic theories and computer technologies correspond to antiquated racial classifications and prove that human racial differences are real and significant. Others are searching for genetic differences among races that can explain staggering inequalities in health and disease as well as variations in drug response.

There has been a corresponding explosion of race-based biotechnologies. In 2005, the U.S. Food and Drug Administration (FDA) approved the first race-specific drug, BiDil, to treat heart failure in black patients. Fertility clinics solicit egg donations on the basis of race and use race in genetic tests to determine which embryos to implant and which to discard. Consumers can send cheek swabs to dozens of online companies to find out not only their genetic ancestry, but also their racial identity. Some law enforcement agencies are using these same forensic tools to identify the race of suspects. And federal and state agents are starting to collect DNA from everyone they arrest, even those never charged with or convicted of a crime, filling ever-expanding government gene banks with compelled samples mostly from black and brown men.

The new science and technology of racial genetics threatens to steer America on a course of social inhumanity that already has begun to dominate politics in this century. Government policies that have drastically slashed social services have been accompanied by particularly brutal forms of regulation of racial minorities: mass imprisonment at rates far exceeding any other place on Earth or any time in the history of the free world; roundup and deportation of undocumented immigrants, often tearing families apart; abuse of children held in juvenile detention centers or locked up in adult prisons, some for the rest of their lives; official and unofficial infliction of torture in police stations and prison cells; and rampant medical neglect that kills.

Today, many Americans believe that the election of Barack Obama as president ushered in a new "postracial" society of equality, harmony, and opportunity. How can the perception of increasing racial fairness coexist with the reality of increasing racialized brutality? At the very moment that race consciousness is intensifying at the molecular level, it is fading at the social level. Genomic science is reinforcing the concept of race as a biological category even as Americans ignore the devastating effects of racial inequality on our society. I argue that this paradox reflects the primary impact of the new biopolitics of race: the seemingly color-blind regime of coercive surveillance imposed on poor communities of color will seem more acceptable to a majority of Americans as their belief in intrinsic racial differences is validated by genomic science and technologies. The new racial biopolitics obscures this modern form of state brutality at a time when the United States claims to have moved beyond violent enforcement of racial hierarchies. As biological theories of race have always accomplished, the redefinition of race as a genetic category and the technologies it is generating make racial inequality, as well as the punitive apparatus that maintains it, seem perfectly natural.

Despite claims that genomic scientists are reconfiguring race in a more precise and accurate way, I have found rampant confusion about this new racial science among friends and colleagues trying to understand where it fits in their conceptions of humanity and social equality. What is the value and impact of modernizing racial classifications with new genomic technology? Will this new understanding of race provide the answers we need to solve the problem of persistent racial inequality in our society?

When I first began reading about scientific discoveries that purported to validate the biological reality of race, I wondered how the claims could be true. My parents, to whom this book is dedicated, taught me that there is only one human race. This has meant more than words to me since childhood. As a scholar and activist, I have studied and seen firsthand the injustices made possible by separating people into different kinds of human beings. I undertook this book project as a personal challenge to my conviction that race is a political system. Would my own racial beliefs withstand my interviews with genomic researchers, the enthusiasm of many African Americans for tracing their genetic ancestry, and arguments by political activists who believe race-based biotechnologies are promoting health?

After five years of intense research and soul-searching, I found not one shred of evidence to counter my belief in the political nature of race. In fact, my journey only strengthened my understanding of our common humanity and the dehumanizing consequences of believing in innate racial differences. The answer to the problem of race will not be found in our genes. Yes, human beings are remarkably similar at the genetic level. But what should link us together is not our genetic unity; we should be bound by a common struggle for the equal dignity of all of humankind.

PART I

Believing in Race in the Genomic Age

1

The Invention of Race

It is virtually impossible to pass a day in the United States without making use of race. Race is the main characteristic most Americans use to classify each other. It is the first or second thing we notice about a stranger we pass on the street or a new acquaintance approaching to shake our hand. Race determines which church most Americans attend, where they buy a house, the persons they choose to marry, whom they vote for, and the music they listen to. Race is evident in the color of inner-city and suburban schools, prison populations, the U.S. Senate, and Fortune 500 boardrooms. Race rears up every time a police officer is shot or an unarmed suspect is shot by the police. Yet most Americans are hard-pressed to define what race means.

Imagine walking into a room filled with people displaying a wide range of skin colors, hair textures, and facial features. It is likely that the very first thing you do, either consciously or unconsciously, is to identify the race of every single person in that room. Americans are so used to filtering our impressions of people through a racial lens that we engage in this exercise automatically—as if we were merely putting a label on people to match their innate racial identities. But the *only* way we know which racial designation to assign each person is by referring to the invented rules we have been taught since we were infants. And the *only* reason we engage in this exercise is the enormous social consequences of classifying people in this way. So we force the mélange of physical features and social clues into a code that tells us how to categorize each person—so as to know where each person fits in our society.

Aristotle claimed that he could tell if a man was a citizen by looking at him. Most intelligent people today would think it preposterous to judge someone's citizenship status by physical appearance. Citizenship is a political category, not a biological one. Citizenship doesn't describe a person's intrinsic characteristics; it defines her relationship to a nation's government, to the other people who are citizens, and to noncitizens. Whether someone qualifies as a citizen isn't inscribed in her body; citizenship is determined by the requirements set forth in the country's constitution. Even if the constitution stated that only people of a certain height qualified for citizenship, we would still understand it as a political category, a relationship among people that has to do with governing them.

Like citizenship, race is a political system that governs people by sorting them into social groupings based on invented biological demarcations. Race is not only interpreted according to invented rules, but, more important, race *itself* is an invented political grouping. Race is not a biological category that is politically charged. It is a political category that has been disguised as a biological one.

This distinction is important because many people misinterpret the phrase "race is socially constructed" to mean that the biological category of race has a social meaning, so that each society interprets differently what it means to belong to a biological race. According to this view, first we are born into a race, and then our society determines the consequences of this natural inheritance. There is, then, no contradiction between seeing race as both biological and socially constructed. Among liberal Americans, recognizing race as a social construction translates into an admonition against racial bias. Some say it is fine to acknowledge the existence of biological races as long as we do not construct them as socially unequal. The problem with this interpretation of race as a social construction is that it ignores its political—and not biological—origin. The very first step of creating race, dividing human beings into these categories, is a political practice.

How do we know race is a political and not a natural grouping?

For a start, human beings do not fit the zoological definition of race. A biological race is a population of organisms that can be distinguished from other populations in the same species based on differences in inherited traits.[1] There are no human populations with such a high degree of genetic differentiation that they objectively fall into races. There is only one human race. As Duke geneticist Charmaine Royal explained to me, in a subspecies or race, "all the entities in that group are the same and different from the entities in the other subspecies." But, she pointed out, humans aren't like this: "Chimpanzees have races; honeybees have races; we don't have races."[2]

Royal makes a good point: human beings are not divided into races in the biological sense of the term. But we are divided into races in the political sense. In contrast to the title of the excellent PBS documentary refuting the biological basis for race, *Race: The Power of an Illusion*, race is not imaginary. Race is very real as a political grouping of human beings and has actual consequences for people's health, wealth, social status, reputation, and opportunities in life. The fact that dividing people into races has biological *effects* does not change the fact that this division is a political exercise. The distinction between the two meanings of race—as a biological versus a political grouping—is monumentally important. If race is a natural division, it is easy to dismiss the glaring differences in people's welfare as fair and even insurmountable; even liberals could feel comfortable with the current pace of racial progress, which leaves huge gaps between white and nonwhite well-being. But if race is a political system, then we must use political means to end its harmful impact on our society. So we cannot ditch the concept of race altogether. Paying attention to race as a political system—which is what it really is—is essential to fighting racism.

More important than the biological evidence is the political evidence of the meaning of race. We know race is a political grouping because it has political roots in slavery and colonialism, it has served a political function over the four hundred years since its inception, and its boundary lines—how many races there are and who belongs to each one—have shifted over time and across nations to suit those political purposes. Who qualifies as white, black, and Indian has been the matter of countless rule changes and judicial decisions. These racial reclassifications did not occur in response to scientific advances in human biology, but in response to sociopolitical imperatives. They reveal that what is being defined, organized, and interpreted is a political relationship and not an innate classification.

Where Does Race Come From?

The use of the term *race* to describe distinct categories of people is surprisingly recent. In 1508, William Dunbar, a Scottish member of King James IV's court, wrote a poem called "The Dance of Sevin Deidly Sins." One of the verses listed among those guilty of Envy, "bakbyttaris of sindry racis"—backbiters of sundry races. Some scholars believe this is the earliest use of the word "race" in the English language.[3] Dunbar employed race to mean family lineage—kinship groups descended from the male line. He probably borrowed *racis* from the Spanish word *raza*, which the Spanish applied to breeds of horses and dogs.

The political origins of race are similarly an artifact of the fifteenth and sixteenth centuries, as Europeans increasingly tried to impose order on the movement of peoples between Christian and Islamic territories and the "Old" and "New" worlds. The Roman Catholic Church, in particular, found a need to distinguish between believers and infidels; the latter were deemed fit for conquest, capture, and enslavement.[4] In 1455, Pope Nicholas V issued a papal directive authorizing the Portuguese to "attack, subject, and reduce to perpetual slavery" all "enemies of Christ" along the west coast of Africa. The authorization included the condition that the captives would be converted to Christianity and then set free. Spanish, Venetian, and Portuguese royalty vastly expanded the importation of dark-skinned slaves and, as the rationale of Christian conversion no longer suited the growing enterprise of perpetual forced black labor, began to evade the papal mandate to emancipate converted captives by describing them as less than human.

It has been argued in defense of a natural concept of race that human classification schemes existed prior to the period of European imperialism. Many millennia before the European age of exploration, for instance, the ancient Egyptians recorded in tomb paintings and hieroglyphs the physical differences between themselves and the lighter-skinned Syrians and the darker-skinned Nubians they conquered. The ancient Greek physician Hippocrates, known as the father of Western medicine, separated the Scythians, Asians, and Greeks into different body types explained by the topography and climate of the regions they lived in.[5] The Bible offers the story of Noah's sons, Ham, Shem, and Japheth, as an explanation for the geographical origins of different groups of people, one that would later be interpreted by white Americans as evidence of a divine justification for slavery.

Believing in the uniqueness and superiority of one's own group may be universal, but it is not equivalent to race.[6] These ancient attempts to describe and explain differences among peoples did not partition all human beings into a relatively small number of innately separable types. Nor did they treat these differences as markers of immutable distinctions that determine each group's permanent social value. Unlike race, these observed differences did not create a set of discrete categories that every single person in the world must fit into from the day he or she is born. The Spanish and the Portuguese had experienced close contact with North Africa as well as domination by highly cultured Arab invaders for centuries. The Moors conquered the Iberian Peninsula in 711, bringing with them several hundred years of cultural influence and ethnic intermixing. Spanish and Portuguese traders did not automatically classify the people from the west coast of Africa as an innately inferior group.[7]

Historian Winthrop Jordan argues that, in contrast, English travelers immediately distinguished themselves from the dark-skinned people they encountered. "From the first, Englishmen tended to set Negroes over against themselves," Jordan writes, "to stress what they conceived to be radically contrasting qualities of color, religion, and style of life, as well as animality and a peculiarly potent sexuality." He goes on to note, however, that this attention to difference acquired a distinctive aspect as the English expanded the slave trade: "What Englishmen did not at first fully realize was that Negroes were potentially subjects for a *special kind* of obedience and subordination which was to arise as adventurous Englishmen sought to possess for themselves and their children one of the most bountiful dominions of the earth" (my emphasis).[8]

A Special Kind of Subjugation

What was this special kind of subjugation? It was not the enslavement of Africans alone, for human servitude was far from a novel practice. Societies across the globe had relied on compelled labor for thousands of years to generate wealth for those in power. All captives in the ancient Egyptian, Greek, and Roman empires could be enslaved, regardless of their geographic origins. People were also enslaved as payment for a debt or as punishment for a crime, and desperately poor parents in ancient Rome made money by selling their children into slavery. In addition, the inhabitants of regions we now call Europe had been enslaved by foreigners for centuries.[9] The word "slave" comes from Slavs, who were held in bondage from as early as the ninth century. The ancestors of people now considered white, who think of themselves as the slaveholding race, were once held as slaves themselves.

Even in the New World, "slave" did not automatically mean "black." The vast majority of people compelled to work in the fields of the American colonies were vagrant children, convicts, and indentured servants shipped from Britain. As Nell Irvin Painter notes in her history of white people, "before the eighteenth-century boom in the African slave trade, between one-half and two-thirds of all early white immigrants to the British colonies in the Western Hemisphere came as unfree laborers, some 300,000 to 400,000 people."[10] At first, European settlers in the American colonies gave the blacks they shipped from Africa and the indigenous people they captured the same status as white indentured servants.[11] Many worked alongside each other for a term of service regardless of color, sharing everyday life together and sometimes forming families. European, African, and Indian servants also joined ranks in a series of revolts demanding better food rations, less arduous working conditions, and rights to property. The few Africans who were able to gain their freedom and purchase land seem to have been treated as equals to other landowners.

By 1700, however, Africans were treated as a distinctly different kind of slave: they were made into the actual property of their masters, a lifelong bondage that passed down to their children. In contrast, the status of white indentured servants was neither permanent nor inherited; whites could work off their bond. Then came Bacon's Rebellion. In 1676, European and African servants in colonial Virginia joined forces to demand that the royal governor, William Berkeley, move more aggressively against Indian tribes to settle lands on the frontier. Led by wealthy planter Nathaniel Bacon, five hundred black and white men, united by enmity against the Indians, marched into the colonial capital, Jamestown, and burned it to the ground. The uprising continued for several months before Bacon died suddenly and Berkeley was able to restore order, later hanging twenty-three of the rebels.

After Bacon's Rebellion and similar revolts, it was imperative for European landowners to prevent future interracial solidarity by driving an impenetrable wedge between African and European laborers.[12] The elite class feared black and white servants joining ranks against it more than it desired a brotherhood of servants fighting against Indians. The colonists increased the importation of African slaves, relying more and more on compelled black labor to produce profits. There was an insufficient supply of white servants making the transatlantic migration, and Indian tribes, with a home territory advantage, had mounted a formidable resistance to wholesale capture. Captive Africans, whose skills at farming, carpentry, and metalworking proved immensely valuable to European capitalists, filled the labor gap. But more significant than the numerical shift from white to black exploitation was a monumental legislative effort to differentiate the status of blacks and whites. As officials split white indenture from black enslavement and established "white," "Negro," and "Indian" as distinct legal categories, race was literally manufactured by law.

The emerging legal regime set *all* blacks clearly apart from *all* whites in terms of power and privilege. Gradually, lawmakers degraded the form of bondage assigned exclusively to Africans. Under a 1705 Virginia statute, for example, white servants received freedom dues, including a musket, money, and bushels of corn, when they completed a term of indentured servitude. Black servants were not entitled to freedom and were banned from carrying firearms. The Virginia racial laws not only treated enslaved Africans worse than white servants, they gave propertyless whites special rights over slaves. Historian David Roediger writes:

> Pass laws restricted the movement of slaves; poor whites patrolled to enforce the laws. Laws required public, often naked, whippings of recalcitrant black slaves but set limits on penal violence against even indentured whites. Beatings awaited any "negroe or other [Indian] slave" attacking "any Christian," according to a 1680 law, putting state power firmly on the side of Europeans in altercations among servants. The term "Christian" increasingly meant "white." Africans would not be defined as such.[13]

A South Carolina statute followed Virginia's example, giving every white man "absolute power and authority over his negro slave," regardless of religion. Converting to Christianity no longer provided Africans an escape from their condition as the lifelong chattel of their masters.

Laws regulating sex and marriage hardened the lines between the emerging racial categories. Virginia had already outlawed interracial sex in 1662, when the legislature amended its prohibition of all fornication to impose heavier penalties if the guilty parties were "negroes" and "Christians." In 1691, the Virginia Assembly beefed up its laws against racial mixing by making it a crime for Negro, mulatto, and Indian men to marry or "accompany" a white woman. By shielding white women from marriage by all nonwhites, Virginia aimed to preserve white racial purity; the statute left Negroes, mulattos, and

Indians free to intermix and marry each other. White people were held out as a privileged race that should be protected from contamination by inferior races.

Antimiscegenation laws also ensured that black men, women, and children would not benefit from the privileges of legal marriage to a white person. As W.J. Cash explained in *The Mind of the South* in 1941, whites passed these laws to protect "the right of their sons in the legitimate line, through all the generations to come, to be born to the great heritage of the white race."[14] The Virginia law, later revised to prohibit all marriages between "a white person and a colored person," remained in force until 1967, when the U.S. Supreme Court ruled it, along with antimiscegenation laws still existing in sixteen other states, unconstitutional in *Loving v. Virginia*. South Carolina did not overturn its ban on interracial marriage until 1998, and even then 38 percent of voters opposed the referendum.

Just as significant as the peculiar form of black slavery was the peculiar form of black freedom. In addition to destroying political unity between white and black servants, white landowners had to make absolutely certain that free blacks would never become socially equal to whites or encourage slaves to revolt in order to join their ranks. Being black had to be equivalent to slavery, regardless of blacks' status as actual property. Free blacks created a particularly vexing dilemma even for abolitionists: if no longer enslaved, would a black person be entitled to the same rights and privileges as a white person? "Suddenly, the country became obsessed with racial distinctions and the problem of freed blacks," writes Gordon S. Wood in *Empire of Liberty*, his recent history of the early republic.[15]

Eighteenth-century lawmakers subjected free blacks and mulattoes to a permanent status of legal degradation that resembled the bondage of their brothers and sisters. "Like slaves, free Negroes were generally without political rights, were unable to move freely, were prohibited from testifying against whites, and were often punished with the lash," writes historian Ira Berlin in his classic *Slaves Without Masters*.[16] Southern statutes provided multiple ways for free blacks to be put in bondage; according to a Virginia law passed in 1782, defaulting on their taxes was sufficient. Many states implemented registration requirements for free blacks, while others barred their entry altogether or banished them from the state. Blacks in the North, too, occupied an inferior position, held in place by a slew of legal restrictions on their mobility, civil rights, and economic opportunities. "Even in the North," writes

Wood, "the liberal atmosphere of the immediate post-Revolutionary years evaporated, and whites began to react against the increasing numbers of freed blacks."[17] By the early 1800s, Northern states tried to keep free blacks from entering their borders and restricted those who lived there to distinct neighborhoods as well as to segregated sections of restaurants, churches, and theaters. Blacks hoping to move to Ohio had to post a $500 bond guaranteeing their good behavior and produce court documents proving they were free. Interracial marriage was outlawed in Massachusetts, just as it was in Mississippi. The New York State legislature, dominated by Republicans who supported universal manhood suffrage without property qualifications, voted to strip blacks of the franchise.[18] Extending greater freedoms to white men necessitated depriving blacks of theirs because it was inconceivable that the two races could share equal political strength.

As a result of the new racial regime, "Negro" and "slave" were becoming synonymous as free and enslaved blacks were grouped together at the bottom rung of society. Combining Africans into a single race eventually obliterated the physical, linguistic, and cultural distinctions that had existed among thousands of ethnic groups on the African continent. The New World Europeans also began to perceive themselves as a different kind of master. English, Scottish, German, French, and Dutch settlers were uniting under the general label "white." W.E.B. Du Bois observed that "the discovery of personal whiteness among the world's peoples is a very modern thing."[19] It was only with the slave trade, Indian conquests, and a legal regime that installed a racial order that Europeans assumed whiteness as a personal identity and possession that naturally entitled them to a privileged social position. White identity *itself*, apart from individual wealth or achievement, "conferred tangible and economically valuable benefits," effectively creating what legal scholar Cheryl Harris calls "whiteness as property."[20] By giving poor white laborers legal dominion over all blacks, enslaved or free, wealthy landowners secured their racial loyalty. Poor whites would cherish their privileged status over blacks—what Du Bois called their "psychological wage"—rather than joining with blacks to fight for a more equitable society.

By 1857, the racial order in the United States was firmly in place. When Dred Scott, an African American slave who had lived on free soil in Illinois and Wisconsin, sued for his freedom in Missouri, the U.S. Supreme Court could point to a coherent system of discrimination against blacks that proved they were not full citizens of the United States. Black Americans, Chief

Justice Roger B. Taney declared, possessed "no rights which the white man was bound to respect." Dred Scott remained a slave, and all blacks, whether in chains or free, were recognized as a separate race that was politically subordinate to all whites.

This is why the Emancipation Proclamation of 1863 did not grant true freedom to black citizens or change the racial definition of citizenship. Despite the passage of the Reconstruction amendments to extend citizenship rights to freed slaves, blacks remained members of an inferior race, and an official regime of segregation, disenfranchisement, and terror effectively reduced them to their former status.[21] Soon after the Civil War, Frederick Douglass observed that the same racial ideas employed in defense of slavery were "employed as a justification of the fraud and violence by which colored men are divested of citizenship, and robbed of their constitutional rights."[22] In addition to the grotesque lynchings that terrorized blacks throughout the South, an especially brutal form of reenslavement was the false imprisonment of thousands of black men who were then leased to white farmers, entrepreneurs, and corporations as a source of cheap labor.

It is in this acute distinction between the *political* status of whites and blacks, this way of governing the power relationship between them, that we find the origins of race. Colonial landowners inherited slavery as an ancient practice, but they invented race as a modern system of power. They employed Aristotle's concept of natural slaves and natural rulers to define permanent features of black and white people. Race separated human beings into two fundamentally distinct groups: those who were indelibly born to be lifelong servants and those who were born to be their masters. Race radically transformed not only what it meant to be enslaved, but also what it meant to be free.

Racial Identity on Trial

One of the most sensational trials in New York in 1925 involved the annulment sought by wealthy socialite Leonard Kip Rhinelander of his marriage to Alice Beatrice Jones. It wasn't the couple's fortune that made tabloid headlines and packed the courtroom with curious spectators. It was the scandal of Alice's race: Leonard sued Alice to end their union on the grounds that she had fooled him into believing she was white when she actually was black. If the jurors found that Alice had perpetrated this racial fraud and that

Leonard would not have married her had he known the truth, their vows would be null and void. It was rumored that Leonard was madly in love with Alice and brought the lawsuit under pressure from his millionaire father, who disapproved of the marriage.[23]

At trial, numerous witnesses painted a confused portrait of the evidence of Alice's race. Neighbors testified that Alice's family was "generally known" as "colored," but that she and her sisters did not "associate with Negroes."[24] Although her teacher said she recognized Alice as colored, her classmates thought she was Spanish. Alice had been admitted into respectable hotels with Leonard, but Dr. Caesar McClendon, a colored doctor, testified that Alice was one of his patients; if Alice were white, she would have been treated by a doctor of her own race.[25] In her defense, Alice surprised Leonard and his attorneys by choosing not to prove she was white. She admitted that she had some "colored blood" but insisted that Leonard—who prior to the wedding had met her family, spent many nights with her at the Hotel Marie Antoinette, and even bathed her naked body—was well aware of her true identity. The climax of the trial occurred when Alice was compelled to provide the key evidence in her defense. The judge, lawyers, jurors, and Leonard waited in the jury room while Alice stripped to her underwear, covered by a coat, in an adjoining lavatory. At her lawyer's direction, a weeping Alice let down the coat before the white male jurors so they could see for themselves how easy it was to tell that she was black. Her lawyer rested his case with one question to Leonard: "Your wife's body is the same shade as when you saw her in the Marie Antoinette with all of her clothing removed?" Leonard replied, "Yes."[26]

The jury found in favor of Alice: they believed that Leonard knew that Alice was colored before he married her. But the verdict was hardly a victory for Alice, who paid a dear price in the humiliating act of disrobing in the public courtroom and, in the end, lost her husband because of her race. W.E.B. Du Bois represented the outrage of the black community when he wrote, "If Rhinelander had used this girl as concubine or prostitute, white America would have raised no word of protest. . . . It is when he legally and decently marries the girl that Hell breaks loose and literally tears the pair apart."[27] Alice and Leonard never reunited after the trial.

Though more dramatic than most, the Rhinelander trial was not unique in its task—to establish the race of a litigant. Trials to determine an individual's racial identity were commonplace in courts throughout the country from the late eighteenth century until well into the twentieth century.[28] Before the Civil War, slaves suing their masters for their freedom and people charged with crimes reserved for blacks, such as traveling without a pass, claimed that they were really white to avoid the disabilities the law imposed solely on blacks. At first, during the colonial era, these racial trials pivoted on documenting the claimant's legal status as slave or free. But as whites justified the slavery and Jim Crow systems on more explicitly racial grounds, juries were given wide latitude to decide a litigant's racial identity based on "performance, reputation, and common sense."[29]

The legal separation of whites from blacks and other people of color created a myriad of reasons to contest racial identity. As legal historian Ariela J. Gross chronicles, "family members tried to have heirs disinherited by alleging that they were of mixed race and therefore the product of an illegitimate marriage. Marriage bans gave rise to criminal prosecutions against husbands and wives, who defended themselves by claiming that the state had mistaken their racial identities. And sometimes husbands sought to avoid the obligations of marriage by claiming they had been tricked into an interracial union."[30] In addition to these family disputes, parents sued school districts for admission by asserting their children were white, feuding neighbors accused their rivals of being black, and Native Americans had their authenticity adjudicated to claim federal land allotments.

In her book *What Blood Won't Tell,* Gross emphasizes that racial identity trials did not always hinge on documentary evidence, physical markers of race, or scientific fact. Although judges and legislators specified a legal definition of racial categories, jury verdicts depended more on social considerations than on the precise fraction of colored blood in the claimants' veins.[31] When a South Carolina judge declared in 1835 that "a slave cannot be a white man," he made clear that racial identity was not a biological fact that could be ascertained with scientific proofs, but rather "a socially and legally defined status that rested on a deeper ideological commitment to race, in which white equaled free (civic, responsible, manly) and black equaled slave (degraded, irresponsible, unfit for manly duties)."[32] Whites pretended that race was inherited biologically but used a mixed bag of legal tools to determine racial identity in order to maintain race as a political system for governing relationships between enslaved and free people.

Another set of racial cases involved litigation over the legal question, Who is white? The Naturalization Act of 1790, the nation's first definition of

American citizenship, restricted eligibility to free white immigrants. Until this racial requirement was abolished in 1952, being either a "white person" or (after the Civil War) a person of "African nativity, or African descent" was a prerequisite of becoming a citizen. Determining which groups of immigrants met the whiteness test for naturalization became a vital legal issue for almost a century. Between 1878 and 1952, state and federal judges issued decisions in fifty-two racial prerequisite cases, including two argued before the U.S. Supreme Court in the 1920s.[33] In these cases, judges ruled that Chinese, Japanese, Koreans, Filipinos, Hawaiians, Afghanis, Native Americans, and anyone of mixed ancestry were not white. Arabs, Syrians, and Asian Indians were considered white by some judges and not by others. Armenians were more successful at claiming whiteness, despite their geographic origins east of the Bosporus Strait, which separates Europe from Asia.[34]

Reviewing these cases, legal scholar Ian Haney Lopez discerned two chief rationales advanced by judges to justify their racial designations: common knowledge and scientific evidence. At times, courts relied on both widely held understandings about race and scientific expertise on the subject to determine which groups qualified as white. The very first case, decided by a California judge in 1878, arose when Ah Yup, "a native and citizen of the empire of China," petitioned in writing to be admitted as a citizen of the United States, raising the central question, "Is a person of the Mongolian race a 'white person' within the meaning of the [naturalization] statute?"[35] Judge Sawyer dismissed the argument that the category "white person" was indefinite, noting that "these words in this country, at least, have undoubtedly acquired a well settled meaning in common popular speech." The phrase "white person," the court stated, ordinarily referred to someone of the Caucasian race. The judge then turned for guidance to the racial typologies developed by European naturalists Johann Friedrich Blumenbach, George Louis de Buffon, Carl Linnaeus, and Georges Cuvier, observing that all grouped Caucasians separately from Mongolians.

The final blow to Ah Yup's claim to whiteness was evidence of Congress's intent to exclude Chinese from citizenship. In the wake of the Civil War and the Fourteenth Amendment, Congress amended the law in 1870 to extend naturalization to persons of "African nativity, or African descent," while deliberately denying Chinese immigrants that right on the grounds they posed a risk to American morals and jobs. Whether to retain the "white person" prerequisite hinged on a heated debate on the Senate floor over the influx of low-paid Chinese laborers to the West Coast. Judge Sawyer quoted Indiana Republican Oliver Perry Morton's tirade against the "Chinese problem," which threatened "a possible immigration of many millions, involving another civilization; involving labor problems that no intellect can solve without study and time."[36]

Congress moved more directly to stanch the "Mongolian invasion" with the Chinese Immigration Act of 1882, barring entry to Chinese workers for ten years, including the wives and families of immigrants already in the country. Subsequent laws passed in 1917, 1924, and 1934 extended the exclusion to immigrants from Japan, India, and the Philippines. The supposedly biological category "Asian" commonly employed today was solidified by the series of statutes and court decisions that classified immigrants from each nation as nonwhite.[37] The racial question was ultimately a political question about which groups the federal government deemed qualified for citizenship.

Further evidence of the political nature of the prerequisite cases is the judges' willingness to repudiate scientific racial classifications when they clashed with popular notions of whiteness. The U.S. Supreme Court was quick to reject scientific evidence on race when the prevailing anthropological classification of Asian Indians as Caucasians rather than Mongolians would have meant citizenship rights to thousands of immigrants from the Indian subcontinent.[38] Bhagat Singh Thind was born to a high-caste family in the Punjab region of India. At age twenty-three, he arrived in Oregon on July 4, 1913, as part of a new wave of Asian Indian immigrants that numbered more than six thousand by 1920.[39] A federal judge initially granted him citizenship in 1920 on the grounds that, because Asian Indians were scientifically classified as Caucasians, they were therefore legally white.

But by this time, a campaign to formally deny all Asians entrance to U.S. shores had gained momentum. Congress had recently passed the Immigration Act of 1917, which prohibited immigration from an "Asiatic barred zone" that stretched beyond China to include much of eastern Asia, the Pacific Islands, and India. The subsequent Oriental Exclusion Act of 1924 put a halt to all immigration from Asia, including Japan. What remained was to deny citizenship to those Asian immigrants like Thind who had entered prior to these exclusionary laws. The federal government challenged Thind's naturalization in the Ninth Circuit, which asked the Supreme Court for instruction on the question: "Is a high caste Hindu of full Indian blood, born at Amrit Sar, Punjab, India, a white person?" Only three months before, in

Ozawa v. United States, the Court unanimously held that Japanese immigrants did not qualify as "white persons" for purposes of naturalization.[40] Relying on "numerous scientific authorities," the opinion by Justice Sutherland determined that the Japanese did not belong to the Caucasian race, despite their fair skin color.

Based on the Court's equation of Caucasian and white, Thind mounted what should have been an airtight case. In addition to deploying numerous anthropological texts that classified Asian Indians as Caucasian, he emphasized that experts placed inhabitants of the Punjab in the Aryan race. To this scientific evidence, Thind added an insightful political proof. High-caste Aryans like himself were white, he argued, because they occupied a superior position in the Hindu caste system comparable to white people's supremacy in the United States. "The high-class Hindu regards the aboriginal Indian Mongoloid in the same manner as the American regards the negro, speaking from a marriage standpoint," he wrote. "The caste system prevails in India to a degree unsurpassed elsewhere." According to Thind, the Indian caste system maintained Aryan racial purity more effectively than did Jim Crow laws in the United States.[41] He was therefore as pure white as any white American.

Despite Thind's anthropological and political evidence, a unanimous Court (in an opinion also written by Justice Sutherland) rejected its precedent in *Ozawa*. Conceding the anthropological classification of Asian Indians as Caucasian, the Court nonetheless dismissed the authority of science as the arbiter of race in favor of "familiar observation and knowledge." "It may be true that a blond Scandinavian and the brown Hindu have a common ancestor in the dim reaches of antiquity," wrote the Court, "but the average man knows perfectly well that there are unmistakable and profound differences between them today." Discrepancies in the leading racial typologies and their inclusion of a range of peoples in the Caucasian category further proved to the Court their unreliability: "We venture to think that the average well informed white American would learn with some degree of astonishment that the race to which he belongs is made up of such heterogeneous elements."[42]

In the 1897 case *In re: Rodriguez*, a federal judge in San Antonio, Texas, also disregarded the prevailing racial classification scheme and legal precedent in determining the eligibility of Mexican immigrants to be naturalized as U.S. citizens. Thirty-seven-year-old Ricardo Rodriguez, a Mexican national, had been living in San Antonio for ten years when his application for U.S.

citizenship was denied because he wasn't white. On appeal, the government's attorney argued that Rodriguez was "one of the six million [Mexican] Indians of unmixed blood" and belonged to the same Mongoloid race that the California court in *In re: Ah Yup* had declared ineligible for U.S. citizenship. The Texas judge conceded that, based on his appearance, Rodriguez would be grouped with the "copper-colored or red men": "If the strict scientific classification of the anthropologist should be adopted, he would probably not be classed as white." Nevertheless, the judge ruled that the same state and federal laws that gave political rights to Mexicans living in Texas when it became a republic and was later annexed as a state also bestowed white status on Mexicans, at least for purposes of naturalization.[43]

Even certain European groups whose whiteness seems unquestionable today were not considered full members of the white race one hundred years ago. In the late 1800s, Irish immigrants were considered to be closer to Africans than to the English and were often portrayed as apelike in caricatures. Italian newcomers were called Guineas, an epithet originally reserved for African Americans and derived from their motherland along the west coast of Africa. Scholars in a number of fields have puzzled over the question, How did the Irish, Italians, Slavs, and Jews "become white"?[44] The wave of immigrants who arrived from southern and eastern Europe from the 1840s to the 1930s were included among the undesirables targeted by the exclusionary immigration laws passed by Congress in 1924. They threw off their native customs, accents, and names not only to become assimilated into American culture, but also to be granted entrance to the privileged rank of whiteness. Obviously, their ancestry did not change to make them whiter; their racial transformation occurred as a result of changing political qualifications for inclusion in the ruling class. There is no biological test for whiteness. *White* means belonging to the group of people who are entitled to claim white privilege.

American judges and legislatures have repeatedly amended the legal contours of racial groups. Before the Civil War, some Southern states had not yet implemented rules specifying the amount of African ancestry that made someone Negro. In South Carolina, for example, until the 1840s, mulattoes could establish they were white through their behavior, reputation, and social standing in order to marry a white person.[45] The Jim Crow era ushered in a regime of official segregation that required stricter enforcement of the borders delimiting whiteness. Laws prohibiting interracial marriage were rife with varying tests for what made someone black enough to be banned from marrying a white person. The infamous one-drop rule, passed in Tennessee in 1910 and mimicked by other Southern states, defined a person as Negro if he or she had any Negro ancestor—or one drop of Negro blood. Other states specified varying fractions of Negro ancestry to qualify as a Negro. Indiana, for example, made it a crime for a white person to marry anyone "possessed of one-eighth or more of negro blood." Oregon used a one-fourth test. The Utah legislature divided blacks into mulattos, quadroons, and octoroons. Only politics and not nature (or genetic science) can explain why in some states "a single black great-grandparent was sufficient to establish a person as 'black,' while seven white great-grandparents were insufficient to establish one as 'white,'" notes anthropologist Jonathan Marks.[46] Only politics and not nature can explain why what legal scholar and sociologist Laura Gómez calls a "reverse one-drop rule" applied to Mexicans, whose fraction of Spanish ancestry enabled them in some circumstances to be deemed "white enough" to acquire certain political rights.[47]

The legal significance of race extends far beyond official definitions of racial categories. Laws have governed not only the formal racial classification of citizens and immigrants but also the material consequences of being classified in one race or another. As Ariela Gross reminds us about racial identity tests, "a lot was at stake in these cases: personal freedom or enslavement, the right to marry the person of one's choice, the right to be a citizen."[48] Colonists first began establishing the distinction between whites, Negroes, and Indians when they passed statutes that specified the privileges and disabilities that accrued to each group. The effort to legislate the political status of whites and nonwhites necessitated legal specifications for these categories. State laws banning interracial marriage, for example, had to stipulate a test for Negroes, who were barred from marrying whites. In other words, the legal construction of racial categories was a means of implementing the white supremacist regime.

Today, the government's identification of racial categories is less directly tied to discrimination, yet it remains part of a complex legal apparatus that enforces racial inequality in the education, economic, political, criminal justice, and health care systems. Dividing people by race facilitates, and is in turn reinforced by, institutions that treat people unequally depending on their racial identity. Reviewing the history of official racial classifications reminds us that these categories are not natural—and neither are the institutional inequities that race undergirds. But when Americans see black and brown people doing most of the menial jobs, dying younger from most diseases, and filling most of the prison cells, it is easy for many to see race and believe it must be part of nature.

The Instability of Race

Perhaps the most compelling evidence that race is a political category is its instability. Since its invention to manage the expansion of European enslavement and the colonization of other peoples, the definitions, criteria, and boundary lines that determine racial categories have constantly shifted over the course of U.S. history. As sociologists Michael Omi and Howard Winant write in *Racial Formation in the United States*, "racial categories are created, inhabited, transformed, and destroyed" in this sociohistorical process.[49] If races are fixed biological groupings, how can the test defining who belongs in each group change? If a person's race is inscribed in her genes, how can a judge officially assign (and reassign) it according to a legal classification system? If race is written in nature, how can people rewrite the rules?

The changes in defining America's racial groups are not just a matter of adding or subtracting options. Rather, they reflect shifts in the rules the government employs to categorize people by race. The U.S. Constitution mandates an enumeration of the nation's inhabitants every ten years to apportion representation in the House of Representatives. The very first U.S. census began on August 2, 1790, a year after the inauguration of President George Washington. Census takers in 1790 counted the number of persons in each household according to the following categories: free white males sixteen years and older, free white males under sixteen years, free white females, all other free persons, and slaves. Since then, every U.S. census has sorted people by race—but the racial groupings have changed twenty-four times over the last two hundred years. In the second census, taken in 1800, Indians were specified as a separate category of free persons. Chinese were added to the 1870 census. In 1920, race had become even more complicated. That census included ten racial categories: white, black, mulatto, Indian, Chinese, Japanese, Filipino, Hindu, Korean, and other. By the end of the twentieth century, the racial groupings were consolidated into five main choices: American Indian or Alaska native, Asian, black or African American, native Hawaiian or other Pacific Islander, and white.

The treatment of Hispanics or Latinos on the census also changed dramatically during the twentieth century. After the Treaty of Guadalupe Hidalgo, which ended the Mexican-American War in 1848, Mexicans were granted U.S. citizenship, effectively conferring on them the legal status of "free white persons." Mexican was included among the racial categories only in the 1930 census. The census indirectly identified Hispanics as a separate category by collecting data on Persons of Spanish Mother Tongue in 1940 and on Persons of Spanish Surname in 1950 and 1960 in five southwestern states.[50] The 1970 census was the first to introduce a full-fledged Hispanic category by asking a 5 percent sample of households a separate question on Hispanic heritage, then requiring them to choose among Mexican, Puerto Rican, Cuban, Central or South American, or Other Spanish.

The 1980 census made it clear for the first time that Hispanic or Latino is considered an ethnicity that is added to a person's race. People were supposed to select both whether or not they are ethnically Hispanic or Latino and also which race they belong to. So someone might be a white Hispanic, a black Hispanic, an American Indian Hispanic, and so forth. Despite the official designation of Latinos as an ethnicity and not a race, almost half of them declined to select white, black, American Indian, Asian, or Pacific Islander on the 2000 census race question and chose "some other race" instead. This refusal to check one of the five racial boxes suggests that many Latinos see themselves as having a separate *racial* identity.[51]

The 2010 census provides fifteen racial categories to choose from, as well as spaces to write in a racial identity not listed on the form. Curiously, the census contains one category called white; one called black, African American, or Negro; one called American Indian or Alaska Native; and seven Asian categories: Asian Indian, Chinese, Filipino, Japanese, Korean, Vietnamese, and Other Asian. Similarly, the form includes four choices for Pacific Islanders: Native Hawaiian, Guamanian or Chamorro, Samoan, and Other Pacific Islander. This classification scheme suggests that there is one white race, one black race, one American Indian/Alaska Native race, but an unspecified number of Asian and Pacific Islander sub-races. The 2010 census emphasized the distinction between ethnicity and race by including the sentence, "For this census, Hispanic origins are not races." It also added Spanish Origin to Hispanic and Latino as the ethnicities individuals could select in addition to race.

The census forms illustrate another American assumption about race: that each person belongs to a clearly defined racial category. Do people whose parents are from different races fall into a distinct racial box? Beginning with the census of 1850, people of African descent were divided into blacks and mulattoes, those with a mixture of both African and recognized European lineage. Every census between 1850 and 1920 except for 1880 and 1900 included a separate category for mulattoes. Indians filling out the 1930 census were asked for the first time to disclose whether they were "full" or "mixed" blood. Despite these prior examples of mixed-race categories, the 2000 census sparked a fierce controversy when it added an option for people of some "other" race. Although prior enumerations contained "mixed blood" and "other" options, individuals were forced to designate only one racial identity. The 2000 census, by contrast, permitted a person to select multiple races—creating the possibility of sixty-three different racial permutations. Despite this range of options, only about 2 percent of Americans chose more than one race. According to a White House spokesperson, President Obama checked only the box for black/African American when he filled out the 2010 questionnaire.[52]

Racial boundary lines also change when we traverse national borders. My birth certificate, issued in Chicago in 1956, states, "Mother—Negro; Father—white." It does not designate my race. Because of long-standing legal and social rules in the United States, it was an unstated given that I was born a Negro. Although a mulatto category was officially recognized until 1920, the system of Jim Crow segregation had settled the rule that one drop of black blood makes you black, even in Chicago. If I had been born in South Africa, I would have fallen into the Coloured category. The Population Registration Act, passed in 1950 during the apartheid era, divided the South African population into three groups—White, Native, and Coloured—and required that every citizen be assigned a single racial classification. Government classifiers based their decisions on both physical traits, such as skin color and hair texture, and social standing, such as the citizen's job, acquaintances, neighborhood, and lifestyle. There was the infamous comb test, which helped to determine race by how easily a comb passed through an individual's hair. The act defined a coloured person as anyone who was not white or a native member of an indigenous African tribe. Whites and natives were racially pure, while coloureds were the product of racial mixing.[53]

What if I were born in Brazil? Brazilian society recognizes an even wider range of identities for people who are neither white (*branco*) nor black (*preto*). In the 1950s, anthropologist Harry Hutchinson found eight in-between categories in the community of Reconcavo, located in northeastern Brazil, ranging from *Cabo verde* ("lighter than the *preto* but still quite dark, but with straight hair, thin lips, and narrow, straight nose") to *Moreno* ("light skin with straight hair, but not viewed as white").[54] I probably would have been classified as *pardo*, designating mulattoes who are the children of the union of *pretos* and *brancos*.

Of course, my genetic makeup remains the same no matter where I was born. But my race, along with all the privileges and disadvantages that go with it, differs depending on which country I am born in or travel to, because race is a political category that is defined according to invented rules.

Making Race Seem Biological

In a respected text, *Anthropology: Biology and Race* (first published in 1923 and reprinted in 1948 and 1963), the influential anthropologist Alfred Kroeber stated the nature of race as a matter of hereditary fact: "To the question why a Louisiana Negro is black and longheaded, the answer is ready. He was born so. As cows produce calves, and lions, cubs, so Negro springs from Negro and Caucasian from Caucasian. We call the force at work *heredity*."[55] As an inherited status, race seems to be passed down simply through the biological process of procreation. But because race is a political category, reproducing it has never been left to heredity alone. One of America's very first laws, enacted in Virginia in 1662, concerned the inherited status of children. "Whereas some doubts have arisen whether children got by an Englishman upon a negro woman should be slave or free," the statute clarified the matter by declaring that the children's status followed the condition of their mother.[56] This ensured that children born from white men's rape of enslaved African women were treated as property.

Making race revolve around biology constructed it as an innate, permanent, and inescapable status. No matter how someone changes her appearance or mannerisms, she remains deep down a member of the race to which she was born. At the time of the Civil War, white Southerners were haunted by the specter of mistaking a Negro for a white person: "What if people of African descent were lurking unknown in their midst, enjoying all the privileges of whiteness despite their hidden black essence?"[57] writes Ariela Gross. This anxiety over telling which race is which may seem like the vestige of a prior era, yet even today most Americans are anxious to know each other's race and remain unsettled until they discover the true identity of any racially ambiguous person.[58]

Making race a biological concept served an important ideological function in revolutionary America. Biological difference was essential to justifying the enslavement of Africans in a nation founded on a radical commitment to liberty, equality, and natural rights. White Americans had to explain black subjugation as a natural condition, not one they imposed by brute force for the nation's economic profit. Treating race as biology constituted the only suitable "moral apology," as Swedish sociologist Gunnar Myrdal put it, for slavery in a society that claimed equality as its most cherished ideal.[59]

While race is not imaginary—it is a very real way our society categorizes people—its intrinsic origin in biology is. Race is not an illusion. Rather, the belief in intrinsic racial difference is a delusion. The diabolical genius of making this political system seem biological is that the very unequal conditions it produces become an excuse for racial injustice. Whites pointed to sickness, fatigue, and illiteracy among the Africans they exploited as evidence of black biological inferiority rather than white inhumanity. So it appears that nature determines one's race and race determines one's nature. In his classic analysis of racial pathology, E. Franklin Frazier incisively describes the delusional rationalization of black inequality that resists all objective evidence:

> The insane support their delusions by the same mechanism of rationalization that normal people employ to support beliefs having a nonrational origin. The delusions of the insane, however, have a greater imperviousness to objective fact. The delusions of the white man under the Negro-complex show the same imperviousness to objective facts concerning the Negro. We have heard lately an intelligent Southern woman insisting that nine-tenths of all Negroes have syphilis, in spite of statistical and other authoritative evidence to the contrary. . . . When the lunatic is met with ideas incompatible with his delusion he distorts facts by rationalization to preserve the inner consistency of his delusions. Of a similar nature is the argument of the white man who de-

> clares that white blood is responsible for character and genius in mixed Negroes, and at the same time that white blood harms the Negroes![60]

To this day, the delusion that race is a biological inheritance rather than a political relationship leads plenty of intelligent people to make the most ludicrous statements about black biological traits. Worse yet, this delusion permits a majority of Americans to live in perfect comfort with a host of barbaric practices and conditions that befall blacks primarily—infant deaths at numbers worse than in developing countries, locking up children in adult prisons for life, the highest incarceration rate in the history of the free world—and still view their country as a bastion of freedom and equality for all.

Some people argue that continuing to treat race as a biological category in genetic research, medicine, and technology is not harmful by itself, as long as safeguards are in place to prevent its abuse by bigoted people. Understanding race as a political classification that supports racism exposes the flaw in this view. British sociologist Paul Gilroy underscores this key premise of the relationship between race and racism: "For me, 'race' refers primarily to an impersonal, discursive arrangement, the brutal result of the raciological ordering of the world, not its cause."[61] In less academic terms: race is the product of racism; racism is not the product of race.

2

Separating Racial Science from Racism

No sooner had the Human Genome Project determined that human beings are 99.9 percent genetically alike than many scientists shifted their focus from human genetic commonality to the 0.1 percent of human genetic difference. This difference is increasingly seen as encompassing race. On the heels of the gene map's completion in 2000, science journalist Nicholas Wade described genomic researchers' follow-up mission precisely in these racial terms. "Scientists planning the next phase of the human genome project are being forced to confront a treacherous issue: the genetic differences between human races," he wrote in a July 20, 2001, article, "For Genome Mappers, the Tricky Terrain of Race Requires Some Careful Navigating." The question was no longer whether or not racial differences exist at the molecular level, but how to go about discovering them.

The speedy resuscitation of biological concepts of race seems less surprising if we consider the intimate marriage of race and science that has lasted more than three centuries. Over this period, the function race serves as a political classification system has remained the same, but scientists have discovered new ways of identifying, justifying, and proving race as a biological category. As evidence contesting the biological basis of race emerges, scientists have reconfigured race without abandoning it. What links racial science from one generation to the next is the quest to update the theories and methods for dividing human beings into a handful of groups to provide a biological explanation for their differences—from health outcomes to intelligence to incarceration rates.

Every modern era has had a science of race. Scientists were instrumental in inventing the concept of biological races, in specifying their demarcations, and in justifying the social inequities between them. Scientists created the classification systems that placed human beings in distinct racial categories. Scientists elaborated the philosophies that explained why human races differ. Scientists made race seem like a natural condition they had discovered about human beings rather than a system of governance imposed on human beings. As Harvard science historian Evelynn Hammonds observes, "The appeal of a story that links race to medical and scientific progress is in the way in which it naturalizes the social order in a racially stratified society such as ours."[1] Science is the most effective tool for giving claims about human difference the stamp of legitimacy. And once scientists were committed to understanding human beings as divided into races, they believed that human biology could not be studied without attention to race.

It would be a mistake to think of this work as confined to "scientific racism." The term implies an exceptional use of science to support racist ideas. Calling it scientific racism identifies the problem as scientists' corrupt misuse of race rather than their support of race itself. By rejecting only specific instances of extreme scientific abuse, scientists proceed to reinforce race in novel ways that are supposedly free of past biases. When the worst abuses of racial science are revealed, the next generation of scientists disavows the "scientific racism" of its predecessors or conveniently forgets that scientists used to think that way at all. Many people, for example, point to the Nazis' eugenicist theories as the prime example of scientific racism.[2] Seeing scientific racism as restricted to extreme cases like Nazi genocide mirrors the view of racism in general as an extremist position that falls outside enlightened Western thinking. But as I have shown in the previous chapter, race and racism emerged as integral aspects of the American republic, not at all in opposition to it.

There is a similar problem with calling the racial science of prior eras pseudoscience. In hindsight, we see the flaws in bizarre means of measuring racial difference, such as craniometry, which anatomists used a century ago to determine intelligence by calculating skull volume, and brand these methods a ridiculous pretense at the scientific method. Scientists today can then claim that it was pseudoscience that fell victim to racial prejudice, not *real* science, which studies racial difference objectively. But what we call racial pseudoscience today was considered the vanguard of scientific progress at the time it was practiced, and those who practiced it were admired by the scientific community and the public as pioneering geniuses. Could it be that our grandchildren will brand as pseudoscience today's racial classifications generated by computerized genome scans?

To understand the role science has played in perpetuating the biological meaning of race—a role that is currently intensifying—it is helpful to trace racial science from its emergence during the European Enlightenment to its practice by genomic scientists today. The belief that race is natural has always been validated by mainstream—not aberrational—scientific theories and methods, and the most advanced science of human nature has always been shaped by current political contests over racial equality. The burning scientific questions of each period have been framed and answered in terms of race not because rational scientific inquiry compelled it, but because race was *presumed* to be an essential biological category. This is not a happy-ending tale of science overcoming racism. To the contrary, this history reveals how scientists have continually rehabilitated a biological understanding of race throughout the scientific and political upheavals of the last three centuries. It also suggests that we should be just as critical of its latest incarnation in genomic science.

How Scientists Created a Racial Order

The way we think about race today is the product of historical coincidence. The concept of race as a natural category arose from the convergence of two transformations in eighteenth-century Europe that intertwined science and politics: the scientific revolution and the age of colonialism.[3] The expansion of the slave trade in the 1700s, which necessitated a racial system of governance, coincided with the shift among European intellectuals from theological to biological thinking, giving the institution of science ultimate authority over truth and knowledge. Enlightenment biologists were preoccupied with classifying all earthly creations, whether plants, insects, or animals, into a natural hierarchy. Their chief scientific method was taxonomy: observing, naming, and ordering the world by partitioning living things into biologically different types. Applying this method to human bodies, naturalists made race an object of scientific study and made European conquest and enslavement of foreign peoples seem in line with nature.[4] The insistence on finding differences among people so they can be categorized governs the study of human biology to this day.

Race was first used as a category for scientifically classifying human beings by the French physician François Bernier, who penned a 1684 essay titled "A New Division of the Earth, According to the Different Species or Races of Men Who Inhabit It." Bernier organized human beings into five types based on their physical characteristics, grouping the people of Europe, North Africa, the Near East, and India together, but separating sub-Saharan Africans and "the blacks of the Cape of Good Hope" into two distinct species. Asians comprised the fourth large grouping. To these, Bernier added the Lapps of Norway, described as "little stunted creatures" and "wretched animals."[5]

The major groundwork for modern biological classification was laid by the writings of Carl Linnaeus, a Swedish botanist and physician born in 1707. Twelve editions of his catalogue of living things, *Systema Naturae* (published between 1735 and his death in 1778), constituted the first classification system that included human beings along with animals and plants.[6] It was in the tenth edition that Linneaus turned to the human species.

Linnaeus divided the genus *Homo* into two species, *Homo sapiens* (man) and *Homo troglodytes* (ape), and divided *Homo sapiens* into four natural varieties—*H. sapiens americanus, H. sapiens europaeus, H. sapiens asiaticus,* and *H. sapiens afer*—linked to the four known regions of the world, America, Europe, Asia, and Africa. He color-coded the subspecies red, white, yellow, and black and assigned each a set of physical, personality, cultural, and social traits. Linnaeus was influenced by the classical concept of the Great Chain of Being described by Saint Thomas Aquinas, which placed everything in the universe—from stones to angels—in a grand hierarchy established by God.

At the pinnacle of beauty and intelligence Linnaeus placed *H. sapiens europaeus*: "Vigorous, muscular. Flowing blond hair. Blue eyes. Very smart, inventive. Covered by tight clothing. Ruled by law." *H. sapiens americanus*, according to Linnaeus, was "Ill-tempered, impassive. Thick straight black hair; wide nostrils; harsh face; beardless. Stubborn, contented, free. Paints himself with red lines. Ruled by custom." Linnaeus described *H. sapiens asiaticus* as "Melancholy, stern. Black hair; dark eyes. Strict, haughty, greedy. Covered by loose garments. Ruled by opinion." And at the bottom, he placed

H. sapiens afer: "Sluggish, lazy. Black kinky hair; silky skin; flat nose; thick lips; females with genital flap and elongated breasts. Crafty, slow, careless. Covered by grease. Ruled by caprice." Here lies the origin of the color scheme for mankind American children still sing about in Sunday school: red, yellow, black, and white.

Linnaeus obviously had moved beyond empirical description to opine about an ideal physical type for each group (such as the flowing blond hair of Europeans), as well as their personality traits (the stubborn and contented Native Americans, for example). In theorizing that these subspecies descended from different ancestors and developed along distinct paths, he started a new debate—did humans have one origin or many?—that would preoccupy biologists for several centuries.

French anatomist Georges Cuvier (1769–1832) applied the Linnaean method to construct "a great catalogue in which all created beings have suitable names, may be recognized by distinctive characters, and [are] arranged in divisions and sub-divisions."[7] He divided human beings into three main groups, which he labeled Caucasian, Mongolian, and Ethiopian. Cuvier's key contribution to the Linnaean classification system was the theory of "racial typology," which held that humans were divided into permanent biological kinds suited to the particular regions on Earth where they originated and differing permanently in their abilities.[8] Meanwhile, in 1795, Johann Friedrich Blumenbach, a medical professor in Germany, published the third edition of his influential *On the Natural Varieties of Mankind*, classifying human beings into five races: Caucasian, Mongolian, Ethiopian, American, and Malay. Although German philosopher Christophe Meiners had divided humans into two branches, Caucasian and Mongolian, in his 1785 work *Outline of the History of Humanity*, Blumenbach was the first to lend scientific credibility to the racial designation Caucasian, derived from the people in the Caucasus Mountain region.[9] Placing all five varieties of humankind in a single species, Blumenbach theorized that Caucasians, the designation for white peoples of Europe and contiguous regions, were the ideal from which the others "degenerated." By 1800, European naturalists had produced a scientific definition of race as a "fundamental taxonomic division of the human species," with their own kind always at the top.[10]

Eighteenth-century racial science did not fall entirely on the shoulders of biological classifiers. Philosophers were also central to the Enlightenment project of using reason to dominate nature, and they joined naturalists in conceiving race by giving it a philosophy as well as a typological structure. In *On the Different Races of Man* (1775), the German philosopher Immanuel Kant (1724–1804) explained the reason for differences in skin color and natural disposition that distinguished the four main races, which he identified as whites, Negroes, Hindustanic, and Kalmuck, the nomadic Mongols of Central Asia.[11] Kant subscribed to monogenism, locating the origin of all humanity in a common ancestor, and defined race as "the hereditary differences of animals belonging to a single stock." "Negroes and Whites are not different species of humans (for they belong presumably to one stock)," he wrote, "but they are different *races*, for each perpetuates itself in every area, and they generate between them children that are necessarily hybrid, or blendings (mulattoes)."[12] The French philosopher Voltaire, on the other hand, believed in polygenesis, the idea that the races of mankind descended from distinct origins. "Only the blind could doubt that the Whites, the Blacks, the Albinos, the Hottentots, the Laplanders, the Chinese, the Americans, are entirely different races," he declared in a 1764 essay.[13]

The disagreement between Kant and Voltaire on the origins of races presaged a question that plagued nineteenth-century typologists. Did the differences that naturalists were cataloging constitute racial variation within the human species (monogenism) or mark completely distinct species that descended from separate creations (polygenism)? Did Africans, for instance, originate from the same ancestor as Europeans and then develop into an inferior type, or did they spring from an entirely different lineage located between human beings and the lower primates?

A major problem with the separate-origin theory was that it breached the main criterion for identifying species. In a 1753 essay on the ass, French naturalist George-Louis Leclerc, Comte de Buffon, offered an influential definition of species as "a constant succession of similar individuals that can reproduce together."[14] Scientists began to adopt "Buffon's rule" that two animals that can procreate together belong to the same species if their offspring can also procreate. Put another way, the mark of separation between one species and another is the sterility of their offspring. Following this rule, all dogs belonged to the same species even though they were grouped into different breeds. Mutts can bear puppies. But a horse and a donkey are separate species because their progeny, a mule, is sterile. Because the children of two people from different races were fertile, Buffon's rule meant that all human beings belonged to the same species. Some scientists nevertheless defended their belief that races were separate species by predicting that the offspring of mulattoes would be born less fertile or sterile.

In the United States, two pioneering scientists were especially instrumental in promoting the view that human beings were split into separate species. Harvard professor Louis Agassiz lent intellectual firepower to the scientific theory of polygenism, while empiricist Samuel Morton developed the scientific technology to validate it. Agassiz, born in Switzerland in 1807, studied comparative anatomy under Cuvier in Paris and traveled to the United States to do research at Harvard, where he became a professor at age forty-one. A tireless lobbyist for public investment in science, Agassiz helped to found the National Academy of Sciences and was appointed a regent of the Smithsonian Institution in 1863. "No man did more to establish and enhance the prestige of American biology during the nineteenth century," wrote the renowned biologist Stephen Jay Gould, who held the Agassiz chair at Harvard.[15]

Agassiz was haunted by his encounters with blacks when he arrived in the United States. His very first experience meeting black people as servants in a Philadelphia hotel in 1846 was so unsettling that he wrote at length about it in a letter to his mother. "I can scarcely express to you the painful impression that I received, especially since the feeling that they inspired in me is contrary to all our ideas about the confraternity of the human type and the unique origin of our species," he wrote. Describing blacks as "a degraded and degenerate race," Agassiz confessed, "it is impossible for me to repress the feeling that they are not of the same blood as us."[16] Agassiz soon turned this repulsion toward blacks into a full-fledged theory that they had descended from a distinct ancestor. He argued that it was wrong to link all living things to "one common centre of origin"; rather, races were created as separate species that adapted to different zoological zones along with other animals and plants peculiar to each region.

Like the European taxonomists, Agassiz professed a scientific duty to discern a natural order among the races. The fact that "there are upon earth different races of men, inhabiting different parts of its surface, which have different physical characters," Agassiz wrote, imposed on scientists "the obligation to settle the relative rank among these races, the relative value of the characters peculiar to each, in a scientific point of view." It should come as no surprise that he placed blacks at the bottom of his scientific ranking system, strenuously arguing that "human affairs with reference to the colored races" should be "guided by a full consciousness of the real difference existing between us and them." In four ardent letters written in 1863 to Lincoln's Civil War commission, Agassiz warned that incorporating blacks as equals in the reunited nation would contaminate the white race both socially and biologically. "Social equality I deem at all time impracticable," Agassiz intoned. "It is a natural impossibility flowing from the very character of the negro race," which he described as "indolent, playful, sensuous, imitative, subservient," and incapable of living on equal footing with whites "without being an element of social disorder." Especially dramatic was his alarm at the specter of racial interbreeding: "How shall we eradicate the stigma of a lower race when its blood has once been allowed to flow freely into that of our children?"[17] Agassiz's predictions that pure blacks would not survive the northern climate and mulattoes would die off from "their sickly physique and their impaired fecundity" were disproved in time. But his construction of a scientific scaffolding for black subordination enjoyed a lasting legacy.

Taxonomies rely on collecting, measuring, and sorting specimens. Physician and anatomy professor Samuel George Morton, born in Philadelphia in 1799, used this scientific method to study racial difference and validate the separate species theory. Morton was well known as president of the Academy of Natural Sciences in Philadelphia and a founder of invertebrate paleontology, having published an acclaimed book describing the fossils collected by the Lewis and Clark expedition.[18] Morton amassed more than one thousand human skulls, sent to him from colleagues around the globe, into the largest collection in the world, known as the American Golgotha. Morton's mission was to test the hypothesis that racial rankings could be established empirically using brain size. To do this, Morton measured the cranial volume of the skulls by filling them with birdseed or lead pellets and pouring the contents into a cylinder. He then methodically calculated, recorded, and compared the skull sizes of people from different races. In his 1839 magnum opus, *Crania Americana*, Morton presented his meticulous measurements of the cranial capacity of Caucasians, Mongolians, Malays, indigenous Americans, and Negroes.

Mimicking the racial ordering of prior taxonomies, Morton concluded that, based on cranial capacity, whites had the "highest intellectual endowments"; Mongolians of Asia were "ingenious, imitative, and highly susceptible of cultivation"; American Indians "averse to cultivation, and slow in acquiring knowledge; restless, revengeful, and fond of war"; and the Ethiopians "joyous, flexible and indolent," representing "the lowest grade of humanity." Morton also defended polygenism. He conceded that the mulatto children of black and white unions were fertile, but wrote that mulatto women had difficulty bearing children. If one of these women mated with another mulatto, he argued, her children would be even less fertile. Eventually, the progeny would die out. Like Agassiz's separate-species theory, Morton's racial methodology proved deeply flawed. But at the time of his death in 1851, the *New York Tribune* wrote that "probably no scientific man in America enjoyed a higher reputation among scholars throughout the world than Dr. Morton."[19]

The publication of Charles Darwin's *Origin of Species* in 1859 forced a sea change in thinking about how the human species developed. Darwin's earth-shaking theory proposed that all biological organisms are products of evolution, not an original and immutable hierarchy designed by a creator. As British sociologist Michael Banton puts it, "Darwin cut the ground from under the feet of the typologists by demonstrating that there were no permanent forms in nature."[20] Darwin addressed variation within the human species in a subsequent 1871 book, *The Descent of Man, and Selection in Relation to Sex*. Refuting the creationist ladder that placed Africans between Europeans and apes, Darwin argued that all humans and apes shared a common ancestor. Next, Darwin emphasized the biological unity of the human species by pointing to the high fertility of mulattoes, which contradicted the expected sterility of true species hybrids. According to Darwin, human evolution had not produced essential types, as earlier scientists thought, but was driven by individual variation of organisms within the species.

Darwin's radical view that the human species comprises varieties that are changeable and impossible to demarcate might have undermined the typological understanding of race. But the implications of Darwin's theory were too much for British and American academics to stomach. "Go back umpteen generations and would blacks and whites find a common ancestor? Itself the descendent of an ape?" English geologist Charles Lyell, a close friend and supporter of Darwin's, remarked. "The very idea would give a shock to . . . nearly all men. No university would sanction it; even teaching it would ensure the expulsion of a professor already installed."[21] Instead, most scientists adapted Darwin's theories to preexisting views of racial types.

Evolutionary reasoning proved useful to racial science in two key respects. First, Darwin's approach made theoretical sense of a race concept that positioned one group of human beings at a higher level of development than others. Darwin's claim that human beings evolved from lower animals fortified the long-held racial hierarchy that put whites at the pinnacle of human advancement and blacks at the bottom, one step away from animals.[22] Scientific studies comparing the facial angles, limb length, and cranial structure of Negroes and whites claimed to offer physical proof of the races' evolutionary parting. Second, Darwin's view of natural selection as resulting from the struggle for existence validated a racist perspective on current political struggles. Darwin wrote at a time of tremendous racial upheaval in the United States. The nation had just survived a bloody Civil War between the slaveholding Southern states and the Union. Four million slaves were freed between 1864 and 1870.[23] Fledgling efforts by black Americans to claim the franchise they were entitled to were brutally repressed by Ku Klux Klan terror, followed by state imposition of a Jim Crow segregation regime, already in place in Northern states, that denied freed blacks the rights enjoyed by white citizens.[24] Meanwhile, the federal government continued to wage brutal conflicts with American Indian tribes aimed at taking their lands and forcibly relocating them to reservations.[25]

At the end of the nineteenth century, American anthropologists, anatomists, and statisticians eagerly embraced the Darwinian notion of survival of the fittest to explain the slaughter of American Indians and forced regression of blacks to a servant class.[26] Social Darwinists filled scientific journals with studies claiming to prove that these inherently primitive peoples were falling victim to a degenerative evolutionary process. "Will the Negro race be eliminated, and his place taken by the white man as a survival of the fittest?" became a key question for scientists to answer.[27] A popular theory held that the defective bodies and minds of "savage races" would gradually generate their own extinction.

After subjecting black people to slave labor in America for three centuries, whites now viewed their emancipation as the cause of their mental and physical deterioration. Noting that the Negro "luxuriates in the parching rays of an equatorial sun," an 1861 article in the *Medical and Surgical Reporter* concluded that it was impossible for blacks to adjust to the North American climate. Although members of a misplaced group might survive for some time, "yet physiological incompatibility will, ere a generation expires, be evident, and failure in propagation of the race will soon bring about its extinction."[28] Statistician Frederick Hoffman claimed to validate the racial extinction hypothesis with scientific data in *Race Traits and Tendencies of the American Negro*, published by the American Economic Association in 1896, compiling extensive disease, mortality, and birth-rate statistics for white and colored populations in the United States. And with the advent of the new science of eugenics, Darwin's theory of natural selection would take racial science in a dangerous new direction.

The Progressive Science of Eugenics

At the turn of the twentieth century, American scientists embraced a theory known as eugenics, which held that intelligence and other personality traits are genetically determined and therefore inherited. This scientific theory about heredity, coupled with Progressive Era reforms, produced a campaign to remedy America's social problems by stemming biological degeneracy. The eugenicists advocated the rational control of reproduction in order to improve society's mental, moral, and physical health through selective breeding. In reality, eugenics enforced social judgments about race, class, and gender cloaked in scientific terms.

The founder of eugenics, English scientist Sir Francis Galton (Darwin's half cousin), became interested in heredity after reading *The Origin of Species*.[29] Although the idea of improving the quality of humans, as well as plants and animals, through selective breeding was not new, Galton was the first to popularize an actual eugenics program. In 1883, Galton coined the word *eugenics*—from a Greek root meaning "good in birth"—to "express the science of improving stock" by giving "the more suitable races or strains of blood a better chance of prevailing speedily over the less suitable than they otherwise would have had."[30] Galton replaced the Darwinian reliance on natural selection for the inevitable extinction of inferior groups with an argument for affirmative state intervention in the evolutionary process. "What Nature does blindly, slowly, and ruthlessly, man may do providently, quickly, and kindly," he proposed.[31]

Galton's theories were grounded in a belief in the genetic distinctions among races as well as individuals. His eugenicist approach adopted the prevailing scientific model that man was divided into different races marked by distinctive physical and psychological features: "The Mongolians, Jews, Negroes, Gipsies, and American Indians severally propagate their kinds; and each kind differs in character and intellect, as well as in colour and shape, from the other four."[32] Like most scientists of the time, Galton reserved the most disparaging description for black people, emphasizing their "strong impulsive passions, and neither patience, reticence, nor dignity,"[33] as well as their intellectual inferiority. "The number among the negroes of those whom we should call half-witted men, is very large," Galton wrote in his 1892 treatise, *Hereditary Genius*.[34]

Galton's eugenicist ideas found fertile ground in America. In the early 1900s, the descendants of northern European settlers were consolidating their hold on commercial and political power in America. They sought to maintain control over an exploited workforce of Southern black sharecroppers and urban factory workers from southern and eastern Europe. Believing that the new immigrants were reproducing faster than they were, native Anglo-Saxons became gripped by a fear of "race suicide." Now, far from believing that blacks and other inferior races were biologically destined to become extinct, whites feared their growing numbers.

Whites were also pathologically obsessed with preserving their racial purity, claiming that sexual relations with blacks would produce a degenerate mongrel race. An intense nativism led to vicious race riots across the country. These attacks, primarily of whites against blacks and natives against immigrants, often ended in dozens of deaths. Thirty-eight people were killed in a race riot in Chicago in 1919. Meanwhile, gruesome lynchings terrorized black citizens in the South.

Eugenics provided a forward-thinking scientific framework to justify whites' efforts to preserve social order in the midst of this turmoil. Concepts about race, in turn, provided a theoretical framework for eugenicist logic. European and American scientists had for over two centuries developed an understanding of races as biologically distinct groups, marked by inherited attributes of inferiority and superiority. Racial science predisposed Americans to accept the theory that social characteristics were heritable and social structure was biologically determined. Eugenicists easily latched on to race as an integral element of their ideology. Moreover, the chattel slavery and Jim Crow systems that violently enforced racial classifications paved the way for the dehumanizing programs that implemented eugenicist ideology. Forced sterilizations, eugenicists' favorite remedy for social problems, were an extension of the brutality inflicted on black Americans. Slaveholders' total dominion over the bodies of enslaved Africans—including ownership of enslaved women's wombs, which they exploited for profit—provided an early model of reproductive control.[35]

Imposing sterilization as a solution for antisocial behavior had long been practiced in the castration of black men as punishment for rape, beginning during slavery. The territorial legislature of Kansas passed a law in 1855 making castration the penalty for any Negro or mulatto who was convicted of rape, attempted rape, or kidnapping of any white woman.[36] In 1913, the Kansas legislature authorized sterilization of state inmates. Around that time, a Texas physician, Dr. Gideon Lincecum, advocated castration as a deterrent to crime by using an anecdote about a "vicious, disobedient, drunken Negro" who was suspected of rape: "After discovering that he had impregnated an idiot white girl, three men went into the field where he worked and castrated him. Less than two years later I heard his mistress say that he had become a model servant."[37] The mutilation of lynching victims' bodies by white mobs, often involving castration of black men, served as entertainment for many residents of Southern states. This long history of sanctioned torture inflicted on black bodies prepared Americans to accept the compelled sexual surgeries performed by state officials on people deemed genetically unfit.

Eugenics was mainstream, and it was financed by the nation's wealthiest entrepreneurs, including the Carnegie, Harriman, and Kellogg dynasties. Johns Hopkins biologist Raymond Pearl observed that, by 1912, "eugenics was catching on to an extraordinary degree with radical and conservative alike."[38] The growth of eugenics as legitimate science owes much to the efforts of Harvard-trained biologist Charles Davenport. As an associate professor at the University of Chicago, he convinced the Carnegie Institute to establish a center for the experimental investigation of eugenics in Cold Spring Harbor, New York, in 1904. With the financial backing of railroad heiress Mrs. E.H. Harriman, Davenport added a Eugenics Record Office to his research station six years later. In an October 10, 1910, letter, Davenport reported to Francis Galton, "There has been started here a Record Office in Eugenics; so you see that the seed sown by you is still sprouting in distant countries."[39]

Darwinian biology and Mendelian genetics, which understood heredity as a matter of quantitative probability, had generated a concept of race as a geographical population rather than a human essence. Eighteenth-century racial typologists grouped human beings into natural kinds that could be distinguished by observing their outward traits. Scientists at the turn of the twentieth century modernized the concept of biological race by defining it as groupings of geographical populations that differed according to their likelihood of inheriting certain physical and behavioral traits. Eugenicists used race to group European populations according to ability, intelligence, and moral character. In the early 1900s, southern and eastern Europeans were considered a racial threat by whites of Anglo Saxon origin. Dressing up racial typologies in the new scientific garb of heredity, Davenport assigned particular behavioral traits to different races of immigrants. He observed that Poles were "independent and self-reliant though clannish"; Italians were prone to commit "crimes of personal violence"; and "Hebrews" fell "intermediate between the slovenly Serbians and Greeks and the tidy Swedes, Germans, and Bohemians."[40] Davenport advocated preventing the reproduction of bad stock through a selective immigration policy, discriminating marriages, and state-compelled sterilizations.

Davenport's Cold Spring Harbor project supplied the burgeoning American eugenics movement with scientific support. It trained and dispersed over 250 field workers, published the *Eugenical News*, and disseminated bulletins and books about the reduction of hereditary degeneracy.[41] Davenport reported his early findings in his widely read book *Heredity in Relation to Eugenics* (1911). As Davenport conducted scientific research, eugenics became the vogue across the county. Ordinary Americans attended lectures, read articles in popular magazines, and participated in Better Babies contests. Those devoted to propagating eugenics joined organizations such as the American Eugenics Society, the American Genetics Association, and the Human Betterment Association. The *Reader's Guide to Periodical Literature* listed 122 articles under Eugenics between 1910 and 1915, making it one of the most referenced topics in the index.[42] At most American colleges, courses on eugenics were well attended by students eager to learn how to apply biology to human affairs.

Meanwhile, Harry Hamilton Laughlin, superintendent of the Eugenics Record Office and an active public lobbyist for the movement, turned biological theory into public policy. In 1914, he prepared a two-volume report that proposed a schedule for sterilizing 15 million people over the next two generations, as well as a model sterilization law to accomplish this plan.[43] The defective "10 percent of our population," Laughlin claimed, "are an economic and moral burden on the 90 percent and a constant source of danger to the national and racial life." It is estimated that 65,000 persons were involuntarily sterilized under eugenicist statutes passed in thirty states.[44]

Like the racial typologists of prior centuries, eugenicists developed scientific methods for testing their claims. Paralleling the development of eugenic theory was the acceptance of intelligence as the primary indicator of human value. Eugenicists claimed that the IQ test could quantify innate intellectual ability in a single measurement, despite the objections of its creator, Alfred Binet.[45] The introduction of "mental tests" at the turn of the century to measure intelligence replaced physical measurements, such as cranial capacity, as the means of determining human inferiority and superiority. Intelligence testing was the ideal instrument for distinguishing the fitness of stocks because it provided "a seemingly objective, quantifiable measure that could be used to rank genetically transmitted ability," writes Rutgers psychology professor William H. Tucker.[46] Psychologists used the tests to demonstrate that blacks and recent immigrants from southern and eastern Europe were intellectually inferior to Americans of Anglo Saxon or Scandinavian descent.

Eugenic science went hand in hand with Jim Crow laws, official disenfranchisement of blacks, and the entrenchment of formal racial segregation—all Progressive reforms intended to strengthen the social order.[47] On a single day in March 1924, the Virginia legislature passed two laws that jointly promoted the state's eugenicist and racist agendas. The first, "An Act to provide for the sexual sterilization of inmates of state institutions in certain cases," authorized the forced sterilization of people confined to government asylums because they were deemed to be "feebleminded." The second half of Virginia's eugenicist scheme was the Racial Integrity Act, which banned interracial marriage. Both focused on controlling reproduction as a means of maintaining the racial order, and both acts would eventually end up in the U.S. Supreme Court.

Six months after the passage of the sterilization statute, the Virginia Colony for Epileptics and Feebleminded approved the sterilization of a seventeen-year-old girl named Carrie Buck. Buck, the daughter of an allegedly feebleminded woman, was committed to the colony by her foster parents when she became pregnant as a result of rape by their nephew.[48] Her court-appointed guardian, in cooperation with the colony's superintendent, appealed the order as a test case challenging the new law. The case made its way to the U.S. Supreme Court. Noting that her sexual depravity was "a typical picture of the low-grade moron," Harry Laughlin testified in a deposition that Buck belonged to the "shiftless, ignorant, and worthless class of anti-social whites of the South." The colony also submitted testimony that her daughter, Vivian, was mentally below average. In its 1927 decision *Buck v. Bell*, the Supreme Court upheld the statute.[49] Rejecting arguments that the Virginia sterilization law violated Buck's equal protection and due process rights, Justice Oliver Wendell Holmes explained the state's interest in preemptively sterilizing people with hereditary defects: "It is better for all the world if, instead of waiting to execute degenerate offspring for crime, or let them starve for their imbecility, society can prevent those who are manifestly unfit from continuing their kind." Holmes, himself an ardent eugenicist, gave eugenic theory the imprimatur of constitutional law in his infamous declaration, "Three generations of imbeciles are enough." He imposed his social judgment condemning Carrie Buck for being poor and pregnant in a disguise of scientific authority.

The Racial Integrity Act required the racial classification of every person at birth and made marriage between whites and anyone with a trace of Negro ancestry a crime. It expressed eugenicists' worry that sexual intermingling between blacks and whites would deteriorate the white race. Over half the papers presented at the Second International Congress of Eugenics in 1921 concerned the biological and social harms caused by marriages between people from different racial groups. Walter Ashby Plecker, the Virginia registrar of vital statistics, was charged with maintaining racial integrity by zealously enforcing the antimiscegenation law. In a letter to Laughlin dated June 18, 1931, Plecker expressed his fears about the genetic contamination caused by intermarriage: "I would feel somewhat easier about the matter if I thought that these near-whites would not produce children with negroid characteristics. I have never felt justified in believing that in some instances the children of mulattoes are really white under Mendel's Law."[50] Plecker selectively rejected scientific tenets that conflicted with his beliefs about racial difference. In 1967, the Virginia statute—a lingering artifact of eugenics—would be struck down by the landmark Supreme Court decision *Loving v. Virginia*.

Laughlin's 1922 survey, *Analysis of America's Melting Pot*, studied the ethnic background of the institutionalized population in order to demonstrate that recent immigrants made up a disproportionate share of the nation's socially degenerate members. Laughlin's conclusion that "the recent immigrants (largely from southern and eastern Europe), as a whole, present a higher percentage of inborn socially inadequate qualities than do the older stocks," helped persuade Congress to pass a 1924 law severely restricting immigration.[51] The Johnson-Reed Act, which set numeric quotas for immigration from nations worldwide, not only introduced the concept of the illegal alien, who was denied any legal rights in the United States, but also defined this disqualification in racial terms.[52]

A year after Laughlin's survey was issued, a new edition of the popular book *The Passing of the Great Race*, by New York eugenicist Madison Grant, appeared. Grant, resident anthropologist of the American Museum of Natural History, similarly warned that the Nordic stock in America was threatened by racial intermixture with blacks and inferior immigrant groups, which inevitably produced children of the "lower" type. Portraying inferior stocks as public enemies, he described racial intermarriage as a "social and racial crime of the first magnitude."[53] *The Passing of the Great Race* was a bestseller, with four editions and numerous reprints published between 1916 and 1923. The *Saturday Evening Post* praised its reflection of "recent advances in the study of hereditary and other life sciences" and recommended it as a book that "every American should read."[54] Legislators quoted passages from the book during congressional debates on immigration restrictions, and President Theodore Roosevelt commended it as "the work of an American scholar and gentleman," stating that "all Americans should be immensely grateful to [Grant] for writing it."[55] Grant was regarded as an important scientist, while his detractors were labeled "Bolsheviks and Jews" who were biased against scholarly investigation of racial difference.

Grant intended readers to learn the central political lesson of eugenic science: that egalitarian social programs are incapable of improving society. As Ellsworth Huntington concluded in his commentary in the *Yale Review*, Grant demonstrated a "lesson of biology . . . that America is seriously endangering her future by making fetishes of equality, democracy, and universal education."[56] Eugenicists' motto was "Nature knows no equality."[57] These scientists argued that inequality is natural: the unequal conditions in which people live are dictated by nature and trying to change them is folly. So, too, 1950s segregationists described integration as a political ideology that disrupted the natural separation of races and threatened white citizens' freedoms. "Integration represents darkness, regimentation, totalitarianism, communism and destruction," declared Robert "Tut" Patterson, founder of the Citizens Council Association, an organization of white businessmen dedicated to preserving white supremacy.[58] Nobel Prize–winning scientist James Watson expressed this sentiment when he told a London newspaper in 2007 that he was "inherently gloomy about the prospect of Africa" because "all our social policies are based on the fact that their intelligence is the same as ours—whereas all the testing says not really."[59]

Watson's remarks were considerably less socially acceptable in 2007 than they would have been in 1925. Shortly after his comments were made public, Watson was asked to resign his position as chancellor of Cold Spring Harbor Laboratory—the same laboratory that had once housed the Eugenics Record Office. By the 1940s, eugenics had been discredited both as bad science and as an excuse for racial hatred. American eugenicists who had initially supported the German sterilization law were shamed by its connection to the Nazi Holocaust.[60] After World War II, a central project of racial scientists became to explain why their investigations of intrinsic racial difference had nothing to do with racism.

The End of Racial Science?

The Carnegie Institute rescinded its support for eugenic studies at Cold Spring Harbor in 1939, and Harry Laughlin resigned as secretary of the Eugenics Record Office in 1941, marking the end of eugenics as an official social program in most of the United States. Beyond U.S. eugenics, the horrific outcome of Nazi racial science also generated a powerful challenge to the belief in the biological basis of race. Renowned scholars such as Columbia anthropologist Franz Boas and his prestigious cast of students, including Margaret Mead, Otto Klineberg, Ruth Benedict, Ashley Montagu, and Melville Herskovits, had demonstrated scientific errors in eugenicists' theories about inherited traits. Raised in a German Jewish home, Boas began teaching at Columbia University in 1896. Within three decades, he and his students had established anthropology as a respected discipline focused on studying culture instead of race.

On July 18, 1950, the *New York Times* ran a front-page story, "No Scientific Basis for Race Bias Found by World Panel of Experts."[61] That day, the United Nations Educational, Scientific, and Cultural Organization (UNESCO) issued a landmark "Statement on Race," drafted by an international committee of experts, declaring that race "is not so much a biological phenomenon as a social myth." The committee emphasized that human populations share most of their traits, meaning that "the likenesses among men are far greater than their differences." Rejecting the Nazi idea that some races are superior to others, the UNESCO statement concluded that biological science demonstrated "the unity of mankind."

The UNESCO committee included scholars from multiple disciplines who were well known for their classic books contesting traditional views on race: Columbia University anthropologist Ashley Montagu, author of *Man's Most Dangerous Myth: The Fallacy of Race*; black sociologist E. Franklin Frazier, author of *Black Bourgeoisie* and *The Negro in the United States*; English humanist Aldous Huxley, author of *Brave New World*; Swedish economist Gunnar Myrdal, author of *An American Dilemma: The Negro Problem and Modern Democracy*; and geneticists Theodosius Dobzhansky and Leslie Dunn, whose 1946 book *Heredity, Race and Society* exposed the genetic fallacies underlying racial classifications.[62] In the 1930s, Dobzhansky and Dunn had joined five other Columbia professors to found the University Federation of Democracy and Intellectual Freedom, dedicated to ridding science of Nazi ideas about race.

The first UNESCO statement was met with a firestorm of criticism from physical anthropologists and geneticists who accused the drafters of allowing their political agenda to distort their scientific claims about race.[63] The following year, a modified UNESCO committee—this time staffed entirely by physical anthropologists and geneticists—issued a revised "Statement on the Nature of Race and Race Differences." Although the second committee unanimously rejected "the racialist position regarding purity of the races and the hierarchy of inferior and superior races to which this leads," it also made clear that scientists could continue to study innate differences between races without being racists. While the original committee insisted that there were no racial differences in mental traits, the 1951 statement conceded that "it is possible, though not proved, that some types of innate capacity for intellectual and emotional response are commoner in one human group than another." In addition, the 1951 statement repudiated the link the first committee made between biological theories of race and social inequality: "We wish to emphasize that equality of opportunity and equality in law in no way depend, as ethical principles, upon the assertion that human beings are in fact equal in endowment." This distinction enables scientists to research inherent racial inequalities while avowing allegiance to the legal and moral principles of human equality. Both UNESCO statements disclaimed the practice of ranking races. But neither document abandoned the concept of biological race altogether. Instead, both statements took issue with race as an *ideological* doctrine of inferiority that was responsible for deadly social conflicts. They distinguished the Nazis' ideological use of race for repressive purposes from the scientific use of race for legitimate research.[64]

At the same time UNESCO issued its statements on race, the NAACP was waging a legal campaign to overturn segregationist state policies. The federal government, now engaged in international competition with the Soviet Union over the spread of communism, felt pressure to reduce blatant racial discrimination in order to gain allegiance from developing nations breaking free from colonial rulers.[65] In the 1954 landmark case *Brown v. Board of Education*, a unanimous U.S. Supreme Court held that public school segregation violated the Constitution because "separate educational facilities are inherently unequal." In rejecting the long-held separate-but-equal doctrine, the Court relied heavily on empirical data collected by the NAACP's expert, Kenneth Clark, showing that segregation caused psychological harm to black children. Clark's heartbreaking experiments that showed black children a white doll and a black doll and asked them, "Which is the nice doll?" no doubt ranks with Newton's apple as one of the best-known scientific studies.

Clark's use of science in the service of racial equality and the gains in black freedom of the 1960s were met by a political backlash that claimed scientific legitimacy. Carleton Putnam, longtime chairman of the board of Delta Airlines, published *Race and Reason: A Yankee View* in 1961, opposing the *Brown* decision on both anthropological and ideological grounds, with a fair share of hysteria about the debilitating effects on white civilization of mixing with inferior Negro genes. Endorsed by a lead review in the Books for Lawyers section of the American Bar Association Journal, *Race and Reason* kept open the debate about the scientific basis for integration accepted by the Supreme Court.[66] The following year, *The Origin of Races*, by Carleton Coon, a University of Pennsylvania professor and the president of the American Association of Physical Anthropologists, claimed once more to provide definitive proof that human races had separate origins. Coon identified five major races in hominid fossils and, based on cranial capacity, theorized that each had evolved toward modern *Homo sapiens* separately and at different rates—Caucasoids (Europeans) and Mongoloids (Asians) advanced the fastest, while Congoids (Africans), Capoids (the Bushmen of South Africa), and Australoids (Australian aborigines and nearby island populations) fell behind. "The length of time a subspecies has been in the sapiens state," Coon postulated, determined "the levels of civilization attained by some of its populations."[67] (With today's more advanced knowledge of human ancestry, this theory would position Africans as the superior subspecies.)

Coon fought an academic battle opposing the revolution in anthropology instigated by the likes of Boas and Montagu. His theory on the origin of races was circulated widely by the media and provided fodder for Putnam and other segregationists. Nonetheless, when the physical anthropologists called a special meeting at their convention in 1961 to censure Putnam's book, Coon resigned his post as president.

Later, Berkeley psychologist Arthur Jensen, Harvard psychologist Richard Herrnstein, and Nobel laureate William Shockley, a Stanford physicist, argued that since genes determined intelligence and intelligence determined social achievement, racial inequality resulted from blacks' cognitive inferiority. The combination of black people's lower IQ scores and higher birth rates, they warned, rendered public programs not only futile for improving blacks' socioeconomic status but a threat to the nation's welfare. "Is there a danger that current welfare policies, unaided by eugenical foresight, could lead to the genetic enslavement of a substantial segment of our total population?" Jensen asked in his 1969 *Harvard Educational Review* article, "How Much Can We Boost IQ and Scholastic Achievement?" Twenty-five years later, Herrnstein and Charles Murray of the American Enterprise Institute rehashed the claim that social disparities stem from the higher fertility rates of genetically less intelligent groups in their controversial bestseller *The Bell Curve*.

Despite some scientists' use of genetic science to promote segregationist ideals, the horrors of Nazi science, the civil rights movement, and the emerging science of population genetics all raised hopes for a scientific consensus rejecting the typological meaning of race. In 1964, *The Concept of Race*, edited by Ashley Montagu, gave a platform for ten of the nation's preeminent biologists and anthropologists to propose replacing scientifically useless racial taxonomies with a more accurate account of human variation. One of the contributions, "On the Nonexistence of Human Races," by University of Michigan anthropologist Frank B. Livingstone, optimistically predicted:

> Just as Galileo's measurements and experiments paved the way for Newton's laws of motion, which totally replaced the Aristotelian laws of motion concerned as they were with describing the nature of bodies and their "essences," our newer genetic knowledge and the measure-

ment of gene frequencies will replace the studies on the nature or essence of race and the mathematical theory of population genetics will replace the Linnaean system of nomenclature.[68]

Although Livingstone adhered to the nonexistence of human races, many biologists and anthropologists were interested in using population genetics to study race, not invalidate it. These scientists "merely believed that race needed to be reformed and refined in the wake of abuses by the eugenicists, white supremacists, and Nazis," observes sociologist Jenny Reardon.[69] By demonstrating the utility of population analysis to understanding how races were formed, leading evolutionary biologists convinced their peers to accept population genetics as the discipline's cutting-edge method. At a 1950 gathering of 129 scientists in Cold Spring Harbor, Joseph Birdsell, a physical anthropologist at UCLA, argued that "problems of human evolution and racial differentiation are essentially population problems. And their solutions will be advanced by borrowing techniques of analysis from the vigorous field of population genetics."[70] A growing view within biology, anthropology, and genetics was that the old racial categories derived from externally visible traits could be upgraded with scientific research on genetic differences that lay beneath the skin.

For many scientists, then, the emerging civil rights ethos did not make racial science untenable. Rather, it made it imperative for scientists to detach their study of biological race from societal racism. It was not the blatantly racist theorizing of Carleton Coon and his ilk that left its mark on today's racial science. Scientists who openly spew racial defamation risk being swiftly ushered out of the limelight, like James Watson. The imaginary wall more liberal-minded scientists erected, separating racial science from racial politics, became critical to the scientific validation of race in a post–civil rights culture that espouses racial equality.

Nevertheless, it is important to note that eugenics never really went away. While mainstream geneticists increasingly condemned what they referred to as "eugenics' excesses," North Carolina's eugenic sterilization program actually *expanded* after World War II and continued until 1974.[71] An investigation by the *Winston-Salem Journal* in the 1990s uncovered records documenting decisions by the Eugenics Board of North Carolina, a panel of five bureaucrats who enforced the state's 1929 eugenics law (revised in 1933) authorizing compulsory sterilization for epilepsy, disability, and feeblemindedness. The program gained momentum when the North Carolina Human Betterment League, formed in 1947 by hosiery magnate James G. Hanes and others in Winston-Salem's elite, joined with Dr. Clarence Gamble, Bostonian heir to the Proctor & Gamble fortune, and local members of the medical community to launch a massive publicity campaign promoting sterilization to improve the state's mental fitness.

Scientists at Wake Forest University School of Medicine also played a key role. Dr. C. Nash Herndon, chair of the department of medical genetics at the university's medical school, helped to promote the state's involuntary sterilization program by serving as president of the Human Betterment League and issuing academic reports advocating eugenics. He also conducted his own eugenic studies with funding from Wickliffe Draper, a wealthy segregationist. Draper was so committed to the merger of Jim Crow and eugenics that he traveled to Germany in 1935 to attend a Nazi eugenics conference. After supporting Herndon's research with two grants in the 1950s, Draper donated $100,000 to the school, solicited by the medical school dean, on the condition that the school promised not to advocate interracial marriage and consider teaching students about therapeutic sterilization. Other faculty members performed involuntary sterilizations on patients at state hospitals, with and without authorization from the Eugenics Board.[72]

In all, the program counted close to eight thousand victims. During the eugenics era, a majority of those sterilized were white. But in the program's final decade, the target shifted to poor black women. Social workers supervising welfare caseloads were empowered to petition the Eugenics Board for orders sterilizing their clients. The board approved 95 percent of the petitions it received, acting largely out of social prejudice against the women it ordered to be sterilized. In 1965, a social worker told Nial Cox, a pregnant teenager living at home, that her family would lose government aid unless she was sterilized. A doctor performed the surgery because she was black, on welfare, and had had a baby out of wedlock. The program ordered sterilizations of two thousand children, including a ten-year-old boy who was castrated.

Governor Mike Easley and Dr. William Applegate, dean of Wake Forest's medical school, apologized after the *Journal*'s revelations in 2003. But eugenicist philosophy continues to make its way into state and federal reproductive, welfare, and immigration policies.[73] Republican congressman Mark

Kirk of Illinois (elected in 2010 to fill the Senate seat vacated by Barack Obama) stated on the House floor during deliberations over a 2007 foreign family-planning provision that slowing the rate of growth of Mexico's population would "reduce the long-term illegal immigration pressure on America's borders."[74] Like eugenicists of old, Kirk argued for fixing a social problem by reducing the fertility rates of people believed to embody it. In reality, birth rates in Mexico have dropped over the last twenty-five years to 2.5 children per couple, very close to the 2.1 birth rate in the United States. But what really matters was the lawmaker's message: America can solve its immigration crisis by decreasing births of undocumented immigrants.

The Final Chapter?

The Human Genome Project was supposed to finally put an end to the biological definition of race. The idea of creating a map locating all the genes in the human body began circulating as early as 1985. By October 1990, the Human Genome Project was official. The National Institutes of Health (NIH) and the Department of Energy authorized $3 billion to be spent over fifteen years on an international venture involving Great Britain, France, Japan, China, and Germany to sequence the entire human genome. The U.S. government's research was headed by a fitting national figure—the ubiquitous James Watson, the geneticist who won a Nobel Prize for discovering DNA's double helix structure, who was then serving as director of the National Human Genome Research Institute. When Watson abruptly resigned his post in 1993, Francis Collins, an accomplished physician and molecular geneticist, accepted President Bill Clinton's invitation to assume leadership of the research institute and the genome project. The genome initiative became a race to the finish in 1998 when Craig Venter, who had defected from a government post to become chief scientist at the private firm Celera Genomics, threw down the gauntlet by promising to complete the genome map and release it to the public in three years.

To understand what it means to map the entire human genome, it is helpful to start with the biological unit being mapped: our genes. Genes are conceived as segments of a six-foot-long molecule called deoxyribonucleic acid, or DNA, that encodes instructions for making proteins. Each nucleus of the trillions of cells in the human body contains two copies of twenty-three

DNA strands, one from each parent, composing forty-six chromosomes. Although each cell has a complete set of DNA, only the portions that are important to the cell's particular function are activated.

How does one gene differ from another? DNA is composed of four kinds of units called nucleotides—adenine, cystosine, guanine, and thymine (A,C,G, and T)—that are strung together like beads on a necklace in different combinations. The sequence of these nucleotides functions like a code using four different letters that are arranged in an infinite variety of spellings. The cell uses part of the DNA code to help determine what sequence of amino acids to build into a protein and another part to help determine when and where to turn on or off a protein. Each person's genome—the complete set of DNA—is distinguished from someone else's by the precise chemical composition of the DNA making up their genes. The Human Genome Project identified the 20,000 to 25,000 genes in human DNA, sequenced the 3 billion base pairs that make up human DNA, and stored all this genetic information in databases.

On June 26, 2000, President Bill Clinton, flanked by Francis Collins and Craig Venter, marched into the East Room of the White House to unveil a working draft of the genomic survey. "Without a doubt, this is the most important, most wondrous map ever produced by humankind," Clinton proclaimed. The genome project was cheered as the Holy Grail that would disclose all the secrets of human life. It was supposed to reveal what it means to be human. Perhaps the most pressing question to be answered was, Is race real? During the 1990s, debates on race had once again permeated scientific circles. Every major scientific journal featured research studies, special issues, editorials, and commentaries on the scientific validity of race. What would mapping every human gene tell us about the biological basis for racial categories? Would scientists discover that race is written in our DNA?

That all three of the chief figures at the White House ceremony pronounced on the issue of race shows the intensity of the angst surrounding these questions. Their conclusion was unanimous: the Human Genome Project revealed that the human species cannot be divided into biological races. President Clinton famously announced, "I believe one of the great truths to emerge from this triumphant expedition inside the human genome is that in genetic terms, all human beings, regardless of race, are more than 99.9 percent the same."[75] Collins ended his remarks by saying, "I'm happy that today the only race that we are talking about is the human race." Venter reported that Celera Genomics had sequenced the genomes of three women and two men who identified as Hispanic, Asian, Caucasian, and African American and found that "there's no way to tell one ethnicity from another." He bluntly declared, "Race has no genetic or scientific basis." The Human Genome Project confirmed what many leading biologists and anthropologists had concluded several decades earlier: race is a social and not a biological category. The project validated a shared humanity that many people had long known without the need for genetic evidence.

Clinton's announcement that human beings, regardless of race, are 99.9 percent genetically identical suggests an obvious question. Can't the 0.1 percent of genetic makeup that people do not share encompass important racial differences? "After all, dogs and wolves are nearly identical at the genetic level, but the difference between a dog and a wolf is huge," a man pointed out at a conference I attended not long afterward.

This 99.9 percent sound bite, commonly deployed to debunk the biological meaning of race, is not enough to prove that race is not biological. Human beings share a majority of their estimated 22,000 genes with other mammals. As fellow primates, chimpanzees are extremely close to humans: we share 98.7 percent of our genes. Humans are also 90 percent genetically the same as mice. So bearing genetic similarity does not mean there are no important differences. With more than 3 billion nucleotide pairs in the human genome, even a 0.1 percent difference can be consequential. Still, as species go, *Homo sapiens* stand out as remarkably homogeneous. There is less genetic variation in the entire human race than in a typical wild population of chimpanzees. This means that what matters to understanding human genetic variation is not only the amount of genetic difference but what the differences are and how they are expressed and distributed.

While it is true that the 0.1 percent of human genetic difference is meaningful, it does not mean that human genetic difference is organized by race. The genetic variation found in the human species is not grouped in discrete, genetically distinct units scientists can identify as races. To the contrary, human genetic diversity occurs in a continuum that cannot be partitioned by clear boundaries and that crosses what are commonly considered racial lines. Consider how, in the United States, people with any amount of African ancestry are grouped together in one race. Yet the greatest amount of genetic diversity in the world exists in sub-Saharan Africa. *Homo sapiens* originated in Africa and remained there for almost 200,000 years before spreading relatively recently throughout the rest of the globe, about 70,000 or 80,000 years ago. African populations vary the most because most of human genetic diversity evolved in Africa, and groups living there had more time to accumulate genetic differences. The people who migrated from Africa and dispersed throughout the globe carried in their genomes only a portion of variants found in the ancestral inhabitants. "From a genetic perspective," writes anthropologist Deborah Bolnick, "non-Africans are essentially a subset of Africans."[76]

In fact, the entire range of human variation for some genetic traits can be found on the African continent.[77] A person from the Congo, a person from South Africa, and a person from Ethiopia are more genetically different from each other than from a person from France.[78] This seems astonishing because we are so used to focusing on a tiny set of physical features, especially skin color, to assign people to racial categories. It turns out that the genes contributing to these phenotypic differences represent a minute and relatively insignificant fraction of our genotypes and do not reflect the total picture of genetic variation among groups.[79] What's more, these phenotypic differences do not even fall neatly into the categories known as races. Rather, the physical features are "discordant" among groups—they are assorted randomly and do not come assembled in racial packages. "Sub-Saharan Africa is home to both the tallest (Maasai) and the shortest (pygmies) people, and dark skin is found in all equatorial populations, not just in the 'Black race' as defined in the United States," writes Richard S. Cooper, a physician epidemiologist at Loyola University.[80] And most genetic variation is found *within* any human population.[81]

How can it be that there is more genetic difference among people within each race than between races? Imagine a white congregation worshipping one Sunday morning at a Lutheran church while a black congregation worships at a Baptist church across town. Our racial habits might lead us at first to view everyone within each church as alike and the two congregations as very different. But think more carefully: each and every person at the Lutheran service is unique and genetically different from every other congregant, especially if they come from different families. Any two unrelated people in the pews have millions of genetic differences, contributing to (though not determining) their looks, personalities, and health.[82] The same is true for every single person at the Baptist church. And those millions of genetic differences that set each parishioner apart as a unique human being, whether black or white, are greater than the differences that separate the black and white congregations. The more significant differences between the two churches—in overall health and wealth, in the neighborhood where each is located, in the songs sung during service—stem from social, not biological, causes.

By the late 1990s, many American biologists, led by Harvard evolutionary geneticist Richard Lewontin, agreed that "no justification can be offered" for continuing the biological concept of race.[83] The American Anthropological Association endorsed this view in a 1997 statement: "genetic data also show that, no matter how racial groups are defined, two people from the same racial group are about as different from each other as two people from any two different racial groups."[84] But this pronouncement came under attack when new genomic technologies permitted analysis of gene clustering on a global scale. In 2003, British statistician and evolutionary biologist A.W.F. Edwards faulted Lewontin for basing his conclusions on simple comparisons of individual genes rather than a more complex structure of correlated gene frequencies. While Lewontin compared a relatively small number of specific points on the DNA molecule, or loci, in different populations, Edwards recommended looking at small frequency differences across many loci to detect population structure. Lewontin, Edwards charged, had mounted "an unjustified assault on human classification, which he deplored for social reasons."[85]

Edwards was proposing that geneticists should examine the question of human genetic difference at the level of gene clusters instead of in terms of individual genes. But Edwards did not refute Lewontin's main claim: that there is more genetic variation within populations than between them, especially when it comes to races. Lewontin did not ignore biology to support his social ideology; he was not making an ideological claim. To the contrary, he argued that there is no biological support for the ideological project of racial classification.

Another reason why human genetic difference defies racial classification is that there are no genetic boundary lines that mark off a handful of large, discrete groups. The genetic differences that exist among populations are characterized by gradual changes across geographic regions, not sharp, categorical distinctions.[86] Groups of people across the globe have varying frequencies of polymorphic genes, which are genes with any of several differing nucleotide sequences. Molecular biologists can use the frequency of alternate alleles, or versions of genes, to trace the geographic ancestry of populations. But they can't draw a line where one set of gene frequencies stops and another begins. Furthermore, these differences are a matter of frequencies of certain gene mutations, not the absence or presence of genotypes.[87] There is no such thing as a set of genes that belongs exclusively to one group and not another. The *clinal*, or gradually changing, nature of geographical genetic difference is complicated further by the migration and mixing that human groups have engaged in since prehistoric times. Race collapses infinite diversity into a few discrete categories that in reality cannot be demarcated genetically.

Did the portrait of genetic unity and diversity unveiled by the Human Genome Project finally put an end to the biological concept of race once and for all? Would genomic science usher in an understanding of human variation devoid of antiquated racial classifications? The story I tell next reveals just the opposite: scientists are using advanced genomic theories and technologies to create a new racial science that claims to divide the human species into natural groups without the taint of racism.

PART II

The New Racial Science

3

Redefining Race in Genetic Terms

At a contentious 2008 meeting at the National Human Genome Research Institute (NHGRI) in Rockville, Maryland, forty scientists and bioethicists debated the best way to talk about the flood of genetic variation research emanating from laboratories across the country and the world. The discussion soon focused on how to handle the pesky term *race*. According to *Science* reporter Constance Holden, "everyone at the meeting agreed on the need for non-'fraught' terminology—'geographic ancestry,' for example, instead of 'race.'"[1] Rejecting race as a valid genetic category creates a vacuum in the scientific vocabulary: how will scientists describe genetic differences between human populations without resorting to the racial categories they have used for centuries?[2]

This reconfiguration of race for the genomic age hinges on applying two key concepts to genetic information: statistical probability and geographic ancestry. With the advent of worldwide genomic population studies, many scientists are using statistical estimates of gene frequencies that differ among geographic populations as a more objective, scientific, and politically palatable alternative to race. Instead of grouping people by race for purposes of scientific studies, why not group them by statistical genomic similarities? A second, related strategy turns to geographic ancestry. But these approaches, as we shall see, tend to merely repackage race as a genetic category rather than replace it.

Statistical Race

Some genomics researchers see race as a statistical grouping based on genetic similarity. Because genes are inherited, biologically related individuals are more likely than unrelated individuals to share genetic variants. A growing branch of population genomics treats race basically as a large family with the same "very distant relatives."[3] According to this theory, individuals belonging to the same race share more of their recent ancestry and therefore are more genetically similar to each other than to those of other races. It is important to note, however, that population biologists using a statistical approach do not actually trace the ancestry of particular individuals to place them in racial categories. Rather, they infer groupings from the statistical frequencies of particular DNA sequences sampled from distinct populations around the globe. Unlike racial typologists who classified people into natural kinds based on outward appearance, modern-day racial scientists classify people according to statistical probabilities based on huge genetic data sets.[4]

But does the statistical race concept really prove the biological nature of race any better than existing racial typologies? How can it when the entire enterprise from beginning to end—identifying populations to enter into data sets, determining which and how many genetic clusters matter, and applying the findings to our everyday lives—inescapably depends on preconceived notions of race? Genomic scientists have not discovered race in our genomes. They are taking already accepted racial categories and telling us a new way, based on computer-generated genetic differences, to verify them scientifically.

To understand how modern genomics reproduces traditional ideas of race, it is helpful to take a closer look at the assumptions behind some particularly influential scientific studies. First, another science lesson. Population genomics leverages the tiny percentage of genetic variation in the human species to identify differences in the frequency of certain alleles, or versions of genes, among groups with different geographic origins. To understand this process, we have to start with SNPs and microsatellites. Single nucleotide polymorphisms, or SNPs (pronounced "snips"), are particular points where the genomes of different individuals vary by a single DNA base pair.[5] SNPs are caused by random mutations that are then passed on to offspring and disseminate slowly away from the group in which they originally occurred.[6]

In some cases, the SNPs produce phenotypical differences; in other cases, the differences are imperceptible. Either way, scientists can infer ancestral relationships from the frequency with which these alternate spellings occur in different groups. A similar principle is at work with short segments of DNA called microsatellites that have a repeating sequence of nucleotides that varies between individuals.

Because the scale of this research is so vast (the human genome includes over 3 billion base pairs), scientists use computer software to infer a population structure from genotype data using multiple loci sampled from a number of groups. With a popular software program known as Structure, a researcher indicates how many genomic clusters the data should be grouped into. The program then allocates the individuals whose DNA was sampled into the predetermined number of clusters based on their genetic similarity. For any given number fed into the program, explains anthropologist Deborah A. Bolnick, "Structure searches for the most probable way to divide the sampled individuals into that pre-defined number of clusters based on their genotype." It uses a mathematically sophisticated algorithm to maximize the chances that the genotypes of individuals in the cluster will match.[7]

Researchers began attempting to group human populations based on their genes soon after the human genome was sequenced. A particularly high-profile project was led by Noah Rosenberg, a former high school math whiz who became a computational biologist at the University of Southern California. His team included genetic scientists from Stanford, Yale, and the University of Chicago, as well as institutes in Paris and Moscow. Unlike in previous studies, Rosenberg's team used computer software to detect clusters of genetic similarity in unidentified DNA sampled from people across the globe. Could the researchers scramble the groups' genetic signatures and put them back together with the aid of computer technology?

Rosenberg's team fed into the computer genetic information from 1,056 individuals representing fifty-two global populations. Then, following a specially designed algorithm, the computer went to work. In a landmark article, published in *Science* in 2002, the researchers announced that they had "identified six main genetic clusters, five of which correspond to major geographic regions." Although race was not mentioned, the "major geographic regions" that matched the genetic structure they discovered—Africa, Eurasia, East Asia, Oceania, and America—were quickly translated into traditional racial divisions. The researchers also concluded that "self-reported population ancestry likely provides a suitable proxy for genetic ancestry" when evaluating individuals for disease risk.[8]

Had high-tech genomic research really confirmed eighteenth-century racial typologies? Closer inspection of the Rosenberg team's findings reveals that they do not verify five classic racial groups at all. Instead, the study's overall results confirmed the basic rule of human genetic unity: within-group genetic variation is much greater than between-group variation. Genetic differences among people within the populations they studied accounted for 93 to 95 percent of all the genetic variation the computer uncovered. Only about 5 percent of the variation found existed between groups. In fact, the distinctions between populations were so minuscule that it took a highly advanced statistical computing program surveying many accumulated differences to make reliable guesses about the geographic origin of the people sampled.

What's more, the numbers of genetic clusters they identified were arbitrary. Although the researchers emphasized six main genetic clusters in reporting their results, they actually told the computer to analyze the DNA data set using a range of numbers, not just six. Their theory was that any statistically significant clusters reflected genetic divisions of the human species and thus the natural structure of human populations. But remember, the number of genetic clusters is dictated by the computer user, not the computer program. Their article presented the results of using two to six predetermined clusters. Rosenberg later revealed that his team also analyzed the data set using six to twenty clusters, "but did not publish those results because Structure identified multiple ways to divide the sampled individuals" when the number was larger than six.[9] The larger number of clusters identified by the study could just as easily have been highlighted to demonstrate the difficulty of dividing human beings into genetic races. There is nothing in the team's findings to suggest that six clusters represent human population structure better than ten, or fifteen, or twenty.[10]

Instructed to find two clusters, the computer divided human beings into groups anchored by Africa and by the Americas. This reflects the portrait of migration that evolutionary biologists have already painted: Native Americans traveled the most genetic distance from our original ancestors in Africa. When researchers told the computer to form five clusters, they were able to divide the human species into groups that matched the indigenous peoples sampled from five continents (Africa, Eurasia, East Asia, Oceania, and the

Americas). Adding another cluster separated out an additional group made up entirely of the Kalesh, a group in the mountains of Northern Pakistan who speak an Indo-European language and whose inhabitants claim to be descendants of Greek soldiers who invaded the Indian subcontinent with Alexander the Great in 327–323 B.C.[11]

Rosenberg's study was touted in media accounts, including a front-page story by the respected *New York Times* science writer Nicholas Wade, as having proved the biological reality of race. The researchers identified "five main human populations," which, in turn, "broadly correspond with popular notions of race," Wade wrote.[12] But the study actually showed that there are many ways to slice the expansive range of human genetic variation. In a 2005 article, Rosenberg and his colleagues acknowledged that the way a genomic study is designed determines what it says about human population structure.[13] Based on their analysis of how changing key variables influenced the outcomes, they reported that the number of loci, the sample size, the number of clusters, the geographic dispersion of the samples, and assumptions about allele-frequency correlations all had an effect on clustering. Although they reiterated their earlier finding that, with a large enough worldwide dataset, "individuals can be partitioned into genetic clusters that match major geographic subdivisions of the globe," they stated this finding "should not be taken as evidence of our support of any particular concept of 'biological race'" and agreed that genetic diversity also consists of clines—differences in allele frequencies that occur gradually across regions. The original Rosenberg study itself had contained the caveat that "genetic differences among human populations derive mainly from gradations in allele frequencies rather than from distinctive 'diagnostic' genotypes."[14]

Understanding human population structure in terms of discrete genetic clusters also misrepresents the path that produced diverse human populations that diverged from shared ancestors in Africa. "Ironically, by ignoring the way population history actually works as one process from a common origin rather than as a string of creation events, structure analysis that seems to present variation in Darwinian evolutionary terms is fundamentally non-Darwinian," Penn State anthropologists Kenneth Weiss and Brian Lambert point out.[15]

One population geneticist called genomics "the computer-assisted comprehensive study of all genes."[16] New genomic technologies dumped in researchers' laps a gigantic array of unorganized genetic data to sort through.

Like the natural world that Enlightenment biologists put in order, the information derived from the human genome beckons molecular biologists to catalog it. And like their predecessors, modern-day scientists are utilizing race to make sense of new genomic discoveries. "Race has rapidly become a prominent 'search tool,'" note the editors of the anthology *Revisiting Race in a Genomic Age*.[17] Today's scientists, however, claim that their focus on "genetic clusters" has removed the political aspects of race from their research. In response to my question asking why Rosenberg's article focused on the clusters that most closely matched the five major continents and therefore our historical ideas about race, a member of his research team told me that "people who share the same continental origin are genetically similar." He went on to explain, "Race has got all these loaded connotations, so we've dropped that term from our work."[18] What this scientist failed to acknowledge was that his own acceptance of racial categories may have influenced the decision to emphasize five genetic clusters, despite his team's attempt to expunge any explicit reference to race.

Is Geographic Ancestry the New Race?

The idea of replacing a typological notion of race with a geographic category traces back to the 1937 text *Genetics and the Origin of Species* by Theodosius Dobzhansky, the founder of population genetics and one of the signatories of the 1950 UNESCO statement on race. Dobzhansky proposed that evolutionary biologists adopt as their unit of analysis "geographical race," which he defined as "populations of species that differ in the frequencies of one or more genetic variants, gene alleles or chromosomal structures."[19] Dobzhansky's population approach was a redefinition of biological race, not a rejection of it. While contrasting his concept of natural populations with Carlton Coon's racist ideas in *The Origin of Races*, Dobzhansky still maintained that "most biological species are composed of races, and *Homo sapiens* is no exception."[20] But how much genetic difference is enough to create a race? Any genetic measure requires applying some a priori concept of race in order to package human genetic variation in a limited number of biological groupings. Informed by genetics, zoologists discarded race as a useful way to divide up animals within a species because so many races were distinguished by only one or two genes: genetic testing revealed that "two animals born in the same litter could belong to different 'races,'" notes Richard Lewontin.[21]

Some genomic scientists I interviewed advocated an absolute rejection of race as applied to the human species. Charmaine Royal, a Duke University geneticist who used to work at the National Human Genome Center at Howard, says she prefers to use ancestry in her research because race simply does not apply to human beings. "So what we are talking about, call them ancestral groups, call them ethnic groups, call them something else," she told me.[22] Her former colleague at Howard, Charles Rotimi, now the director of the NIH Center for Research on Genomics and Global Health, takes a similar tack. When I spoke with him about the center's research initiatives, I noticed that he had not used the word *race* during our entire conversation. "That actually is quite deliberate," he replied. "What I was describing has nothing to do with race. I don't use race because I know the people I study are not races; they are ethnic groups."[23]

Using *ancestry* can also be a way to acknowledge that individuals inherit traits from groups whose members share genetic similarities, while reserving *race* to designate a political category. "People are born with ancestry that comes from their parents but are assigned a race" is how Camara Jones, a research director at the Centers for Disease Control (CDC), explains it.[24] Ancestry is a far more accurate tool than race for describing human genotypes because it can reflect the true nature of individual and group heterogeneity. An individual can have ancestors from multiple geographic regions (as opposed to belonging to one race), and the regions can be defined in multiple ways, from small local areas to entire continents.[25] In terms of genetics, ancestry gives a better account than race, for example, of someone whose Irish-descended mother was born in Wichita, Kansas, and whose father came from the Luo group in Nyangoma-Kogelo, Kenya.

But while some scientists reject race altogether and others distinguish between (biological) ancestry and (social) race, an increasingly prominent trend is to redefine race *as* genetic ancestry. Concerned about how the Human Genome Project should deal with the subject of race, Robert Cook-Deegan, policy advisor to HGP director James Watson, wrote a letter to population geneticist Luca Cavalli-Sforza in 1989 asking him "what genetic research revealed about the reality of race and whether human genomics could lead to a new racism."[26] In a long and detailed reply, Cavalli-Sforza agreed that it was difficult to distinguish between groups based on genetic traits: "[W]hy classify races if the result is arbitrary and uncertain?" But he stopped short of refuting the existence of races at the genetic level or denying the validity of using the concept of race in genomic science. What he objected to were the "social" constructions of race popular with the "man on the street" based on visible physical traits, such as skin color and hair texture. Far from rejecting its scientific validity altogether, Cavalli-Sforza replaced the popular conception of race with a more scientific one—a "genetic definition of race" discovered at the molecular level using advanced scientific methods.

In 2002, genetic scientists Neil Risch and Esteban Burchard went further, erasing any distinction between race and ancestry.[27] Instead of distinguishing between a "man on the street" concept and a genetic concept of race, as Cavalli-Sforza had, they defined race "on the basis of the primary continent of origin." Ancestry, on the other hand, "refers to the race/ethnicity of an individual's ancestors, whatever the individual's current affiliation."[28] In other words, race is where one's ancestors come from, and ancestry is the race of one's ancestors. Their concession that "migrations have blurred the strict continental boundaries" did not dissuade them from associating race with continental origin, nor did their observation that Ethiopians and Somalis of East Africa, as well as North Africans, are "intermediate between sub-Saharan Africans and Caucasians" owing to their genetic resemblance to Caucasians. "The existence of such intermediate groups should not, however, overshadow the fact that the greatest genetic structure that exists in the human population occurs at the racial level," they stated, never explaining why we should simply ignore the blurred boundaries and intermediate groupings to uphold an equation between race and ancestry.[29] The definition of race in a 2010 article in the *Pharmacogenomics Journal* as "population clusters based on genetic differences due to evolutionary pressure" that is "often used to imply geographic or genetic ancestry" is increasingly common.[30] These scientists are treating *social* categories, determined by law, custom, and political affiliation, as if they are *biological* ones.

Studies that seek to discover natural groupings of human beings are only as informative as the populations they sample. For their part, Risch and Burchard relied on genetic material from two or three indigenous groups from each of the five continents, which together were supposed to represent the entire human race.[31] In another study, three sub-Saharan populations—two pygmy groups and the Lisongo—stood in for Africa; while Chinese, Japanese, and Cambodians represented East Asia; and Northern Europeans and Northern Italians were the Caucasians. Sampling a handful of ethnic groups to symbolize an entire continent mimics a basic tenet of racial thinking: that because races are composed of uniform individuals, any one can represent the whole group.[32] "Even our view of the Big Few might change were it not for our curious convenience of overlooking places such as India," wrote medical geneticist Rick Kittles and biological anthropologist Kenneth M. Weiss.[33] People from India don't fall neatly into an "Asian" genetic cluster. A 2003 study of fifty-eight DNA markers from many Indian populations, for example, traced their ancestral lineages to Africa, Central Asia, southern China, and Europe.[34]

This flawed sampling method is now built into the infrastructure of genomics research. A major initiative to document human genetic variation, the Human Genome Diversity Project (HGDP), relied on samples drawn from groups assumed to be geographically separate and isolated.[35] Based on the fear that many indigenous tribes were on the brink of extinction, HGDP scientists collected DNA from such groups around the globe to preserve this "precious genetic information" before it vanished. Led by Cavalli-Sforza, who launched the project in 1991, research teams descended on more than seven hundred indigenous communities worldwide to take blood from dozens of their members. In addition to "immortalizing" indigenous genes, the HGDP analyzed the genetic data to compare variation among indigenous groups in order to "facilitate studies of the genetic geography and history of our species," Cavalli-Sforza wrote.[36] The HGDP met an ignominious and unexpected end when tribal leaders accused the scientists of biocolonialism, for exploiting native genetic information in the same way that European colonizers had exploited their ancestors' natural resources.[37] The cell lines derived from the samples live on, however. Most of the research on human population structure, including the Rosenberg study cited above, used the HGDP samples for their data sets.

The relatively small number of indigenous populations sampled for the HGDP archives do not represent humankind's genetic diversity, nor do they paint an accurate portrait of the migrations and intermixing that contributed to most contemporary groups. Similarly, geographic areas with high levels of intermixture—North Africa, Spain, the Middle East, and the Balkans, for example—are rarely included in genomic studies.[38] Northern and eastern Africans are never selected to represent the continent because they do not fit the profile of "black" Africans—they have mixed too much with Europeans, Arabs, and other non-Africans. Even assuming the isolated indigenous groups sampled by the HGDP are genetically "pure," their unusual purity is all the more reason they cannot stand in for all the other populations of the world that are marked by intermixture from migration, commerce, and conquest.

A more accurate study of human genetic variation would use an objective sampling method. It would select the populations randomly and systematically across the globe, including those that reflect historic intermingling, instead of cherry-picking groups that best fit a priori racial classifications.[39] If researchers collected DNA samples continuously from region to region throughout the world, they would find it impossible to infer neat boundaries between large geographical groups that look like races.

When they do not attempt to fit findings into predetermined boxes, genomic population studies have discovered that (1) many of the individuals sampled fit in more than one cluster, and (2) the clusters nonetheless leave out whole groups of people who do not fit anywhere. Almost all of the Mozabites from Algeria, for example, belong to both Eurasian and African clusters. A 2008 *Science* study conducted by Cavalli-Sforza and Marcus Feldman analyzed a data set of 938 individuals from fifty-one populations at 650,000 common SNPs with an enhanced computer program called Frappe. Although the scientists were able to segregate the populations into five continental groups, their more significant findings challenged this simple breakdown, revealing mixtures among many groups such as Palestinians, Druze, and Bedouins, who have ancestral contributions from the Middle East, Europe, and South/Central Asia.

The expanded analysis also detected "finer substructures" when individual regions were examined separately. The East Asian populations divided into a "north–south genetic gradient," Europeans separated into eight populations, and those from the Middle East divided into four. Similarly, when Michael Bamshad of the University of Washington ran Structure using DNA from several continents, he found two separate sub-Saharan clusters: one consisted of the Mbuti, one of the indigenous pygmy groups in the Congo, plus three stray individuals; the other consisted of all sub-Saharan Africans *except* the Mbuti and the three other individuals.[40]

While the computer-generated findings from all of these studies offer greater insight into the genetic unity and diversity of the human species, as well as its ancient migratory history, none support dividing the species into discrete, genetically determined racial categories.[41] In 1994, at the outset of genetic-clustering research, Cavalli-Sforza predicted that classifying clusters as races would prove a "futile exercise" because "every level of clustering would determine a different population and there is no biological reason to prefer a particular one." Unfortunately, this did not stop Cavalli-Sforza from illustrating his work with color-coded world maps showing "four major ethnic regions": African in yellow, Mongoloids in blue, Caucasoids in green, and aboriginal Australians in red.[42]

The way some population geneticists treat the mixing of different groups also illustrates how genomic research is organized by race. If there are no pure races, we should not conceive of people with mixed ancestry as being a combination of two or more pure races. But this is exactly how many genomic scientists describe what they call racial "admixture."[43] An article titled "Reconstructing Genetic Ancestry Blocks in Admixed Individuals," co-authored by Neil Risch, states, "If the admixing occurred recently, we can imagine that each chromosome was assembled by stitching together long segments of DNA from a particular ancestral population." The authors refer to these imagined chromosomal segments that are identified with the component populations as "ancestry blocks."[44] But remember, this is how they *imagine* what their theory of admixing pure populations would look like.

A scientist from the Rosenberg team whose talk I attended used this same building-block imagery to explain how African Americans fit into the continental clusters the team identified. Noting that a large portion of African Americans have both European and African ancestry, he urged the audience to think of African American genomes "as a series of pieces that come from one or the other population." To illustrate this point, he showed a picture of a string of yellow and green blocks, representing an African American individual's chromosome. "As I go along the chromosome," he told the audience, "I can actually estimate which bits come from European ancestry, and which bits come from African." On the screen was projected a color-coded genome, with yellow and green blocks symbolizing "European" and "African" genes.[45] This graphic left the impression that, at the molecular level, African Americans are composed of distinguishable pieces of pure European and pure African ancestry neatly strung together. A 2010 article on admixture in the *Pharmacogenomics Journal* included a similar graphic with blue and red chromosomes representing different ancestral populations and a caption explaining that "the admixed individual's genomes are a mosaic of the two initial ancestral chromosomes."[46] These pictures of color-coded genomes represent what genomic scientists imagine about race, supported by statistical calculations based on racial assumptions, not on discoveries of scientific fact.

From Segregated Gene Banks to Color-Coded Genomes

Another problem with replacing a social conception of race with race based on geographic ancestry is that the distinction gets completely blurred when genetic scientists use race as a variable in laboratory research. These researchers typically abandon the usual rigor applied to scientific studies in order to classify their DNA samples, analyze their data, and report their findings according to race. A young medical anthropologist from Harvard, Duana Fullwiley, trailed scientists in two biopharmaceutical labs to investigate firsthand how they categorized the genetic samples used in their research. During a six-month fieldwork stay at the lab run by Esteban Burchard at University of California at San Francisco's department of biopharmaceutical sciences, Fullwiley interviewed researchers investigating the pharmacogenetics of cell membrane transporters, molecules that are vital to drug delivery. She discovered that race served as an unquestioned organizing principle for the collection, analysis, and reporting of genetic data.[47] Far from carefully scrutinizing the scientific validity of racial classifications, the researchers simply inserted race into their studies.

Fullwiley found that the laboratories practiced an extreme form of racial segregation at the genomic level. To obtain molecular data for their research, the scientists purchased DNA from the Human Genetic Cell Repository located at Coriell Institute for Medical Research, a nonprofit company in Camden, New Jersey, which houses the world's largest collection of human cell lines available for scientific research. Coriell labels samples according to the self-reported race of the donors, so the genetic material arrived at the lab already classified by race. Unsatisfied with Coriell's racial labeling, Burchard applied for a grant to build a genetic database specifically for his research that collected more "racially pure" DNA by "excluding anyone who reported racial mixing in their genealogies for the past three generations."[48] Burchard believed that this sampling method would allow him to segregate the DNA in his lab according to a more accurate test for race. Yet even his own, supposedly more rigorous, criteria still incorporated the DNA donors' own social definitions of race. Thus the concept of biological race is stamped on the very raw materials that go into pharmacogenetic studies, starting researchers on a trajectory that shapes the scientific conclusions they reach.

Once, Fullwiley noticed that two young investigators assigned to analyze the racial breakdown of a particular gene spent days "playing with" the data by applying various statistical tests using two different software programs. When Fullwiley asked about this exercise, one of the researchers confided that they were trying to manipulate the data to make the racial associations appear stronger. "These genotypes are specific to Caucasians, and we know that they are different in minority groups," the researcher explained. "So we want to make that difference stand out, which needs to be done, or else science will never change. People will just keep looking at Caucasian genes." The researcher apparently believed that showing nonwhite, race-specific variations would make minorities more worthy of study.[49]

The lab scientists not only assumed that African American and Caucasian DNA samples had significantly different allele frequencies, but they also perceived each as the other's "*opposite* race." They predicted that black and white DNA would always produce dramatically disparate findings. When researchers found results that were inconsistent with their perception of racial categorization, instead of rethinking their presumptions about racial sameness and difference, they usually "reacted against the data," writes Fullwiley. So when African American allele frequencies turned out to be more similar to Caucasian ones than expected, one scientist concluded the racially labeled samples must have been contaminated.[50]

The idea that blacks and whites represent opposite races is patently unscientific. Aside from the flaws inherent in treating Caucasians and African Americans as biological races in the first place, the pattern of human populations migrating out of their African homeland starting around eighty thousand years ago does not place Europeans and Africans the farthest apart. DNA studies of human evolution, buttressed by fossil and archeological evidence, show that human groups that journeyed out of Africa reached Europe about forty thousand years ago. Evolutionary biologists posit that geographic distance is a good predictor of genetic distance, and parts of Africa and Europe are swimming distance from one another. The intimate intertwining of Europeans and Africans in the ensuing centuries through trade, conquest, enslavement, and migration make it absurd to consider them opposites from a genetic standpoint.

So where does the notion of "opposite race" come from? It is part of an ideology about race that pits blacks against whites as the moral and social antithesis of each other.[51] This fictional opposition expresses the fundamental contrast between the innate character of each group essential to U.S. racial hierarchy. This ideology manifests itself in familiar (though often implicit) racial stereotypes that paint blacks as having a negative trait for every positive trait possessed by whites: blacks are lazy, while whites are industrious; blacks are ugly, while whites are beautiful; blacks are ignorant, while whites are smart; blacks are criminal, while whites are law-abiding; and so on.[52] Taken to its extreme, "opposite race" signifies not only difference but also enmity. "The black race is believed to be the perennial enemy of the white race, against whom all whites must unite," writes theologian George Kelsey. "'Opposite race' thus means 'race in opposition.'"[53] In racist circles, the worst thing you can call a white man is "nigger lover." But the idea of enemy races can also be seen in old claims that blacks are taking white people's jobs, or that Latinos are invading U.S. borders.

It is this ideological opposition, so ingrained in our racial culture—not any genetic evidence—that makes some scientists automatically think of black and white genotypes as being opposites. Of course, the researchers did not have these stereotypes in mind. To the contrary, they seemed to be accentuating assumed racial differences at the genetic level in a misconceived effort to fill a gap in research on minorities. Whatever the motivation, however, their statistical analyses of the genetic data sets were heavily influenced by social ideas about race.

Two anthropologists from Michigan State University, Linda Hunt and Mary Megyesi, found the same lack of scientific rigor when they interviewed thirty genetic scientists about their use of racial classifications in their research. The scientists, who held medical degrees or PhDs in fields ranging from human genetics and molecular biology to biostatistics, were all principal investigators in research projects in which race was a central variable. Most of the researchers categorized the DNA samples they worked with according to the familiar racial categories adopted by the federal government for the census and other administrative purposes: American Indian, Asian/Pacific Islander, black, white, and Hispanic. They simply lumped together people from different ethnic groups and geographic locations into these large social classifications, without any valid biological justification. "For example, samples collected from a relatively isolated village in rural China were described as 'Asian,' as were those taken from individuals of partial Japanese heritage living in suburban Detroit," wrote Hunt and Megyesi in their 2008 article. "Or a person labeled as 'black' in a study in San Francisco was classified as belonging to the same ancestral group as individuals being sampled in Nigeria."[54]

What's more, the researchers did not use any scientific criteria or specialized language to describe the racial categories. Instead, they dredged up the familiar colloquial labels that the average person on the street would use in identifying someone's race: Caucasian, white, Jewish, Hispanic, Mexican, African American, and so forth. Many of the classifications were nonequivalent (juxtaposing skin color with national origin, for example) as well as overlapping, so researchers had to make a subjective decision about where to place some subjects. A medical doctor studying the genetic basis for chronic disease in African Americans developed his own idiosyncratic technique for handling mixed ancestry among his research subjects. "The way we classify people sort of minimizes the admixture of whites," he explained. "You don't get considered 'white' if you look too much black or you look too much phenotypically nonwhite or if you have certain type of hair—you don't get called just plain 'white.'"[55]

An endocrinologist used a computerized list of eight thousand Spanish surnames to classify research subjects as Hispanic. "We took the view that if you have a Hispanic surname, you're Hispanic until proven otherwise," he stated.[56] Who knows how many people with Spanish names who had no Hispanic ancestry (whatever that means) were admitted to the study or how many Hispanics without Spanish names were left out—but treating people with certain last names as a genetic grouping is no sillier or more arbitrary than the other methods scientists use to make race seem like a biological classification.

One wonders how genetic scientists using widely varying, inconsistent, arbitrary, and ambiguous definitions of racial categories can possibly rely on or replicate the results from studies dependent on such classifications (or get them published in respectable journals). Because the definition of race varies across countries, it is even more hazardous to link data from race-based genetic research conducted globally. Added to this confusion is the fact that the research subjects who contribute the genetic data typically identify their own race, without any uniform criteria. Researchers have no way of knowing whether or not (or how) the participants are applying identical racial identity tests, and it is very unlikely they all define race precisely the same way. Many scientists use inconsistent definitions of race even within the same article, shifting from self-identified race to describe their research subjects to a classification of genotypes when discussing their findings.

Published reports of biomedical and genetic studies rarely describe how race was determined or the rationale for analyzing the data on the basis of race. "The lack of disciplinary clarity or consensus with respect to a central term of analysis . . . was not a barrier to publication of thousands of articles evaluating racial differences in a host of medical conditions," reported a survey on the use of race variables in genetic studies.[57] Some medical and scientific journals have addressed these methodological errors by adopting editorial policies that require more rigorous scrutiny of racial variables. In 2000, the editors of *Nature Genetics* declared that the journal would start requiring that "authors explain why they make use of particular ethnic groups or populations, and how classification was achieved," writing, "We hope that this will raise awareness and inspire more rigorous design of genetic and epidemiological studies."[58] Despite these improvements, unscientific gaps remain. As Margaret Winker, deputy editor of the *Journal of the American Medical Association* (*JAMA*), notes, "Still lacking are careful consideration of what has actually been measured when race/ethnicity is described, consistent terminology, hypothesis-driven justification for analyzing race/ethnicity, and a consistent and generalizable measurement of socioeconomic status."[59]

In no other field do scientists routinely use such a poorly defined variable as a critical component of their research. In the field of genetics itself, "genetic and disease variables are carefully defined and systematically classified," Hunt and Megyesi point out.[60] And yet the findings produced by scientists' faulty use of race as a research variable are taken to confirm the very racial categories that are being employed in such a sloppy manner. The public and major media outlets assume that researchers claiming to show racial disparities at the genetic level must have used rigorous scientific methods to define racial classifications, identify the race of research subjects, and group them with others in the same category. Nothing could be further from the truth.

When social scientists or legal scholars point out these flaws, we are often accused of meddling where we have no place. It appears to be a common belief that genomic and biomedical researchers should be left alone to investigate race objectively at the molecular level, while sociologists and their ilk should stick to understanding how race functions in society. "In an unadulterated scientific environment," wrote Rick Carlson, then a clinical professor at the University of Washington School of Public Health, "racial variables would be weighted by the same measures applied to other variables, such as temperature." He bemoaned that, in "addressing questions about race and genetics, social sciences have achieved parity with the 'harder' life sciences." The influence of social approaches to race creates, he argued, "a real peril that lowbrow theories wrapped in tendentious and oily slogans will get the public's ear and gain even footing with scientific proof as worthy of belief," comparing the debate about intelligent design and human evolution to the debate about social science and biological perspectives on race. Here again, we see the refrain that biological studies of race are ipso facto scientifically valid while the mountain of evidence that race is a political category amounts to "social science posturing."[61] This closed-minded faith that racial science must be true helps to shield scientists' flawed methods from public scrutiny.

Although genetic researchers routinely force genetic samples into preexisting racial categories, they often have trouble articulating what race means or even identifying what their own races are. One of the oddest discoveries Fullwiley and Hunt and Megyesi made was that some of the scientists they interviewed could not apply to themselves the racial classifications they applied to the DNA in their labs. When asked to describe his racial/ethnic background, one researcher, who was born in Mozambique, explained that he had one grandparent from Cape Verde and others from Portugal. "But if you go back a few generations, I've got people from all over the place," he elaborated. "So, I usually go for 'Other.'" The researcher revealed that he also considered identifying as Hispanic, but decided against it after noticing that the official definition doesn't include Mozambique or Portugal. Despite his confusion about his own racial category, this researcher was somehow able to categorize his clinical subjects by race.[62]

Fullwiley found a similar disconnect between the presumed reality of genetic race and the fuzziness of social race in the minds of the scientists she studied. One researcher defended the racial categories she used in the lab because she believed "there are ethnic-specific SNPs," but then conceded that she could not apply these same categories to herself because her father was Indian and her mother was part Czech. She concluded that she approached race in two divergent ways: "When I'm doing my genetic type research, I want things very well defined, and in a social setting I don't even want to think about it."[63]

These scientists seem oblivious to their reliance on social assumptions about race in conducting their genetic research. Without realizing it, they import social classifications into their work as if these classifications had biological validity. The problem ultimately lies not only in scientists' shoddy classification methods, but also in the impossible task of classifying people into a few clearly demarcated biological groups called races, even with the most advanced genomic knowledge and statistical computing at their disposal. Yet many of today's genomic scientists have faced this challenge by redefining race to fit twenty-first-century theories and technologies.

Where Does Geography Get Us?

While the scientists I have discussed so far use traditional or redefined racial categories in their research, there is another group of scientists who are trying to eliminate notions of race from their research altogether. Some biomedical geneticists who conduct genome-wide association studies (GWAS) to identify genetic contributions to common complex diseases, such as heart disease, type 2 diabetes, and cancer, are replacing race with the concept of "genetic ancestry" to differentiate populations. They use a population genetics software technology called Eigenstrat that divides DNA samples into clusters on the basis of SNP variation scores. Unlike the Structure program, Eigenstrat doesn't require prespecifying the number of expected clusters; nor does it depend on any presorting of samples using race, ethnicity, ancestry, or a theory of human evolution. Thus, it enables GWAS researchers to "create *categories of genetic difference* that are *not* categories of race," write University of Wisconsin sociologists Joan Fujimura and Ramya Rajagopalan.[64]

Yet Fujimura and Rajagopalan discovered that these medical geneticists tend to translate the Eigenstrat clusters of genetic similarity in terms of "genome geography": the scientists believe that individuals with similar SNP variation scores are likely to have "shared ancestry" that traces to a specific geographic location. Similarity is assumed to mean relatedness, which is assumed to mean a common geographic origin, such Europe, Africa, or Asia. In other words, genetic ancestry is equated with geographic ancestry. Just as happened with the Rosenberg study, the media, the public, and other researchers often read the genetically similar populations identified by a computer program as racial categories.[65]

Geographic ancestry does not solve the problem of race. If you look at a map of the world, you will see that parts of Africa are very close to Europe and the Middle East and other parts are very far from these regions. Because they are closer to the Arab Peninsula, African Somalis are genetically more similar to people in Saudi Arabia than they are to people in western or southern Africa. Likewise, the Saudis are more similar to the Somalis than to Norwegians, who are geographically more distant.[66] Yet molecular geneticists routinely refer to African ancestry as if everyone on the continent is more similar to each other than they are to people of other continents, who may be closer both geographically and genetically.

The same is true for Europe and Asia. We speak about the two "continents" as if they are very far apart. But Europe occupies the same land mass as Asia. England is much closer to Turkey, the nation seen as bridging the two continents, than it is to the eastern edge of Russia. Most of Russia is much closer to China than it is to Germany. The Rosenberg cluster study actually identified Eurasia—comprised of Europe, the Middle East, and Central Asia—as one of the five main continental groups. Yet newspaper coverage of the Rosenberg study conveniently ignored how the Eurasian cluster contravened, rather than confirmed, everyday racial categories.

Many geographic boundary lines are not dictated by natural barriers; they are drawn by political deal-making in the wake of wars, colonialism, and negotiated treaties. Consider the Middle East or the Arab world. The region comprising twenty-two countries in northern Africa and the Arab peninsula stretches across the continents of Africa and Eurasia, from the Atlantic to the Persian Gulf. Sudan, the African country devastated by civil war and the humanitarian crisis in Darfur, is a member of the Arab League despite its large population of black groups in the south. Is Turkey, which is predominantly Muslim and geographically closer to the Middle East than Sudan, part of the Arab subgrouping, or is it part of Asia? Or Europe? Or both?

The misperception of continental populations as natural groupings is grounded in a broader concept of populations as natural, isolated, and static. Populations came to be seen as "bounded units amenable to scientific sampling, analysis, and classification" as a result of Western linguistic and anthropological studies of indigenous peoples in the late nineteenth century and the first half of the twentieth century, historians of science Lundy Braun and Evelynn Hammonds show. They trace the scientific framing of African populations to European missionaries at the end of the nineteenth century who condensed diverse tongues of multiple groups into a single written language in order to introduce these groups to the Bible and convert them to Christianity. Later, colonial administrators foisted unified tribal identities on people who were geographically dispersed, spoke a variety of dialects, and had not previously felt any political allegiance to each other. When social anthropologists conducting extensive fieldwork in Africa in the 1930s and '40s treated tribes as self-contained units of study, they hardened the view of populations as bounded, fixed, and natural entities.[67]

"Once named and entered into international atlases and databases by anthropologists in the U.S.," write Braun and Hammonds, "the existence of populations as distinct, naturally occurring and static formations became self-evident, thus setting the stage for their use in large-scale population genetic studies—and for the reinvigoration of broad claims of human difference based on population identity."[68] In designing the Human Genome Diversity Project, Cavalli-Sforza relied heavily on the comprehensive atlas of the world's people compiled by the Yale anthropologist George Peter Murdock in 1967. These seemingly objective groupings based on geographical ancestry, in turn, are taken as verification of the racial classifications that they appear to mirror.

Population genomics trades the fallacy that races are naturally bounded by biology with the fallacy that populations that map onto races are natural formations that became biologically cohesive. While it was once thought that races are created from a biological essence, it is now thought that populations create a biological essence that mirrors race. But we should challenge genomic scientists who take it for granted that human beings are naturally organized into definable, genetically cohesive populations. If we pause for a moment to examine the political, cultural, and even arbitrary borders that delimit populations and consider how mutable, porous, and continually changing these boundaries are, the scaffolding of population genomics that seems to be supporting race begins to look very wobbly.

It is true that ancestry explains why some groups are more genetically similar than others better than race. Ancestry, at least, is a biological mechanism, whereas race is a political relationship. While categorizing someone by race requires an "other," tracing someone's ancestry is more politically neutral. But scientists run into trouble when they simply substitute geographic ancestry for race in order to dodge controversy or to give race biological legitimacy. Cramming findings about genetic ancestry into predetermined racial categories—or geographic populations that map onto them—turns serious objections to the biological concept of race into a battle over semantics and political correctness, ignoring the grave political consequences of dividing people into races. It appears that many scientists do not even believe this distinction makes a difference: they have concocted a thinly disguised euphemism for race they hope will not stir up as much controversy. Geographic ancestry has not replaced race—it has modernized it.

The Enduring Faith in Race

How should we view the progression of racial science that brings us to the doorstep of a new genomic concept of race? Are the errors of the past—the legend of Ham, Cuvier's racial typology, craniometry, eugenics—the product of flawed research methods that today's scientists have corrected with advanced genomic theories, state-of-the art computing, and giant DNA databases? Did Blumenbach happen to devise an accurate classification of human races in 1795 despite using a faulty technique? The answer: there are no biological races in the human species. Period. That conclusion was confirmed by the most ambitious research project on human biology yet undertaken, the Human Genome Project. A mountain of evidence assembled by historians, anthropologists, and biologists proves that race is not and cannot possibly be a natural division of human beings. Think about the origins of the concept of race, the way racial groupings have been reconfigured over time, or their differing meanings around the world, as I described in part 1. Race *must* be a political category.

Why, then, do most Americans cling to a false belief that biological races really do exist? Why do they latch on to whatever trivial proof they can muster to confirm their misconceptions about race? I am not referring to redneck white supremacists who spout vitriol about the inferiority of colored people. Many of my left-leaning colleagues, for example, balked at my book project. "Of course we should be working toward racial equality," they said, "but what if scientists *are* able to identify races genetically?"

"Racism is a faith." This was the summation of George D. Kelsey, the prominent black theologian who mentored Dr. Martin Luther King Jr. Kelsey argued that racism initially arose as an ideological justification for colonialism and slavery but gradually "heightened and deepened in meaning and value so that it pointed beyond the historical structures of relation, in which it emerged, to human existence itself."[69] The same can be said of race itself, the system of human classification that facilitates racism. Race started as a crude device for parceling people into servant and master classes, whose historical roots and scientific rationales we now reject. Yet race has outlasted its original historical context because it developed into a deeply held belief about the nature of human beings, a belief that continues to be useful in ordering our contemporary society.

Most Americans do not deduce that biological races exist from sound scientific evidence and reasoning. They are inculcated with this belief in the same way a child is raised in a religion. Children in the United States learn to divide all people into racial groups and come to have faith in race as a self-evident truth, like a traditional creation story that explains how the world works. Anthropologists describe the common meaning of race that defies scientific facts as a "folk concept." This is why Ashley Montagu called race "the witchcraft of our time." In 1942, he wrote, "It is the contemporary myth. Man's most dangerous myth."[70] Perhaps the best proof of its power is Montagu's own inability to renounce race definitively in the 1950 UNESCO statement.

According to folklorist Judith Neulander, for a folk story to persist it must contain "elements that can be modified without changing what the tale is about," enabling it "to dodge later discreditation."[71] Science has been responsible for giving racial folklore its superficial plausibility by updating its definitions, measurements, and rationales without changing what the tale is about: once upon a time human beings all over the world were divided into large biological groups called races.

Believing in race can be compared to believing in astrology. People who have faith in astrology find constant confirmation that horoscope predictions are reliable and that astrological signs determine personality types. For the faithful, the twelve divisions of the zodiac are as accurate as Blumenbach's five divisions of human beings. The funny thing is, biostatisticians can find significant medical differences according to astrological signs. In the 1990s, a major randomized clinical trial compared the effectiveness of an intravenous drug, an oral aspirin, and a placebo to treat 17,000 patients who were hospitalized with signs of a heart attack. The study found a huge overall statistical benefit for patients who got the aspirin over the placebo. To test the strength of the outcome, the researchers divided the patients into twelve subgroups by their astrological signs. They found that the zodiac made a difference: their statistical analysis showed that patients born under

Gemini or Libra suffered an adverse effect from aspirin.[72] Unsurprisingly, physicians laughed off this finding because it was more scientifically plausible to interpret the results as an insignificant coincidence. But an astrology enthusiast would take it as proof that zodiac signs determine people's health and drug response.[73]

If race is a faith, like astrology, how can science resolve the debate over its meaning? If scientists could settle the question whether or not race is biological, we would not still be debating it. As we have seen, a generation of scholars has definitively refuted prior versions of racial science by revealing their errors and biases.[74] The work of scientists is invaluable for dispelling misconceptions about the biological definition of race, and scientists should continue to educate the public about its scientific invalidity. I, too, felt it was crucial to lay an empirical groundwork in part 1, to use genetic science to challenge the racial lies still being told about humanity. Many Americans' belief in race depends on sheer ignorance of the scientific evidence about human genetic unity and diversity.

But the resilience of racial science raises serious doubts about the efficacy of fresh efforts to debunk it based on more accurate and less prejudiced scientific methods. Race is a political system that will not be brought down with scientific evidence alone. Race persists neither because it is scientifically valid nor because its invalidity remains to be proven. Race persists because it continues to be *politically useful*. It is therefore imperative to evaluate the political function of race at the present time and wage a political assault against it. I realize this talk of politics will be called unscientific by those invested in preserving the biological view of race. But as many scientists have shown, the science of race has long been riddled with unscientific flaws. And as both history and current practice clearly demonstrate, racial science and politics are inseparable.

Believing in biological races, not only racism, is an irrational moral conviction that scientific evidence alone has been unable to overcome. In fact, over the course of U.S. history, scientists themselves have worked as much to uphold this moral conviction as to defeat it. In November 2010, University of Chicago business professor Richard Thaler, coauthor (with Cass Sunstein) of *Nudge: Improving Decisions About Health, Wealth, and Happiness*, asked contributors to the Internet forum Edge to name their favorite examples of "wrong scientific beliefs that were held for long periods of time." The responses listed dozens of false but durable scientific theories, ranging from "life generates spontaneously" to "stomach ulcers are caused by stress" to "genes are made of protein."[75] I hope one day the scientific theory of biological races will make the list. Scientists' defining mission to test accepted beliefs utilizing empirical evidence remains an important aspect of challenging antiquated views about race. It is the faith in race—the religion of separating human beings into racial groups—that makes it difficult for Americans to think like scientists.

4

Medical Stereotyping

Imagine if every single day a jumbo jet loaded with 230 African American passengers took off into the sky, reached a cruising altitude, then crashed to the ground, killing all aboard. According to former surgeon general David Satcher, this is exactly the impact caused by racial health disparities in the United States. In a 2005 article, he and several other health experts reported that there had been 83,570 "excess" black deaths in 2002. That represents the number of African Americans who would still have been alive that year if their life expectancy were the same as that of whites.[1] The number of excess deaths is closer to 100,000 today.[2] In one generation, between 1940 and 1999, more than 4 million African Americans died prematurely relative to whites.[3] Overall life expectancy is actually declining in some counties where there is a high proportion of African Americans.[4]

In my hometown of Chicago, one third of all black deaths are excess in terms of the black–white mortality gap. In other words, one out of every three black people who died in 2000 would have survived if black and white death rates were equal. Chicago is a very segregated city, so longevity varies geographically. Of Chicago's seventy-seven community areas, the twenty-two with the lowest life expectancies are more than 90 percent black. There is a difference of sixteen years between the white neighborhood with the highest life expectancy and the black neighborhood with the lowest. Blacks are more likely to die prematurely (before the age of sixty-five) from most major illnesses: cancer, stroke, diabetes, kidney disease, AIDS, and coronary heart disease, to name a few. Race matters at the beginning of life as well.

Black infants are almost three times more likely than white infants to die before their first birthday. Blacks also spend a greater portion of their lives in poor health. They experience a rate of preventable hospitalizations more than double that of whites.[5] "That's not a racial health *gap*: it's a *chasm* wider and deeper than a mass grave," says Harriet Washington, whose book *Medical Apartheid* chronicles medical experimentation on black Americans from colonial times to the present.[6]

Similarly dreadful statistics are available for other U.S. minority groups, including Latinos and Native Americans. Recent immigrants from Latin America tend to be in better health than white Americans, despite having higher rates of poverty and less access to health care—a curiosity known as the "Hispanic Paradox."[7] But their health advantage disappears in the generation born in the United States. Latinos and blacks are about twice as likely as whites to have diabetes.[8] In November 2009, President Barack Obama convened the largest gathering of Indian tribal leaders in U.S. history—bringing together representatives from more than five hundred tribes—to discuss their urgent health needs. "Native Americans die of illnesses like tuberculosis, alcoholism, diabetes, pneumonia and influenza at far higher rates," Obama said.[9]

What explains this racial chasm in health? Is there something innate in racial minorities that predisposes them to disease and early death? Or is there something about being a racial minority in America that jeopardizes health? The stakes are high. Which answer you choose determines who you think is responsible for the jumbo jet crashing every day and what should be done about it. Genomic science is increasingly pointing us toward the molecular level for the answer. Before addressing the current debate among scientists about the cause of the shocking disparities, let's consider past scientific understandings of race and disease that influence how we understand the subject today.

Racial Diseases

The notion of "racial diseases"—that people of different races suffer from peculiar diseases and experience common diseases differently—is centuries old. It is tied to the original use of biology in inventing the political category of race. As I showed in chapter 1, white slaveholders explained race in biological terms to demarcate slaves from masters and to provide a moral excuse for slavery. Whites argued that the biological peculiarities of blacks made enslavement the only condition in which blacks could be productive and disciplined. Every perceived physical difference conveniently supported Africans' slave status. Defining disease in racial terms played an essential part in this biological strategy that was enshrined in biomedical research and practice. It was precisely "by locating disease in physiologic difference—be it susceptibility or resistance—that medicine served to mark blacks as deserving of their inferior social status in society," writes Lundy Braun.[10] Medicine has historically promoted a racial construction of disease that in turn perpetuates a biological construction of race.

In 1781, before becoming the third president of the United States, Thomas Jefferson wrote an influential treatise in response to a set of questions about Virginia posed by a visiting French dignitary. *Notes on the State of Virginia* compiles extensive information about the state, as well as Jefferson's most eloquent defense of key constitutional principles from individual liberty to the separation of church and state. A naturalist as well as a political philosopher, Jefferson approached the question of slavery from a biological and political perspective. He both condemned chattel slavery and justified it on grounds that "the real distinctions which nature has made" between blacks and whites made it impossible for them to live together as equal citizens. Even if blacks were freed from bondage, their biological infirmities prevented them from joining the American polity as free individuals. Jefferson opined, for example, that blacks sweat more than whites because of "a difference of structure in the pulmonary apparatus, which a late ingenious experimentalist has discovered to be the principal regulator of animal heat." He also observed that black people were disposed to sleep too much because their minds were empty: "An animal whose body is at rest, and who does not reflect, must be disposed to sleep of course," he wrote.[11]

Jefferson conveniently attributed the physical differences he saw in blacks to their innate weaknesses rather than to their forced toil in the hot Virginia sun. He transformed perceived physiological differences into indicia of character flaws, and ultimately into political disqualifications: "This unfortunate difference in colour, and perhaps of faculty, is a powerful obstacle to the emancipation of these people." It also must have eased his conscience to believe that blacks soon forgot the "griefs" of slavery because "their existence appears to participate more of sensation than reflection."[12] Jefferson was not misusing a biological category of race he had discovered in nature; he was helping to invent it for political reasons. Attributing blacks' poor health to inherent racial difference allowed whites like Jefferson both to ignore how disease is caused by political inequality and to justify an unequal system by pointing to the inherent racial difference that disease supposedly reveals.

Benjamin Banneker, the free black astronomer and mathematician born in 1731, confronted Jefferson, then secretary of state, about the statesman's contradictory position on slavery. On August 19, 1791, Banneker wrote a letter appealing to Jefferson to lay aside his prejudices against black people and live up to the principles he avowed in the Declaration of Independence: "Sir, how pitiable is it to reflect, that although you were so fully convinced of the benevolence of the Father of Mankind, and of his equal and impartial distribution of these rights and privileges, which he hath conferred upon them, that you should at the same time counteract his mercies, in detaining by fraud and violence so numerous a part of my brethren, under groaning captivity and cruel oppression." Banneker presented to Jefferson a copy of his Almanac for 1792, based on his own careful astronomical calculations, to substantiate his equal intelligence.

Jefferson's reply, dated August 30, 1791, is fascinating for its ambivalence on the scientific question of whether blacks' status stemmed from biological or social disadvantage. "Nobody wishes more than I do, to see such proofs as you exhibit, that nature has given to our black brethren talents equal to those of the other colors of men; and that the appearance of the want of them, is owing merely to the degraded condition of their existence, both in Africa and America."[13] Apparently Jefferson was not persuaded. After Banneker died, Jefferson disparaged his talents, suggesting that his almanacs were crafted with the aid of a white mentor. "I have a long letter from Banneker, which shows him to have had a mind of very common stature indeed," Jefferson wrote.[14]

During the Civil War, Union Army physicians reported that their anatomical examinations of black Union soldiers, measuring chest width and lung capacity, showed that blacks had smaller lungs, making them more susceptible to pulmonary disease.[15] Most doctors in the United States believed that although tuberculosis was communicable, a predisposition to it was inherited. They were so wedded to the view that tuberculosis was a hereditary disease that they initially resisted the discovery in 1869 by French physician J.A. Villemin that it was caused by transmissible pathogens. It was not until

1882, when the tubercle bacillus was positively identified, that the medical profession accepted the evidence that there was no internal cause for tuberculosis passed down from parent to child according to race.

After slavery ended, racial differences in disease worked just as well to reinforce black people's continued subordination. Locating blacks' inferior status in their diseased bodies provided a reason for retaining white supremacy and discredited the need for radical social transformation. Instead, whites argued that the best way to improve the condition of emancipated blacks was through either benign neglect or coercive medical intervention.

The most notorious example is the Tuskegee Study of Untreated Syphilis in the Negro Male launched by the U.S. Public Health Service in Macon County, Alabama, in 1932.[16] The study was designed to confirm the long-held view that venereal disease acted differently in blacks than whites. Scientists believed, for example, that syphilis caused the greatest damage to the neurological system in whites but ravaged the cardiovascular system in blacks, bypassing their "underdeveloped" brains.[17] The Tuskegee researchers enrolled six hundred infected black men, withheld treatment from them, and then recorded the progression of symptoms. They tricked the poor and illiterate sharecroppers into participating with false promises of treatment, then let the men die so they could autopsy their bodies and examine the destruction left by the disease. Despite condemning the scientific racism practiced by Nazis, U.S. scientists continued the syphilis study for forty years, until national press coverage prompted the Department of Health, Education, and Welfare to halt it in 1972.

The Syphilis Study is so frightening because it shows the inhumanity that can result from a belief in intrinsic racial difference. The study "had nothing to do with treatment," notes James H. Jones, author of the leading history of the experiment, *Bad Blood*; its ultimate purpose was to document the presumption that syphilis was a different disease in blacks.[18] Far from testing the efficacy of new therapies, the researchers prevented the men from getting existing therapies. Focusing on racial difference in disease rather than on the human beings who were suffering from disease had tragic consequences. Jones calls this the "scientific blind spot to ethical issues."[19] Race has been and continues to be a major cause of blind spots in biomedical research.

An article by Dr. Thomas Murrell, a leading expert on syphilis, published in the *Journal of the American Medical Association* in 1910, represents the idea that syphilis was different in blacks and illuminates the way this idea supported repressive racial policies. "The knowledge of syphilis as affecting the Caucasian, however profound, will not give one an insight into the conditions confronting the negro," Murrell begins the article, titled "Syphilis and the American Negro." He explains that the greater intensity of syphilis in blacks stemmed from the "terrible changes fifty years have wrought"—the physical and moral degeneration of blacks between 1859 and 1909 as a result of their emancipation from slavery. Whereas "by a forced system of hygiene the negro's body, as a piece of property, was not allowed to deteriorate," freedom left blacks unprepared to take care of themselves.[20]

Murrell devotes a section to the "political aspects" of the disease, which he argues warrant an immediate, nationwide response. "If the healthy negro is a political menace, then the diseased one is doubly a social menace, and the invasion of the South by the North forty years ago has brought about an invasion of the North, and that by the man they freed." Although Murrell focuses on the "sociology" of race in identifying the roots of black disease, he favors a strictly biological remedy. "The future of the negro lies more in the research laboratory than in the schools. . . . When diseased he should be registered and forced to take treatment before he offers his diseased mind and body on the altar of academic and professional education," Murrell concludes.[21]

Other medical experts reached the exact same conclusions about the causes and remedies for the high rate of tuberculosis among newly freed blacks. Around the time Dr. Murrell wrote about syphilis in the *Journal of the American Medical Association*, Alabama physician Seale Harris asserted in the same journal that "consumption is almost a scourge to the emancipated negro . . . and the only reason why it was not so with the slaves was that their habits and sanitary surroundings were better than those of many of their masters." Pittsburgh physician W.T. English bluntly told the South Carolina Medical Society that his anatomical studies uncovered in the "body of the negro a mass of minor defects and imperfections from the crown of the head to the soles of the feet." Tuberculosis, he predicted, "threatens to settle the race problem by elimination."[22]

Black intellectuals issued a stunning assault on claims that high rates of disease among blacks stemmed from their biological inferiority and incapacity to cope with freedom—what famed African American sociologist W.E.B. Du Bois described as "an unfortunate attempt to reclothe an old and discredited thesis." Writing in the *Philadelphia Negro* in 1899, Du Bois said, "Particularly with regard to consumption it must be remembered that Negroes are not the first people who have been claimed as its peculiar victims; the Irish were once thought to be doomed by that disease—but that was when Irishmen were unpopular."[23] The National Medical Association, founded as the National Association of Colored Physicians, Dentists, and Pharmacists in 1895, was also an outspoken critic of whites' view of race-based disease.[24] When R.M. Cunningham, a white Alabama prison doctor, reported in an 1894 issue of *New York Medical News* that his research on black inmates confirmed a racial predisposition to pulmonary disease, a black physician named M.V. Ball published a strong rejoinder. Ball took Cunningham to task for multiple methodological errors: the prison sample was not representative of blacks as a whole, and the study failed to control for relevant nonracial variables such as socioeconomic status or previous health history. The racial difference in tuberculosis occurred because "the negro, as is well known, usually occupies the poorest quarters of a town or city," Ball wrote.[25] In Baltimore, as historian of medicine Samuel Roberts has shown, claiming that pulmonary disease was different in blacks justified coercive state intervention in their lives and avoided the glaring need for residential desegregation and housing reform. In fact, viewing blacks as predisposed to tuberculosis provided an excuse for preserving segregation, to protect white families from, in the words of Baltimore mayor James Preston, the "insidious influence of slum conditions into our very midst to defile and destroy."[26]

Meanwhile, Jewish immigrants from Eastern Europe had settled on the Lower East Side of New York City after arriving at Ellis Island in the Upper New York Bay. Between 1881 and 1914, more than 2 million Jews—one third of the Jewish population of Eastern Europe—came to the United States in flight from poverty and persecution. At this point in America's racial history, the Jewish immigrants were not considered white. To the contrary, as medical historian Sander Gilman notes, the "general consensus of the ethnological literature of the late nineteenth century was that the Jews were 'black' or, at least, 'swarthy.'"[27] One anthropologist of the period explained the "predominant mouth of some Jews being the result of the presence of black blood." Most Americans viewed Jews as a biologically inferior race stricken by a host of hereditary diseases that resulted from "inbreeding" and "racial incest."[28] Tay-Sachs disease was highlighted as a racial illness demonstrating that Jews were innately degenerate. Jews were susceptible to developing

Tay-Sachs, known as a "Hebraic debility," because, according to Dr. Isador Coriat, "the Jew possesses certain racial characteristics of organic inferiority through which he differs from the non-Jew."[29] The inferior organ Coriat meant was the Jewish brain.

Doctors described Jewish immigrants as too sickly, weak, and neurotic to cope with urban life in America. U.S. Navy medical director Dr. Manly H. Simons, for example, attributed the immigrants' "increasing degradation, retardation, and extinction to an inability to adapt their genetic constitution to their new circumstances."[30] He placed special blame on their "family diseases . . . which tend to early death or sterility." The stereotype of Jews as a "community of the ill" fit the emerging eugenicist ideology in casting them as a threat to American society. "This great Polish swamp of miserable human beings, terrific in its proportions, threatens to drain itself off into our country as well, unless we restrict its ingress," wrote MIT and Columbia professor William Z. Ripley in 1899. As the vectors of specifically *genetic* disease, Jews were seen as capable of contaminating American racial stock more permanently than immigrants who spread infectious ailments alone.[31]

On the other side of the country, in Los Angeles, city health officials were targeting Chinese and Mexican communities as "rotten spots" that polluted the city's idyllic environment of fresh air and sunshine.[32] Confirming popular stereotypes, public health officials portrayed Chinatown as "dirty, depraved, and disease ridden."[33] A local zoning ordinance passed in 1908 gave city officials a means to shut down Chinese laundries, which constituted half of the laundry business in Los Angeles. The law was patently unconstitutional. The U.S. Supreme Court ruled in the 1886 landmark case of *Yick Wo v. Hopkins* that a similar zoning law passed in San Francisco violated the Equal Protection Clause by discriminating against Chinese launderers on the basis of race. But when Quong Wo was arrested for operating a laundry in a residential section of Los Angeles, the California Supreme Court dismissed his civil rights claim on the grounds that the zoning law did not racially discriminate. The justices upheld the law as necessary "to protect the public health, morals, and safety, and comfort."[34] Public health provided a color-blind excuse for city officials to target Chinese businesses without running afoul of the Constitution.

In 1916, when twenty-two Mexican railroad workers were struck by typhus, an infectious disease transmitted through lice and tick bites, the Los Angeles health department launched a "campaign against filth and lack of personal hygiene" that covered more than one hundred railroad camps. "Every individual hailing from Mexico should be regarded as potentially pathogenic," wrote one doctor in a 1914 issue of the trade magazine *Southern California Practitioner*.[35] The typhus outbreak soon became reason for widespread government inspection of all Mexican residents that extended into private homes and public school classrooms. State efforts focused exclusively on "enforcing bodily cleanliness" instead of requiring railroad and city officials to improve sanitation where Mexicans lived. Similarly, John Pomeroy, who became health department chief in 1915, explained that Mexican babies died at high rates because their mothers were "absolutely ignorant of the fundamentals of hygiene," not because of malnutrition and polluted water in Mexican camps.[36]

The public health campaigns against Chinese and Mexican immigrants provided an excuse for racial discrimination and carved racial boundary lines around eligibility for citizenship. "By systematically associating dirt, disease, and disorder with immigrant status," writes ethnic studies scholar Natalia Molina, Los Angeles health officials "redefined citizenship in racialized and medicalized terms."[37] Diagnosing disease according to race was a powerful means of defining blacks as natural slaves, Jews as a contaminating threat to U.S. welfare, and Chinese and Mexican immigrants as inherently unfit for citizenship.

Just as whites explained the symptoms of black oppression as caused by disease, so they explained black resistance as a symptom of disease. Shortly before the Civil War, the Medical Association of Louisiana charged a committee of four local physicians to investigate "the diseases and physical peculiarities of our negro population." Its chair, Dr. Samuel Cartwright, a University of Pennsylvania Medical School graduate and well-known expert on Negro medicine, stood before members of the association in 1851 to deliver the committee's report. Cartwright argued that slavery was beneficial to blacks for medical reasons. The physical labor that white slaveholders forced upon naturally slothful blacks helped their lungs to "vitalize" blood. "It is the red, vital blood, sent to the brain, that liberates their mind when under the white man's control; and it is the want of a sufficiency of red, vital blood, that chains their mind to ignorance and barbarism, when in freedom."[38] By converting race into biological difference, Cartwright could ingeniously turn enslavement of Africans into a form of freedom and turn black freedom into slavery.

Cartwright's greatest contribution to racial medicine was his catalog of diseases peculiar to Negroes. Cartwright coined the term *drapetomania*—combining Greek words for "runaway slave" and "crazy"—to describe the mental illness that caused blacks to abscond from bondage. Whites brought on the symptoms by "trying to make the negro anything else than *'the submissive knee-bender'* (which the Almighty declared he should be) by trying to raise him to a level with himself, or by putting himself on an equality with the negro." Cartwright diagnosed blacks with a second form of insanity he labeled *dysaesthesia aethiopis*—what overseers called rascality. This disease was characterized by a lack of respect for the master's property that resulted when blacks were not closely monitored by whites. According to Cartwright, it afflicted virtually all free Negroes and attacked those slaves who lived as if they were free by drinking too much and working too little. The cure Cartwright proposed to his white audience was to subject the deranged slaves to hard physical labor as well as harsh corporal punishment "until they fall into that submissive state which it was intended for them to occupy."[39]

In *The Protest Psychosis*, Jonathan Metzl, a professor of psychiatry at the University of Michigan, traces how schizophrenia became a black disease in the 1960s in much the same way that drapetomania became a black disease in the 1860s. Like Dr. Cartwright's use of mental illness to explain black resistance to enslavement, psychiatrists in the civil rights era began to explain black urban unrest as a symptom of mental instability. Unlike drapetomania, which explicitly originated as an illness peculiar to Negroes, however, schizophrenia was first considered a psychological malady that afflicted primarily white middle-class housewives.

But the race of the disease radically morphed as the civil rights movement erupted. "Many leading medical and popular sources suddenly described schizophrenia as an illness manifested not by docility, but by rage," writes Metzl. "Growing numbers of research articles from leading psychiatric journals asserted that schizophrenia was a condition that also afflicted 'Negro men,' and that black forms of the illness were marked by volatility and aggression." Black male accusations against the white power structure were diagnosed as signs of paranoia, delusion, and hostility. Metzl borrowed the title of his book from an article in the prestigious *Archives of General Psychiatry*,

in which psychiatrists Walter Bromberg and Frank Simon described schizophrenia as a "protest psychosis" whereby black men developed

> "hostile and aggressive feelings" and "delusional anti-whiteness" after listening to the words of Malcolm X, joining the Black Muslims, or aligning with groups that preached militant resistance to white society. According to the authors, the men required psychiatric treatment because their symptoms threatened not only their own sanity, but the social order of white America.[40]

Accompanying articles like this in leading psychiatric journals, calling for chemical management of black male antagonism, were advertisements for antipsychotic medications that showed "angry black men with clenched, Black Power fists." The white press popularized the racial diagnosis by warning the public about violent black men crazed by schizophrenia and assigning the label even to leading black activists like Roy Wilkins, executive director of the NAACP. Schizophrenia became a metaphor for any black dissatisfaction with racial inequality. Most shocking, state mental institutions began to fill with young black men arrested for minor offenses who were confined indefinitely when psychiatrists diagnosed them with schizophrenia. The locked wards at Ionia State Hospital for the Criminally Insane outside Detroit, once an asylum for "neurotic" white women, swelled to more than one thousand black "patients" thought to have gone mad from their anger at white people. There, a black man's complaint that his civil rights were being violated or "tough guy attitude towards white authority figures" were noted in medical charts as evidence of mental illness.

The very long history, with many recent examples, of concocted diagnoses reminds us that understanding disease in racial terms inevitably involves social understandings of race and reflects current debates about race relations. Whites have always turned their anxieties about race mixing, immigration, and resistance into medical findings of minority disease.[41] It is when scientists and doctors insist that their use of race is purely biological that we should be most wary.

Racial Profiling in the Doctor's Office

Medical students are taught to take race into account when they treat patients. In fact, race is the second thing about a patient they are trained to see, after age and before sex. A patient's history begins with "This is a [insert patient's age, presumed race, sex] who presents with a chief complaint of. . . ."[42] Sex is confirmed during the physical exam. Doctors guess the patient's race based on their assumptions about how the patient looks, speaks, and acts. Immediately noticing that a patient is a forty-five-year-old Asian woman is supposed to shape the way the doctor diagnoses her complaint of pain in the abdomen.

Sally Satel, a psychiatrist and a senior fellow at the conservative American Enterprise Institute, blew the lid off racialized medical practice with her defiant admission, "I Am a Racially Profiling Doctor," a 2002 cover story in the *New York Times Magazine*. "In practicing medicine, I am not colorblind," Satel began. "I always take note of my patient's race. So do many of my colleagues. We do it because certain diseases and treatment responses cluster by ethnicity. . . . When it comes to practicing medicine, stereotyping often works." She gave the example of prescribing medications on the basis of race at the Washington, D.C., drug clinic where she works as a psychiatrist: "When I prescribe Prozac to a patient who is African-American, I start at a lower dose, 5 or 10 milligrams instead of the usual 10-to-20 milligram dose. I do this in part because clinical experience and pharmacological research show that blacks metabolize antidepressants more slowly than Caucasians and Asians." She also referred to a colleague, an anesthesiologist in a Baltimore-area hospital, who takes race into account when placing a breathing tube down a patient's windpipe. The anesthesiologist gives his black patients a drying agent in advance because, he says, "black patients tend to salivate heavily, which can cause airway complications." He also always starts Asian patients undergoing surgery on a lower dose of narcotics because he believes they have a higher sensitivity to the drugs.[43]

One of the most enduring and disturbing medical beliefs about blacks is that they are impervious to pain. This myth has excused surgical experimentation without anesthesia on blacks, as well as providing blacks with inadequate pain relief for their injuries and illnesses. Nineteenth-century physician Charles White asserted that blacks "bear surgical operations much better than white people and what would be the cause of insupportable pain for white men, a Negro would almost disregard."[44] Some more modern medical experts still believe that tolerance for pain differs by race. A 1983 study conducted at Memorial Sloan-Kettering Cancer Center, published in the medical journal *Pain*, investigated the reasons for variation in the relief of chronic pain in cancer patients receiving morphine.[45] The researchers divided 715 patients by age, race, and sex and then looked at their responses to different amounts of morphine. They reached the astonishing conclusion that "black patients obtained relief comparable to white patients at *half* the morphine dose." Doctors who accept this finding may believe it advisable to give their black cancer patients only half the pain relief they give their white patients.

In fact, that is precisely what many doctors do: they are less likely to prescribe pain medication to black and Latino patients than to white patients for the same injuries. Dr. Knox Todd began documenting how patients' race affects the treatment of pain when he was a doctor in the UCLA Emergency Center in the 1990s.[46] He and colleagues examined the way doctors treated 139 white and Latino patients coming to the emergency room over a two-year period with a single injury—fractures of a long bone in either the arm or leg. Because this type of fracture is extremely painful, there is no medical reason to distinguish between the two groups of patients. Yet the researchers discovered that Latinos were twice as likely as whites to receive no pain medication while in the emergency room.[47] Although it's possible that the Latino patients complained less of pain, the doctors should have been aware of the high degree of pain they suffered, given the nature of their injuries.

When Todd moved to Emory University School of Medicine, he led an Atlanta-based study that confirmed his finding in Los Angeles. This time his research team analyzed medical charts of 217 patients who were treated for long-bone fractures at an inner-city emergency room that served both black and white patients. In a 2000 article in *Annals of Emergency Medicine*, Todd reported that 43 percent of blacks, but only 26 percent of whites, received no pain medication. In this study, Todd took the additional step of documenting whether or not the patients expressed pain to their doctors. By carefully looking at notations in the medical files, he found that black patients were about as likely as whites to complain of pain. Black patients thus received pain medication half as often as whites because doctors did not order it for them, not because blacks do not feel pain or do not want pain relief.[48]

A national study of emergency-room treatment for pain, published in 2003 in the *American Journal of Public Health*, provides at least one clue why doctors are prescribing relief based on race.[49] The study examined the type of pain medication given to people who came to an emergency room for treatment of long-bone fractures, back injuries, and migraine headaches.

Doctors can choose between two types of drugs for the pain associated with these injuries: nonopioids, like Tylenol or Advil, provide minimal relief and are not addicting; opioids, like codeine or Demerol, are far more powerful against pain but can be abused by patients. The study found that blacks were significantly less likely than whites to receive the stronger relief for back pain and migraines. Why are emergency room physicians more reluctant to prescribe opioids to black patients in pain? Some experts in the field suggest that these doctors harbor stereotypes that blacks are more prone to substance abuse and so conclude that blacks are making up pain symptoms to get drugs—or that it would be harmful to give blacks in pain potentially addicting medications.[50]

Racial profiling by doctors can also lead to *over*medication. A 2009 federally funded study found that doctors are four times as likely to prescribe powerful antipsychotic drugs to children covered by Medicaid as they are to children whose parents have private insurance.[51] Prescriptions to Medicaid youth have jumped dramatically over the last several years, far outpacing the modest rise in prescriptions to other children. What's more, children on Medicaid are more likely to be given these medications for less severe conditions such as attention deficit hyperactivity disorder (accounting for more than a third of Medicaid youth on antipsychotic drugs), anxiety, aggression, or "persistent defiance." Prescribing antipsychotics to children has raised safety concerns because they cause metabolic changes that can result in long-term health problems. Some commentators suggest that socioeconomic status explains the disparity: doctors are quicker to prescribe antipsychotics to poor children "because it is deemed the most efficient and cost-effective way to control problems that may be handled much differently for middle-class children," who are more likely to receive family counseling or psychotherapy, writes *New York Times* reporter Duff Wilson.[52]

Given the history of race-based psychiatric diagnosis, it seems likely that the disproportionate number of African American children receiving Medicaid helps to explain the disparity in treatment. Just as doctors invented the "protest psychosis," might they be equating black children's perceived hyperactivity and defiance with mental illness? The deeply entrenched view, either conscious or unconscious, that black diseases stem from innate abnormalities may suggest to doctors the necessity of a chemical therapy for black children's behavior problems. Racial stereotypes held by doctors are then accentuated by the Medicaid system, which pressures them to make quick and low-cost diagnoses.

Most doctors take race into account because they think it is good medical practice. When I was living in Palo Alto, California, in 2008, I got to know a family physician whose daughter was in the same child care program as my son. She had a liberal perspective on the political issues we chatted about and believed in treating everyone equally when it came to marriage and voting. She bragged that she was an Obama supporter. But she had a different attitude about practicing medicine. When I told her about my book project, she looked confused. She found it hard to see race apart from biology. "I would never treat a black patient with heart disease the same as a white patient with heart disease. Heart disease is different in blacks and whites," she told me. This widespread view among doctors who may harbor no ill will can lead them to misdiagnose their patients and to excuse inferior treatment as medically beneficial.

In 2006, four years after publishing "I Am a Racially Profiling Doctor," Satel issued an "Erratum?" on her Web site:

> Since this piece was published, a number of physicians have told me that they were not aware of differences in salivation rate between blacks and whites. This prompted me to look further into the statement by the anesthesiologist I quoted. There is not a big medical literature on this but after reading those studies and querying the experts on salivation, it appears that racial differences in this domain are unremarkable. I am not sure if the anesthesiologist I quote in the piece may have been detecting a differential secretory response to surgical premedication in black and white patients or something else, but it appears that, at baseline, race-related differences are negligible. There are many other intrinsic differences, though, so this does not undercut the overall argument of the article.[53]

Why hadn't Satel checked to see if there was any basis for her statement of racial discrepancy in salivation before publishing it? If this race-based medical practice is unfounded, why not question others? Myths about black people's bodies conjured up during slavery continue to circulate in doctor's offices and in the media. I shudder to think of the black patients who may have been injured by the anesthesiologist Satel quoted and others who believe the salivation myth. I also wonder how Satel and her colleagues determine which of their patients are black. Do they modulate their prescriptions by skin tone or follow a one-drop, one-size-fits-all diagnosis for everyone with any amount of perceived African ancestry?

After a reading by Rebecca Skloot from her book *The Immortal Life of Henrietta Lacks*, about a black woman whose cancer cells were harvested for research without her consent, I chatted with a group of first-year medical students about the role of race in their training. "Things have changed since Henrietta Lacks was a patient at Johns Hopkins," one eager young woman assured me. "We're taking a required course on 'Cultural Dynamics in Medicine' and learning about how racial bias can affect doctors' decisions." They were determined not to repeat the mistakes of the past generation of physicians. When I asked the students if they were taught to take race into account when diagnosing and treating illness, they just as assuredly told me they were. Race, they said, revealed how likely it is that a patient is suffering from one malady rather than another. Could their use of race in making medical judgments be influenced by societal bias? The students vehemently denied that could happen: they used race only as an objective biological contributor to disease and had been warned about allowing racial prejudice to enter their medical decision making. What they did not consider is that treating race as biological difference itself might shape their view of patients in harmful ways. The narrow view of bias as a prejudiced attitude on the part of individual doctors misses the more powerful intertwining of racial stereotypes with their understanding of disease.

The Source of Disparities

In 2002, the health arm of the National Academy of Sciences, the Institute of Medicine (IOM), dropped a bombshell by scientifically documenting widespread racial disparities in health care and suggesting that they stemmed, at least in part, from physician bias. Its 562-page report, *Unequal Treatment: Confronting Racial and Ethnic Disparities in Healthcare*, noted that, although these disparities are associated with socioeconomic status, the majority of studies it surveyed "find that racial and ethnic disparities remain even after adjustment for socioeconomic differences and other healthcare access-related factors." As directed by Congress, the IOM committee defined "disparities" in health care as "racial or ethnic differences in the quality of health care that are not due to access-related factors or clinical needs, preferences, or appropriateness of intervention." *Unequal Treatment* concludes that, after factoring out these access-related differences, remaining disparities can be attributed to discrimination by the medical profession—physician prejudices, biases, or stereotyping of their minority patients.[54]

The IOM report sparked controversy before it was even released. When Bush administration officials read the draft report concluding that racial health disparities are "pervasive in our healthcare system," they tried to censor it. As recounted by Gregg Bloche, a Georgetown law professor who was a member of the IOM committee responsible for the report:

> top officials at the Department of Health and Human Services ordered HHS researchers to strike the term *disparity* from a Congressionally mandated annual report on—"health care disparities." . . . Two days before Christmas, HHS Secretary Tommy Thompson released a neutered rewrite, one that rejected the IOM's findings of racial disparity and dismissed the "implication" that racial "differences" in care "result in adverse health outcomes" or "imply moral error . . . in any way."[55]

Fortunately, the administration's cover-up was disclosed when outraged HHS officials leaked the earlier version. Under pressure from congressional Democrats, Thompson retracted the doctored report and issued the original version he had suppressed.

Conservatives took offense at the report's charge of racial bias. It was unfair, they argued, to suggest that blatant racial prejudice, as was demonstrated in the Tuskegee syphilis experiment, still lingered in contemporary medical care. "I would stress that the attitudes of physicians today have shown a true revolution from those that permeated the generation or two ago," wrote University of Chicago law and economics professor Richard Epstein. "It is a shame to attack so many people of good will on evidence that admits a much more benign interpretation."[56] The "benign" interpretations offered by critics were threefold: racial disparities stem from patient behavior, biological difference, and economic inequality. Each of these explanations sounds familiar to historians of race and health in America.

Michael Byrd and Linda Clayton, physicians who teach at the Harvard School of Public Health, are the nation's leading scholars on the history of racial inequities in medical care. In their painstaking two-volume medical history of African Americans, the couple has shown that the long legacy of treating blacks unequally is deeply ingrained in medicine.[57] When I met them at their home in Boston in fall 2007, they escorted me to an apartment on another floor that houses their vast library on the diagnosis and treatment of African American diseases—what they call "race medicine." "Two hundred forty-six years, or almost two thirds of our experience here in America, we were in slavery," Michael Byrd began. "So, to understand African American health, you've got to know something about slave health." Enslaved Africans were treated by a separate and awful "slave health subsystem" that either neglected their health, treated them so as to benefit their masters, or subjected them to barbaric medical experimentation. The medical profession was intimately involved in perpetuating slavery. "The slave ship surgeons were so important to the slave trade that they were the second highest paid person on a slave ship other than the captain," Byrd told me.[58]

Following the Civil War, black health plummeted further with the collapse of the slave health subsystem and the beginning of Jim Crow–governed medical care. "Now the next hundred years, from 1865 all the way up to 1965, when Medicare and Medicaid were signed, represents a period of open overt health care segregation and discrimination," says Byrd. "That's 346 out of 388 years in the health system. The habits of treating people unequally and treating people in a discriminatory manner were ground into the system from its very beginnings. Slavery established the way we do business in health care in the United States. It might be a reason that we are the last Western industrialized country not to have a universal system of health care."

But most doctors fail to connect race medicine across different periods of history. Cardiologist Jay Cohn, who invented the first race-specific drug approved by the FDA, calls the use of race to distinguish patients "a universal and well-accepted custom in medicine." Like the medical students I talked to, he draws a line between the biased past and the scientific present. "Although the origin of this practice may, in part, reflect past prejudicial attitudes," he writes, "its use today can certainly be defended as a useful means of improving diagnostic and therapeutic efforts." Perhaps a decade ago, doctors employed race to misdiagnose and abuse patients, but now race makes their decisions more precise and efficient, he argues. This is a matter of statistical accuracy, not a bad attitude: "It is not cost-effective to search for sickle cell disease in a white anemic patient or to search for cystic fibrosis in a black patient with lung disease," says Cohn.[59]

But this crude statistical approach, relying on unscientific assumptions about racial difference, can harm patients. Richard Garcia, a California pediatrician, tells the story of his childhood friend, Lela, who was described by doctors as a "2-year-old black female with fever and cough" and later as a "4-year-old black girl with another pneumonia," as she continued to suffer from an unshakable respiratory ailment.[60] Finally, when she was eight, a radiologist who read her chest X-ray without seeing her face-to-face asked, "Who's the kid with cystic fibrosis?" Garcia points out that, had Lela been white, she would have been diagnosed correctly and treated when she was a toddler. But because she was black, her doctors never considered she could be suffering from a "white" disease. They misinterpreted the data showing that black children are less likely to have cystic fibrosis to mean that they *never* have it. Or perhaps they thought it was not "cost effective" to check Lela for cystic fibrosis despite her symptoms.

Applying a sophisticated biostatistics model to several uses of race in medicine, epidemiologists Richard Cooper and Jay Kaufman calculated that differences between racial groups are usually too small to warrant using this variable as a predictive tool or as a factor in clinical decision making. The practice risks "stereotyping and the tendency to misapply quantitative differences between groups as though they were categorical differences," they argue—as happened with the misdiagnosis of Lela.[61] Instead, when a patient presents with a particular set of symptoms, doctors should evaluate the symptoms based on clinical diagnostics and family history, not assumptions about their race. "Doctors are practicing race medicine when they take into account the color of the patient," says Linda Clayton.

Although medical schools may insert a class on racial bias in the curriculum, training future doctors to practice race medicine will only intensify with the expansion of racial genomics. Genetics is becoming a more important subject in medical school as doctors are prepared to incorporate discoveries about DNA in their treatment of patients. Many genetics professors will teach the new racial science that claims to have confirmed that the "principal human races" can be found at the molecular level. A single course on cultural dynamics in medicine can hardly compete with centuries of racialized notions of disease now reinforced by the popular and lucrative science of genetic difference. The University of California in San Francisco recruited a team of medical anthropologists, historians, and clinicians to lead small-group discussions on "Culture, 'Race' and Ethnicity in Cancer" with first-year medical students. Physician and historian of science Warwick Anderson withdrew from the team in 2003 when the lecture he was supposed to discuss told students uncritically about the recent discovery by Bay Area scientists that genetic clustering corresponded with racial groupings. The lesson's objective used to be to "describe the dangers (both moral and scientific) of conflating socially constructed categories of difference such as 'race' with actual genetic or biological difference across the human population." Now it had drastically changed to: "Recognize the ways in which 'race' is often characterized in relation to genetics and culture."[62]

The main "benign interpretation" of the IOM findings that conservatives, including Satel, propose is economic. Instead of blaming racial bias, they argue, "understanding health disparities as an economic problem tied to issues of access to quality care and health literacy, rather than as a civil rights problem borne of overt or unconscious bias on the part of physicians, is a more efficient and rational way to address the problem of differential health outcomes."[63] Critics of the IOM report are right that the roots of health inequities run far deeper and wider than individual cases of prejudice or misdiagnosis. But where they are wrong is in their assumption that racial disparities in health can somehow be separated from racism. Our health care system is structurally designed to give minority patients inferior treatment.

Unequal access to care does not necessarily mean that medical care is completely unavailable to minorities or that it is provided by biased doctors. It typically means that minority patients have access only to inferior medical services and are systematically barred from high-quality care. In an eye-opening study reported in 2004 in the *Journal of the American Medical Association*, Yale epidemiologist Elizabeth Bradley and colleagues investigated why it took much longer for nonwhite heart attack victims to receive lifesaving therapies after they were rushed to the hospital. They learned that African Americans wait the longest of any group to be treated—20 percent longer than whites. (A 2008 *New England Journal of Medicine* study similarly found that defibrillation after in-hospital cardiac arrest was delayed more often with black patients, leading to lower rates of survival.)[64] Bradley discovered that what mattered most is the hospital to which a patient is admitted, not different treatment inside the hospital. Black patients tend to go to hospitals with the worst procedures for treating heart attacks.[65]

But Bradley also found that, even holding the hospital constant, there remained racial disparities in timeliness of treatment. Race affected service within the same hospital independent of differences in patients' clinical characteristics, socioeconomic background, and insurance status. In other words, black heart attack patients are doubly harmed—they are treated in inferior hospitals, and they receive inferior treatment no matter which hospital they go to. The two types of bias are most likely related: the historic devaluation of black lives created and perpetuates a system of inferior care in which physicians' stereotypes about black patients are fostered. Ending the inequitable treatment of black heart attack victims requires both dealing with individual physician bias and making systemic changes that guarantee all patients access to high-quality hospitals.[66]

Another study published in 2004 in the *New England Journal of Medicine* linked the lower-quality health care that black patients usually receive to the clinical training and resources of the physicians who treat them. Researchers at New York's Memorial Sloan-Kettering Cancer Center analyzed 150,391 visits by black Medicare beneficiaries and white Medicare beneficiaries who were seen by 4,355 primary care physicians.[67] The research group immediately noted the striking segregation of medical care. Visits by black patients were concentrated among a small group of physicians who rarely treated white patients. What's more, the doctors who treated blacks were less equipped to provide high-quality care than the doctors who treated whites: physicians visited by black patients were less likely to be board certified in their primary specialty; were more likely to report that they were unable to provide high-quality care to all their patients; and faced greater obstacles in referring their patients to qualified subspecialists, in getting excellent diagnostic imaging, and in nonemergency admission to the hospital. "We have the only organized health care system on Earth in which the sickest are treated by the least skilled and the least qualified," Michael Byrd told me. The expansion of Medicaid in the 2010 federal health care overhaul to cover millions of poor and low-income Americans who cannot afford insurance will help to narrow racial disparities in health. But the evidence of differential treatment despite equal access to state-sponsored programs shows that health care reform must address racial bias at the individual and institutional levels.

The impact of second-class treatment on black people's bodies is devastating. It is manifested not only in the black–white death gap but also in the drastic measures required when chronic disease is left unmanaged. Black patients are less likely than whites to be referred to kidney and liver transplant wait lists and are more likely to die while waiting for a transplant.[68] If they are lucky enough to get a donated kidney or liver, blacks are sicker than whites at the time of transplantation and less likely to survive afterward. "Take a look at all the black amputees," said a caller to a radio show I was speaking on, identifying the remarkable numbers of people with amputated legs you see in poor black communities as a sign of health inequities. According to a 2008 nationwide study of Medicare claims, whites in Louisiana and Mississippi have a higher rate of leg amputation than in other states, but the rate for blacks is five times higher than for whites.[69] An earlier study of Medicare services found that physicians were less likely to treat their black patients with aggressive, curative therapies such as hospitalization for heart disease, coronary artery bypass surgery, coronary angioplasty, and hip-fracture repair.[70] But there were two surgeries that blacks were far more likely to undergo than whites: amputation of a lower limb and removal of the testicles to treat prostate cancer. Blacks are less likely to get desirable medical interventions and more likely to get undesirable interventions that good medical care would avoid.

Difference Under the Skin

Nineteenth-century scientists such as Samuel Morton, who amassed an enormous collection of skulls, studied comparative anatomy to establish physical distinctions between the races. Black body parts became a useful tool for scientists looking for ways to precisely measure racial variation. Ears, pelvises, skulls, fingers, and genitalia "could be easily dissected, stored, counted, and circulated by interested scholars," report historians Evelynn Hammonds and Rebecca Herzig, creating a gruesome traffic in black people's bodies for scientific purposes.[71] Medical researchers became convinced that differences in the shape and size of internal organs and in bodily systems produced racially specific diseases. "It is commonly taken for granted, that the color of the skin constitutes the main and essential difference between the black and white race," wrote the prominent Louisiana physician Samuel Cartwright in 1851. "But there are other differences more deep, durable and indelible, in their anatomy and physiology, than that of mere color."[72] This shift in scholarly focus from skin color to the internal anatomy of blacks was an effort by scientists to dig deeper and deeper in order to find more accurate evidence of racial differences—an ongoing search that has brought scientists today to the molecular level.

Before Henrietta Lacks died of cervical cancer in 1951 in a colored "clinic patient" ward at the Johns Hopkins charity hospital, a doctor cultured the cells from her cervix without her consent. Her cells, known by scientists as HeLa, had the amazing ability to reproduce endlessly and prolifically, providing material for more than sixty thousand scientific studies that contributed to a wealth of medical advances from the polio vaccine to chemotherapy and in vitro fertilization. While the HeLa cells enabled a lucrative biomedical industry, the Lacks children, who were not informed for decades about the fate of their mother's tissue, struggled with inadequate health care. The injustice is captured in the words of Lacks's daughter: "I *would* like some health insurance so I don't got to pay all that money every month for drugs my mother's cells probably helped make."[73] The story of Henrietta Lacks reflects the horrible history of medical exploitation and neglect of African Americans that has been reinforced by the view that their bodies are intrinsically inferior. But her story also defies the belief in inherent racial differences. Her cells, although they came from a black woman, helped to improve the health of human beings the world over and testify to our common humanity.

5

The Allure of Race in Biomedical Research

By the mid-1980s, researchers had again turned to the idea that racial identity was a useful way to distinguish human bodies for medical purposes. Ironically, as sociologist Steven Epstein has chronicled in his book *Inclusion: The Politics of Difference in Medical Research*, this campaign was led by minority and women health advocates in a push to diversify the populations studied in medical research. At the time, most clinical research subjects were middle-aged white men, whose bodies had been defined as the scientific norm. Critics demanded instead that medical research be made more inclusive to attend to the particular health needs of women, children, and racial minorities. Claims about justice in scientific research had shifted from protecting socially disadvantaged subjects from unethical practices toward promoting access to clinical trials and biomedical products.[1]

Inclusion of groups that were previously underrepresented in clinical research meant measuring biological differences across these groups—what Epstein calls the "inclusion-and-difference paradigm." In response to these demands, the federal agencies "ratified a new consensus that biomedical research—now a $94 billion industry in the United States—must become routinely sensitive to human differences, especially sex and gender, race and ethnicity, and age," writes Epstein.[2] Starting in 1986, a series of federal laws, policies, and bureaucratic offices institutionalized the scientific use of racial categories to ensure greater participation of minorities in clinical research and to address health inequities. Any federally supported university scientist performing biomedical research involving human beings or any company seeking approval to market pharmaceuticals is required to include racial and ethnic minorities as research subjects and to analyze their findings by race.

The campaign for inclusion resuscitated federal interest in minority health, but one that focused on biological rather than systemic inequities. Within the space of a decade, the federal research bureaucracy incorporated a new focus on minority health at every level. A task force commissioned by Margaret Heckler, secretary of the Department of Health and Human Services (HHS) from 1983 to 1985, to examine the state of minority health released its pathbreaking report in 1985, documenting "a continuing disparity in the burden of death and illness experienced by blacks and other minority Americans as compared with our nation's population as a whole."[3] In the space of five years, HHS founded an Office of Minority Health, the Centers for Disease Control created the Office of the Associate Director for Minority Health, and the NIH established the Office for Research on Minority Health. Congress also began to require that biomedical researchers receiving government money pay attention to race. The NIH Revitalization Act of 1993 mandates that federally funded clinical studies enroll women and minorities as subjects to "elicit information about individuals" in these groups and, in the case of trials evaluating interventions, "examine differential effects on such groups,"[4] Many researchers interpret this policy as a requirement to break down research findings into racial categories. The act also specifies that researchers must use the racial categories provided in OMB Directive No. 15 for all federal reporting.

There were dissenters to the race-conscious approach to inclusion in biomedical research. Some conservatives argued that the NIH rules imposed unlawful gender and racial quotas on researchers, likening them to affirmative action policies. Some minority doctors dedicated to racial equality in medicine were also worried. Otis Brawley, an African American oncologist then at the NIH National Cancer Institute and now chief medical officer of the American Cancer Society, warned that the NIH Revitalization Act "may eventually do more harm than good for the minority populations that it hopes to benefit. The legislation's emphasis on potential racial differences fosters the racism that its creators want to abrogate by establishing government sponsored research on the basis of the belief that there are significant biological differences among the races."[5]

Race consciousness in federal funding guidelines creates a perplexing paradox. While designed to correct historic neglect of people of color in biomedical research, requiring that biomedical researchers use race as a variable risks reinforcing the very biological definitions of race that have historically supported racial discrimination. Paying attention to racial disparities in health is crucial to eliminating them, but attention to race in biomedical research can also make these disparities seem to be grounded in biological difference rather than social inequality.[6]

The way out of this paradox is to focus on how race is used and defined by the research at hand. Some federal research initiatives that investigate the reasons for race-based disparities and develop programs to address them properly treat race as a social grouping that has consequences for people's health. The NIH guidelines fall into trouble when they import these social categories into research that reaches biological conclusions—as if race were a biological category. As I showed in chapter 3, forcing genetic findings from population and biomedical research into social categories of race threatens to make these categories seem genetically determined. The NIH requirements can easily be interpreted to treat races as biologically distinct populations whose health status and responses to therapies vary for genetic reasons inherent to each group.

Researchers often mechanically break down their findings by race to comply with the NIH guidelines. Even if they are careful to identify study participants in narrower geographic, ethnic, or indigenous terms, they are compelled by NIH rules to "aggregate" their findings into the approved racial categories.[7] They then proceed to report race-based conclusions at the end of the study—even when racial subsamples are too tiny for statistically sound results and even when race was not related to the purpose of the study in the first place. When I asked a scientist who studied genetic contributions to hypertensive heart disease why she reported her findings according to the race of research subjects, she responded simply, "I had to report by race to get NIH funding!" In his interviews with biomedical researchers studying racial differences in health, Steven Epstein "found that researchers consistently were unable to articulate clear definitions of 'race.'"[8] Good science requires a cogent definition of racial variables used in research as well as an intelligent hypothesis of why they are relevant. "I think that requiring these racial categories traps us into continuing to use them without breaking past them and trying to find other ways of actually directly measuring the things that we are really interested in," Stanford bioethicist Mildred Cho told me.[9]

It appears that Congress had not thought through these problems when it passed the NIH Revitalization Act in 1993. Most of the advocacy for diversifying biomedical research came from women's organizations, and the debate centered on the biological differences between the sexes. In fact, the original mandate in the NIH reauthorization bill, drafted by congressional staff in 1990, referred only to inclusion of women as research subjects. It was only after the Congressional Black Caucus belatedly intervened that the phrase "and minorities" was added to the language requiring inclusion of women in NIH-funded research.[10] Congress did not seriously weigh what it meant to treat race, along with sex, as a biologically distinctive category. The law lumps together women and minorities as "second-class citizens" who had been wrongfully excluded from clinical trials without recognizing that race and sex are not parallel kinds of identity. The biological distinctions between men and women (though far more fluid than commonly held) are not mirrored in biological distinctions among races.

For most of U.S. history, people of color were exploited in medical experiments that injured or stigmatized them while they were excluded from clinical trials designed to improve health. Congress was right to correct this injustice but went about it in the wrong way. The purpose of diversifying biomedical research should not be to find innate differences among racial groups. It should be, first, to give patients equal access to the benefits that can accrue simply by participating in a clinical trial and, second, to give scientists a richer resource to investigate the mysteries of human biology—what makes cancer tumors grow in human tissue, why a therapy is effective for some patients and not for others, and how to stop the progression of Alzheimer's, for example.[11] Adding minority patients to the research pool provides a more accurate reflection of human diversity.

Third, diversifying clinical research can aid in investigating how racism harms people's health. Scientists need a political, not a biological, definition of race to accomplish this. If race is treated accurately as a social category, there is nothing wrong with recruiting members of a particular racial group to investigate the causes of illness in the group and the best ways to eliminate them. Through its Racial and Ethnic Approaches to Community Health (REACH) program, the CDC Office of Minority Health sponsors scientific studies that "target" diseases within particular racial and ethnic communities. As chair of the board of directors of the Black Women's Health Imperative, I support the organization's research projects and educational programs (some funded by the CDC) that have addressed black women's greater burden of diabetes, obesity, breast cancer, heart disease, and HIV/AIDS because these efforts tackle the preventable social reasons for these disparities.

In the decade since the federal inclusion policy was launched, little progress has been made, however. A 2009 analysis of fifteen health status indicators found that disparities widened significantly for a third of them: mortality rates for heart disease, breast cancer, diabetes, and suicide, as well as cases of tuberculosis. The disparity in infant deaths remained virtually unchanged. In short, "there was no significant trend toward overall improvement."[12] The 2007 National Healthcare Disparities Report, issued by HHS, had reached an equally dismal conclusion: "Overall, disparities in health care quality and access are not getting smaller."[13] The paradigm that includes minorities in biomedical research to discover their intrinsic biological differences has done nothing to close the racial chasm in health. Yet biomedical researchers are increasingly turning to genetic explanations for racial disparities in health and disease.

Searching for the Gene

Asthma, an inflammatory disorder of the airways, is the most common chronic disease striking children and is the illness that most frequently puts children in the hospital. Asthma symptoms are triggered by allergens or irritants in the environment that cause the airways to narrow. People with asthma have a hard time breathing and are subject to bouts of coughing and wheezing. "I think that asthma's worse for children, though, because play is part of childhood and children cannot play with real abandon when they feel so bad," writes Jonathan Kozol in *Ordinary Resurrections* about children living in the South Bronx, where asthma rates are sky-high. "Even mild asthma weighs their spirits down and makes it hard to smile easily, or to read a book with eagerness, or to jump into conversation with entire spontaneity."[14]

It has been estimated that one third of children living in public housing have allergic asthma.[15] Puerto Rican and African American children have especially high rates of the disease: while 8 percent of white children have asthma, 19 percent of Puerto Rican and 13 percent of black children do. Asthma prevalence and death rates have been increasing in recent decades, especially in inner-city communities, further widening the racial gap.[16]

Many research studies have identified the environmental allergens that trigger asthma. For example, a team of scientists from Boston University School of Medicine traced the cause by exposing mice to dust particles collected from inner-city homes and studying the effects on their lungs. The culprit turned out to be exoskeletons and droppings from cockroaches.[17] From 2002 to 2005, New York University researchers attached air pollution monitors to the backpacks of children with asthma in the South Bronx. They found that the children, who were twice as likely to attend a school near a highway as children in other parts of the city, were exposed to fine-particle pollution from diesel exhaust (a known asthma trigger) that exceeded EPA standards.[18]

It is clear that exposure to pests and air pollution increases the risk of asthma, but why does the risk vary according to race? Why is the rate and severity so high in Puerto Rican and black children in particular? Esteban Gonzalez Burchard, the biopharmaceutical researcher at University of California at San Francisco we met in chapter 3, thinks it has to do with their genes. Burchard calls himself a "physician scientist" who studies the influence of race on the genetic causes of disease. He established the Genetics of Asthma Laboratory in a quest for the unique genetic signature that predisposes children of certain races to get sick with asthma. He also wanted to know why Puerto Rican children respond poorly to albuterol, the top asthma drug. To assist the effort, Burchard pulled together a multidisciplinary team of experts in genetic epidemiology, biostatistics, genomics, clinical asthma, and pulmonary molecular and cell biology. He has collected thousands of genetic samples, stored by race in the university's DNA Bank, to create a database his lab team can scan for genetic clues as to what distinguishes rates of asthma in different racial and ethnic groups. The distinctive genetic variant in Puerto Ricans, he hypothesizes, is related to their recent African ancestry and also explains asthma severity in African Americans.

I sat down with Burchard in June 2008 at his office in the Genetics of Asthma Laboratory to ask him why he was so sure that disparities in asthma had a genetic root. "I'm fascinated as to why disease rates and severity vary across populations and how racial or ethnic background influences that. I'm also impressed by how race modifies risk factors, whether they would be genetic or environmental," Burchard tells me. But he quickly leaves the environmental part to boast about his lab's recent discovery: "We just published a paper that came out two weeks ago in *Human Molecular Genetics* in which we identified an African-specific mutation. Meaning that we screened

Caucasians, Asians, different Hispanic subgroups, and Africans, and this mutation was specific to African origin." Burchard explains that the mutation is involved in regulating a protein that affects smooth muscle tone in the airways, saying, "Long story short, it causes more severe asthma."[19]

"When you say it's African specific, do you mean that this mutation is only found in people of African descent?" I asked. I was surprised because the scientific literature usually speaks of differing allele frequencies, not racially exclusive mutations. Yet Burchard told me the mutation was confined to people with African ancestry. I learned from a later reading of the article Burchard mentioned that this conclusion was based on his lab's screening of research subjects, including twenty-four African Americans, for the polymorphism C818T. "We did not find a single subject heterozygous for the C818T SNP in screening 96 Puerto Rican, 96 Mexican, 86 Caucasian, and 7 Asian asthmatics, implying this SNP is specific to populations of African origin," his research team concluded.[20] Yet, given the research on human genetic diversity, the uncertainties inherent in racial self-identification, and Burchard's small sample, I needed better evidence of this "implied" African gene. It makes no sense for African Americans but not Puerto Ricans to have this supposedly race-specific allele when people in both groups have recent African ancestry.

"Some people would say that by focusing on minority health, you're reinforcing the idea that minorities are biologically different. How you do respond to that?" I asked Burchard.

"I think populations *are* biologically different," he replied. "I mean, whether you are minority or non-minority, I think populations are biologically different, just like males and females are biologically different."[21] He continued, "So, for example, cystic fibrosis, that is a Caucasian mutation, only in Caucasians. We are finding it now in African Americans and Puerto Ricans. And that's because of the intermixing of populations."

But as we talked, Burchard's views of race and genetics became complicated by his understanding of race as a social category. "I identify as being Mexican or Hispanic, knowing that I am like Obama. I had a white father, my mom looks black—she's darker than you. She identifies as being Mexican, and when I told her she was twenty percent African, at first she said no, that's not correct. But in her infinite wisdom, she later said, fine, who cares." I pushed Burchard on the contradiction between his uses of social and biological race. He seemed to separate the race-based genetic research he was conducting from race-based social effects on health.

"That's why we need more research in this area. It's fascinating to think about the interaction between your biologic background and your genetic background and the social forces that are operating on it. Now, we know if we talk socioeconomic status, somehow your social position in life gets internalized. The physiological outcomes of stress translate into high blood pressure, which is translated into kidney disease and the heart disease, which is translated into premature death. We also know that the social discrimination, particularly in United States, somehow gets internalized in African American males in particular." By the end of our conversation, Burchard was emphasizing genetics less. "Personally, I think I have been pretty good at saying that it's not all genetic. People try to paint me as a pure geneticist, but I know that there is an interaction between social and environmental and genetic factors."

"Then why establish a lab devoted to finding the genetic roots of racial differences in asthma?" I asked. "Right now I don't think we know enough about the potential outcomes of genetic testing in specific racial groups. I think in the case of the mutation that we identified, if it did pan out, it could be a novel drug target that we would say, gee, if you are an African American male, here is the drug that you should take. . . . That would be an example of how our sort of work could be directly translated into clinical applications."

The Genetics of Asthma lab is one of countless research projects at universities and biotech firms around the country hunting for the genes that are responsible for health disparities in America. They are supplementing a large body of published studies that claim to show that racial gaps in disease prevalence or mortality are caused by genetic differences. In addition to asthma, disparities in infant mortality, diabetes, cancer, and hypertension have all been attributed in the scientific literature to genetic vulnerability that varies according to race. Most of these studies never even examined the genotypes of research subjects, as Burchard's lab does; they just infer a genetic source of racial differences when they fail to find another explanation. As interest in health disparities converges with the genomic science of race, a new brand of racial stereotyping is gaining hold in biomedical research.

Consider an effort to explain the enduring black–white gap in premature births and low birth weight. A team of obstetric researchers examined all births in Missouri between 1989 and 1997 to test the hypothesis that "black race independent of other factors increases the risk of extreme preterm birth and its frequency of recurrence." The researchers used statistical methods to calculate the independent influence of race, socioeconomic status (whether the mother was a recipient of Medicaid, food stamps, or the WIC program), and maternal medical risk factors such as lack of prenatal care and cigarette smoking. An article published in 2007 in the *American Journal of Obstetrics and Gynecology* reported that black women were not only more likely to deliver preterm babies but also to have preterm births in subsequent pregnancies. Because this overrepresentation occurred even when they controlled for the medical and socioeconomic factors, the researchers concluded that their findings "suggest a probable genetic component that may underlie the public health problem presented by the racial disparity in preterm birth." Although conceding that they may have overlooked "hidden variables" that also contribute, they nevertheless speculated about an unproven genetic mechanism operating in "the black race":

> We postulate that although preterm birth is a detrimental outcome in pregnancy, it may be a result of a selective advantage, conferring inflammatory protection against other disease processes. This selective advantage phenomenon has been well described for diseases afflicting the black race, particularly sickle cell disease, glucose-6-phosphate dehydrogenase deficiency, and nitrous oxide synthase polymorphisms and their effects on the incidence of malaria.

The article ended by downplaying "disparate access to medical care or other environmental factors," arguing that "our data suggest that the proposed genetic component to preterm birth may be a greater etiological contributor than previously recognized"—despite presenting no genetic data whatsoever![22]

Despite its weaknesses, the Missouri birth study was dignified with a published roundtable discussion in which commentators granted that "the genetic link is very strong" and that the disparity "may best be explained by a genetic etiology."[23] The research also led to the headline "Study Points to Genetics in Disparities in Preterm Births" in the *New York Times*, which repeated the totally unsubstantiated conjecture that premature births may provide some evolutionary advantage to black women. Neil Risch, the colleague of Esteban Burchard who conducts genetic research on health disparities, criticized the study's inference of a genetic cause without ever examining genes. "They're inferring something is genetic by elimination of other factors," he told the *New York Times*. "But geneticists believe that to implicate something as genetic requires direct evidence, as opposed to evidence of absence."[24]

In this study and others like it, guesswork about a peculiar black predisposition toward unhealthy births imports an old notion about sickle cell disease "afflicting the black race."[25] Whenever I give a talk on this topic, there is inevitably someone in the audience who invokes the mantra that sickle cell anemia is a black genetic disease and therefore proves that race is a genetic category. This misconception was first popularized in the early twentieth century by hematology experts who believed the capacity to develop sickled cells was uniquely inherent in "Negro blood."[26] Stereotypes about black resistance to malaria and susceptibility to sickle cell justified sending black workers to malaria-infested regions in the first part of the century and later led to discriminatory government, employer, and insurance-testing programs in the 1970s.[27]

The error is easily exposed by looking at two world maps, one highlighting the regions around the globe where malaria is prevalent, the other highlighting areas where sickle cell disease is present. The maps mirror each other perfectly. By comparing them, it is plain to see that malaria and sickle cell aren't restricted to Africa and that much of Africa is unaffected. High frequencies of the trait also occur in parts of Europe, Oceania, India, and the Middle East, all places where there is malaria. In fact, people in the town of Orchomenos in central Greece have double the rate of sickle cell disease reported among African Americans.[28] If frequency of the sickle cell gene determined racial boundaries, it certainly would not prove there is a black race. Instead, as Jared Diamond pointed out in the November 1994 issue of *Discover*, if we grouped together people by the presence or absence of the sickle cell gene, "we'd place Yemenites, Greeks, New Guineans, Thai, and Dinkas in one 'race,' Norwegians and several black African peoples in another."[29] It would be more accurate to call the groups with the sickle cell gene the "anti-mosquito race." Of course, that would be a silly way of grouping people, except for studying the sickle cell gene. But "black race" is an equally silly way of grouping people for identifying genetic contributions to disease.

Another favorite playground for genetic speculation is hypertension. Until recently, virtually every study of hypertension among African Americans accepted the premise that blacks have higher rates of the disease than whites because of inherited susceptibility.[30] In volumes 27 through 30 of the scientific journal *Hypertension*, published in 1996 and 1997, thirty articles hypothesized the existence of innate physiological differences among racial groups.[31] Since then, theories about the precise genetic mechanism behind the hypertension gap are legion. Authors of one study published in the *Journal of Hypertension* in 2000, for example, "postulate that the genetic factor increasing the propensity of black people of sub-Saharan African descent to develop high blood pressure is the relatively high activity of creatine kinase, predominantly in vascular and cardiac muscle tissue."[32]

Even Oprah was familiar with one of the genetic theories for the hypertension gap. On an "Ask Dr. Oz" segment of *Oprah* in 2007, an audience member asked, "Why do I sweat so much?" After citing overactive thyroid, body toxins, and high blood pressure as possible causes, Dr. Mehmet Oz turned to Oprah. "Do you know why African Americans have high blood pressure?" Oprah replied with confidence, "The reason why African Americans have higher blood pressure, Dr. Oz, is because during the Middle Passage, the African Americans who survived were those who could hold more salt into their body."

"I'm off the show, you don't need me anymore—that's perfect!" Dr. Oz cheered.[33]

One of the most popular yarns about black genetic difference is the "slavery hypothesis" for hypertension. Originally spun by Thomas Wilson and Clarence Grim in the 1980s, the theory holds that blacks in America today suffer from higher rates of hypertension because their ancestors survived the brutal transatlantic voyage from Africa by overcoming water deprivation and dehydrating illnesses owing to their genetic predisposition to retain sodium. This hereditary trait, proponents claim, came to dominate the gene pool of enslaved Africans and was passed down to a disproportionate share of present-day African Americans. Their genetically impaired ability to excrete salt expands water volume in the blood vessels, leading to higher rates of hypertension. But many experts—including slavery historian Philip Curtin, on whose work Wilson and Grim had relied; biological anthropologist Fatimah Jackson; and epidemiologist Jay Kaufman—have refuted this conjecture on methodological, evidentiary, and theoretical grounds, while others have provided more plausible social explanations for African American hypertension rates.[34] For one thing, blacks in former slave societies like the West Indies do not have the high hypertension rates of blacks living in the United States.

A landmark study led by Richard Cooper contested the conventional wisdom that blacks have an inherent predisposition to hypertension.[35] Comparing hypertension rates around the world, Cooper analyzed three surveys of blacks from Africa, the Caribbean, and the United States and eight surveys of whites from the United States, Canada, and Europe. Collectively, the studies enrolled 85,000 participants. If African Americans' higher hypertension risk were genetic, we would expect that people of African descent are more likely to have high blood pressure than people of European descent. Instead, after pooling the global data, Cooper found just the opposite. White populations on average have a substantially higher burden of hypertension. Germans have the highest. Nigerians have the lowest. U.S. whites come close to black Nigerians and Jamaicans, while U.S. blacks come close to whites from England and Spain.

Although the slavery hypothesis has been thoroughly debunked, it still holds sway in the popular imagination and even in professional circles. The theory is described in numerous hypertension textbooks without mention of the refutations and "frequently invoked in the medical literature to justify the more general proposition of innate biologic difference in cardiovascular disease risk and treatment efficacy," writes Jay Kaufman.[36] The myth received a shot in the arm when the Harvard economist Roland G. Fryer Jr., an African American, embraced it in his larger research project to "figure out where blacks went wrong."[37] Fryer co-authored a 2005 paper with two white colleagues in the Harvard economics department, Edward Glaeser and David Cutler, attributing the six-year disparity in life expectancy between blacks and whites to blacks' inherited tendency to retain salt.[38] Fryer rehashed the discredited theory on the CNN series *Black in America*. A 2005 *New York Times Magazine* article about Fryer by Stephen Dubner, co-author with Steven Levitt of *Freakonomics*, states that Fryer "came across a period illustration that seemed to show a slave trader in Africa licking the face of a prospective slave," presumably to "try to select, with a lick to the cheek, the 'saltier' Africans." Dubner writes that Glaeser and Cutler appreciated having a black collaborator to circulate the theory: "There's an insulation effect," Glaeser said. "There's no question that working with Roland is somewhat liberating." According to Dubner, Fryer is able to "raise questions that most white scholars wouldn't dare."[39] It is not the case, however, that white scholars are fearful of attributing racial disparities in health to genetic difference; most of the hundreds of articles making precisely these claims are authored by whites and published in prestigious scientific journals. Rather, having a black co-author on a dubious theory about black genetic difference confers seeming legitimacy.

The slavery hypothesis may be particularly egregious, but in fact the whole body of genetic explanations for health disparities is questionable. To begin with, most of these studies suffer from the methodological sloppiness I discussed in chapter 3. They group research subjects into conventional racial categories, fail to explain the relationship between these social categories and genetic traits, and then reach conclusions about genetic difference among the subjects. A survey published in the *Journal of Medical Ethics* in 2006 examined 268 published reports of genetic research that used race as an independent variable.[40] The research team found that 72 percent of the studies failed to explain their methods for assigning race to research subjects. Despite this glaring flaw, 67 percent of the same studies drew conclusions associating genetics, health outcomes, and race.

But there is a far more fundamental defect. Genetic explanations for health disparities are basically implausible. Remember, the issue is not whether genes affect health—of course they do—but whether *genetic difference* explains *racial disparities* in health. If you approached health disparities with a completely open mind, with no preconceived assumption that racial differences must be genetic, it would make perfect sense that social groups that have been systematically deprived for centuries have worse health than social groups that have been systematically privileged. The logical cause is the social distance between them and all the ways societal advantage and disadvantage affect people's experiences, environments, and access to resources, including health care.

Likewise, it would seem strange for a large group of people as genetically diverse as African Americans to have such a concentrated genetic susceptibility to so many common complex diseases. When he came to the United States from his native Nigeria, the geneticist Charles Rotimi was struck by the gulf between white and black health. "It seemed highly unusual to see these disparities," he told me. "I will call it a privileged perspective—I came from a different environment and to see that a group of people in this society were so heavily at risk for multiple conditions was a curious thing. I thought, this cannot be genetics—why would a group of people inherit so many bad things?"[41] A more plausible hypothesis, given the persistence of unequal health outcomes along the social matrix of race, is that they are caused by social factors.

In order to conclude that the cause of health disparities is genetic, scientists must first rule out more logical social explanations. That is a near-impossible task because of the nature of gene–environment interactions: it is very hard to separate a genetic cause from environmental influences. Definitively showing a genetic cause for a racial disparity in disease prevalence or outcome would require the kind of experiment researchers perform on laboratory mice. The standard scientific test required to prove that a phenotypic difference (such as different disease rates) results from a genetic difference is a controlled breeding experiment with rigorously regulated environments that spans at least two generations. Applied to human beings, the study would last sixty years and examine the offspring of men from different races who mated with at least four carefully selected women, so the offspring could be compared. The researcher would also have to dictate "what those children could eat, where they could live, and what exercise regime they could have maintained," points out evolutionary biologist Joseph Graves Jr.[42]

Perhaps it is unfair to expect such a high degree of scientific precision. But studies that conclude health disparities are caused by genetic difference do not even come close. These studies typically control for the socioeconomic status (SES) of the research subjects in an attempt to compare subjects of different races who have the same SES. If there remains a difference in the prevalence or outcome of a disease, the researchers typically attribute the unexplained variation to genetic distinctions between racial groups. But this conclusion suffers from a basic methodological error. The researchers failed to account for many other unmeasured factors, such as the experience of racial discrimination or differences in wealth, not just income, that are related to health outcomes and differ by race. Any one of these unmeasured factors—and not genes—might explain why health outcomes vary by race. Statisticians call this the problem of residual confounding: falsely concluding that there is a causal relationship between two variables (here, genetics and disparate health outcomes) because other variables are not measured.[43]

An important aspect of this problem is that SES measures used in genetic studies are woefully inadequate. The typical measures—occupation, income, and education—do not capture fully the social and economic factors that determine social status. Whites and blacks with the same income and educational levels occupy different rungs on the social ladder and are not interchangeable. Because of their racial privilege, whites earning the exact same salary as blacks tend to have greater wealth—money in the bank, property and investments, and anticipated inheritance. A large federal survey showed that, even after adjusting for SES and household characteristics, blacks were more likely than whites to have experienced economic hardships during the previous year.[44] Blacks who appear to be as poor as poor whites when income alone is measured are at greater risk of being unable to pay the rent, having their utilities shut off, and being evicted. Black poor people experience a more intense poverty than white poor people.[45] Even when black individuals reach the middle class, chances are they have close relatives who are poor, so they bring family financial needs with them into a higher bracket. Their neighborhood conditions also tend to be drastically different. Blacks are more likely to live in all-black neighborhoods with fewer services, more pollution and crime, and higher overall poverty rates. College-educated African Americans applying for jobs routinely "whiten" their résumés, deleting clues to employers that they are black because they fear their race will hurt their chances of getting an interview.[46] Because of a multitude of individual and institutional biases against blacks, the typical measures fail to control adequately for true SES.

Nor do typical SES variables measure a research participant's socioeconomic position across time. The "snapshot" model of SES data collected by biomedical researchers ignores subjects' entire life experiences.[47] Poverty and deprivation early in life may affect a tumor or heart condition or diabetes later in life. As I discuss in the next chapter, a pregnant woman's living conditions shape fetal development in ways that have lasting effects on a child's health into adulthood. This omission is compounded by the distinctive nature of black child poverty: black children are not only three times more likely to be poor than white children, but they are also more likely to be poor for their entire childhoods. So equating a black bank teller with a high school diploma earning $25,000 a year and a white bank teller with a high school diploma earning $25,000 a year may overlook extremely different life circumstances, such as childhood years in poverty, current family wealth, or neighborhood segregation, that can have a huge influence on their health.

Even with better measures of socioeconomic status, there would remain a fatal flaw. Studies testing a genetic hypothesis fail to account for the impact of racism on health at both the individual and societal levels. Genetic studies do not even attempt to measure the health effects of experiencing racism or of inequitable social systems.[48] There is growing evidence that living in a society that devalues your intelligence, character, and beauty, where you encounter discrimination on a daily basis, and in which entire institutions systematically disadvantage the group you belong to, exacts a toll on health that scientists are only beginning to fathom. Researchers cannot resort to genetic causes when they have omitted this crucial variable. "The biology is a fall-back black box that many researchers use when they find racial differences," says Harvard sociologist David Williams, a leading expert on health disparities. "It is a knee-jerk reaction. It is not based on science, but on a deeply held, cultural belief about race that the medical field has a hard time giving up."[49] Leaping to genetic conclusions after failing to account for the impact of racism on health is fundamentally unscientific.

All these methodological problems lead back to a more basic question about research testing the hypothesis that health disparities are caused by genetic differences. Perhaps it is so easy to leap to genetic conclusions, but so hard to prove them scientifically, because the hypothesis itself is faulty.[50] It is founded on a misunderstanding of race as a naturally created biological division instead of a politically invented social division. The belief in natural races despite the evidence obscures the circular logic of studies of race and genetics. Scientists observe racial disparities in health and hypothesize they are caused by biological difference based on an ideological premise that race is a biological category. After collecting data on health disparities, they conclude that unexplained differences between racial groups must be genetic, which they claim proves that races are biologically different.

But this type of research has not *proven* that health disparities stem from innate biological difference. It has simply restated the original observation of health disparities in genetic terms based on an unsubstantiated assumption of biological distinctions among races. Witness the tautological explanation appearing in the 1995 text *Biologic Variation in Health and Illness*: "Human beings are similar; they are of the same species, but belong to several different races; hence, they may differ in several important ways: in growth and development rates, in enzyme systems, in disease susceptibility, and in response to environmental stresses."[51] Because it is assumed that races differ biologically, the differences between them appear to be biological.

There is another fatal flaw in the hypothesis that health disparities are caused by genetic difference. It is not just difficult to isolate genetic causes without the type of experiment I described earlier that carefully breeds the research subjects and raises them in controlled environmental conditions. It is actually *impossible* to separate genetic from environmental contributions to health. We usually talk about genes plus environment, as if one is added to the other and each part can be independently measured and quantified. But any genetic scientist worth her degree knows that DNA's contribution to disease *always* interacts with environment in a dynamic and ongoing process. Genes are not the original foundation for health that is acted upon by the environment. From the moment a pre-embryo is created, its traits are determined *both* by genes and environment, and this interaction continues during every moment of its existence. This is why genetically identical organisms, including human beings, raised under different environmental conditions differ in physiology.[52]

When experts claim that "genes are responsible for fifty percent" of a disease or behavior, it gives the impression not only that genes are more important than they are, but that it is possible to separate genetic and environmental contributions.[53] My eight-year-old son understood this when he realized that if he cloned himself, he could not guarantee his clone would be exactly like him. "He would turn out different if you treated him differently," he observed wisely. We do not need clones to know that identical twins, with the same complement of genes, are different from each other even before they are born because of their positioning within the womb or chemical changes that happened in their cells while they were still gestating. Once they get out into the world, they develop into two distinct individuals, even when raised in the same home, because their environments and experiences are different. Similarities in twins raised apart do not obviate the myriad differences that still exist between them.

The additive model of nature versus nurture misrepresents human biology. Under a more accurate model of interactive effects on health, there is no separable genetic cause that researchers can identify through a process of elimination. As Cooper and Kaufman put it, the question of whether observed racial differences in blood pressure, low birth weight, or asthma are caused by genes "falls properly within the realm of nescience—the unknown and the unknowable."[54] It seems heretical to say that scientists are incapable of knowing everything. Scientists are supposed to speculate about possible causes of observed realities. By putting forth creative hypotheses that can be tested, scientific imagination advances our understanding of human biology. Yet contemporary scientific publishing, which generally tends to ignore studies that *disprove* hypotheses, rarely reports when these hypotheses go wrong. Instead, researchers tend to "postulate" genetic mechanisms for racial disparities that are never proved outright.

Some health disparities research is now focusing on gene–environment interactions rather than trying to isolate genetic causes of disease. As Francis Collins put it in 2004, researchers investigating risk factors for disease "must be *equally* rigorous in their collection of genetic and environmental data. If only genetic factors are considered, only genetic factors will be discovered" (my emphasis).[55] Beyond paying lip service to vaguely defined "nongenetic" factors, however, most genomic scientists are not incorporating rigorous measures of environmental factors—especially social ones—into their health disparities research. What's more, the genetic associations they discover tend to attract more academic and media attention, usually eclipsing the social influences on health altogether. Genes are frequently described as "the cause" of disease, while environmental contributions are merely "triggers," and little attention is paid to how the environmental and the biological actually interact. This emphasis on genetic versus social contributions is reflected in federal research funding. For the years 1995 to 2004, a search of research awards in the National Institutes of Health database using the term *genetics* identified 21,956 new grants (including 181 cross-indexed by the term *race*), while only 44 new grants were indexed by the terms *racism* or *racial discrimination*.[56] When the NIH launched a new center to study population health, it was originally named the Center for Genomics and Health Disparities—the environmental component was completely missing from the title. It has since been reconfigured as the Center for Research on Genomics and Global Health to eliminate the implication that it is studying genomic causes of health disparities.

It is possible that Nigerians, Jamaicans, and African Americans are all genetically prone to high blood pressure, but there is something in the environment that causes elevated rates in this country and lowers rates elsewhere. Or perhaps there is a SNP more prevalent in people with African ancestry that makes them more susceptible to environmental triggers for asthma. But if our goal is eliminating the gap between white and black hypertension or asthma in the United States, our focus should be on the environmental causes of the gap because these are factors that can and should be changed. Continuing to dwell on an unknown genetic component of health disparities only distracts scientists from the more relevant task of identifying and tackling the preventable causes of disease. Spotlighting genetic "causes" as more important than environmental "triggers" steers solutions to health disparities toward gene-targeted therapies rather than toward improving the environment for everyone. It can cause other kinds of trouble—especially when racial stereotypes come into play. A black man in San Diego who developed hypertension because of exposure to toxic chemicals lost half of his disability award after a doctor reported that blacks are genetically prone to hypertension.[57] The failure of racial science to stem the disaster of the racial gap in American health is not surprising, given the flawed ideological, theoretical, and methodological foundation that supports it.

Why, then, do scientists continue to hunt for genetic explanations for race-based health disparities? The faith in biological race is incredibly powerful. Every methodological error or theoretical infirmity is seen not as a reason to question the hypothesis, but as a challenge to look harder for the genetic difference that is presumed to exist. Dissenters are often marginalized, their scientific objections dismissed as "politically correct" or failing to grasp the importance of genetics. Nevertheless, another group of researchers has taken up a more promising line of investigation that demonstrates that racial injustice, and not genes, causes America's glaring inequities in health.

6

Embodying Race

White women in Chicago are slightly more likely than black women to get breast cancer, but black women are twice as likely to die from it. That is a startling statistic by itself. But what is equally as shocking is that in 1980 Chicago's black and white breast cancer mortality rates were identical: black and white women died at the same rate. Over the course of the next twenty-five years, the astounding gap emerged.[1] Consider this additional aspect: the disparity in breast cancer mortality in New York City is only 15 percent. In Chicago, the racial gap is ten times greater than in New York.

It is unlikely that genes explain these numbers. Did something change in white women's DNA between 1980 and 2005 that decreased their likelihood of dying from breast cancer? Is there something genetically distinct about black women in Chicago versus New York that makes breast cancer deadlier? A more logical explanation is that there is something about having breast cancer that changed and that affected black and white women in Chicago differently.

Life and Death in Chicago

In 2006, a group of Chicago breast cancer researchers released their study showing the alarming racial divergence in breast cancer deaths. An article in *Chicago* magazine featured a photo of co-author Steven Whitman, an accomplished epidemiologist with a PhD in biostatistics from Yale who directs the Sinai Urban Health Institute.[2] Whitman and the Institute have been at the forefront of documenting health disparities in Chicago. Whitman is pictured holding a giant graph with two lines representing the white and black mortality rates across the 1980s and 1990s. The lines converge at the beginning and remain fairly similar for the first decade, then begin to move apart in the mid-1990s. The space between them gradually widens over the next decade, until they reach the huge gulf between black and white women in 2005. I met with Dr. Whitman on a hot summer afternoon in 2009 to find out why.

Whitman's office in an aging brick building at Mount Sinai Hospital was small and shabby. There was no air-conditioning; a couple of fans blew hot air at us from opposite ends of the room. Whitman began by describing the hospital's location. Mount Sinai sits in an all-black community called North Lawndale on Chicago's West Side, a block from the border of South Lawndale, which is predominantly Mexican. "The patient population here at Mount Sinai is about half black and half Mexican. Sometimes, if a white person gets hit by a car down the street, they'll bring them here, but almost never do white people walk in here and get treated willingly," Whitman says.[3]

Whitman tells me that before 1950 all of the residents of North Lawndale were white, mostly Jews from Eastern Europe. Then black families began to move into the neighborhood, triggering a white exodus. What followed was a racial metamorphosis in the space of a decade. There were 110,000 white people living in North Lawndale in 1950. By 1960, they were replaced by 110,000 black people. "It's really extraordinary," Whitman says. "That's a migration of 220,000 people, which is larger than most cities in the United States." When Martin Luther King Jr. brought his civil rights campaign to Chicago, he stayed in North Lawndale. Half the neighborhood was burned down in the riots precipitated by the King assassination, and decades of disinvestment followed. North Lawndale claims only 40,000 residents today. The median income is strikingly below that for the city overall, $28,203 in North Lawndale compared to the $46,767 Chicago average; half the adults are unemployed and uninsured.[4]

The fate of Mount Sinai Hospital was tied to the neighborhood's racial transformation. In 1950, the hospitals in Chicago, like the city's neighborhoods, were segregated by race. When whites fled North Lawndale, Mount Sinai stayed behind, losing its support from white philanthropists and politicians. Its finances mirror those of its patients: it is one of the poorest hospitals in the city. Whitman likes to compare the paltry resources at Mount

Sinai to the bounty at fancy university hospitals that serve mainly white patients. "To get an A rating for a bond, a hospital is supposed to have 115 days of cash on hand," Whitman tells me. "Northwestern University Hospital has about 400 days of cash on hand. They probably spend about 5 or 10 million dollars a day. So they have literally a couple billion dollars on hand. Mount Sinai normally is lucky to have one day of cash on hand. The way we measure cash on hand is more often in hours than days. There's virtually no money."

Whitman believes this backdrop is essential to explaining why black women in Chicago are dying of breast cancer at twice the rate of white women. "The way that happened was the black rate hasn't changed at all in twenty-five years while the white rate has halved. The improvements in the white rate began to take place just as we began to figure out how to do early detection with mammography. We also began to learn more about treatment—developing medicines and radiation therapy. White women were able to take advantage of these improvements and black women not at all. So what you have is a stunningly painful observation that in twenty-five years black women have gained nothing, not one iota, in terms of breast cancer mortality from any of our advances."

As Whitman sees it, this is a man-made catastrophe for black women: "One hundred and ten black women die each year from breast cancer simply because the black rate is not the same as the white rate. That's almost half of the black breast cancer deaths. So every week in Chicago, a little more than two black women on average die from breast cancer just because of the disparity. It's literally a matter of life and death."

What blocks black women from getting the cancer care available to white women? One barrier is that black women do not have the same access to mammography. Black neighborhoods have fewer facilities that provide breast cancer screening. The sole mammogram machine in Englewood, a predominantly black area on Chicago's South Side, was broken for months. Women were sent ten miles away to get screened. Even the state-of-the-art John H. Stroger Hospital, which replaced Chicago's aging Cook County Hospital in 2002 and serves many of the city's poor African Americans, ran up a backlog of more than ten thousand women seeking mammograms.[5] Mammograms cost about $150, which can be prohibitive for a woman struggling to feed her children. Medicaid paid only about half of the cost, so many hospitals in Chicago didn't offer mammograms to women on Medicaid. "What does it mean if you have to take three buses to get to a place that gives mammograms, and then when you get there, you say, 'Here is my Medicaid card,' and they say, 'Sorry, we don't take that'?" Whitman asks.

Many private insurers required co-payments for mammograms. A survey of 366,000 women by researchers at Harvard and Brown medical schools published in the *New England Journal of Medicine* in 2008 found that co-pays deter women from getting mammograms.[6] "A small co-payment for a mammogram can lead to a sharp decrease in breast cancer screening rates," says Amal Travedi, the study's lead author.[7] The 2010 Patient Protection and Affordability Act increased women's access to mammograms by requiring health insurance plans to cover the full cost of biennial screening for women aged fifty to seventy-four years, the current recommendation from the U.S. Preventive Services Task Force.

Lack of trust in a health care system that has treated black women badly for generations is another possible deterrent. "If you've been screwed over by the health care system all your life, why would you go?" says Whitman. "It's one thing if you're bleeding; you don't have a choice. But why would you get a preventive test on a stigmatized part of your body if you didn't trust the health care system?"

A second reason is mammogram quality. Even when black women in Chicago get screened, the quality of the mammography tends to be inferior. Compared to white women, black women are far more likely to have their mammograms performed at public facilities and far less likely to be screened at academic institutions. These public facilities usually rely on older equipment and frequently lack digital mammography and trained mammography specialists—both of which are key to diagnosing cancers earlier and more accurately. The opposite is true for most white women.[8] One local facility serving poor women was catching only two cancers per thousand instead of the expected six.[9] "It wouldn't matter if every black woman in Chicago could get mammograms if we miss half the cancers," Whitman said.

A third reason has to do with access to breast cancer care. When cancer is detected in black women, they encounter multiple obstacles to high-quality treatment. Neighborhood segregation is the main one. The best treatment facilities are located far from black neighborhoods. Of the twenty-five Chicago community areas with the highest breast cancer mortality rates, twenty-four are predominantly black. Only one of these has a hospital with a cancer program approved by the American College of Surgeons Commission on Cancer.[10] "They are located the farthest from the women are who are sickest from breast cancer," Whitman explains. "It's almost like someone on purpose put the institutions in a different part of the city. Go find the fancy institutions—they are all in white neighborhoods."

Other studies bear out this racial inequity in breast cancer treatment. In 1987, when researchers had begun to notice the decreased survival rate of black breast cancer patients, a team at the National Cancer Institute linked survival rates to the kind of treatment patients received.[11] White women were more likely to be treated initially with surgery, which gave them a better chance of survival. Black women were more likely to be treated without surgery or to receive no treatment at all. Twenty years later, this finding was confirmed by an Emory University study that analyzed all primary invasive breast cancers diagnosed during 2000–2001 among black and white women living in Atlanta.[12] It found that black women experienced longer treatment delays and were less likely to receive the surgical, radiation, and hormonal treatments recommended by established guidelines.

In short, black women are diagnosed with breast cancer when it has reached a more advanced stage and is harder to treat, and the treatment they receive is inferior. "I think it's totally fair to say the system of breast health care in the city of Chicago is killing 110 black women a year," Whitman concludes.

Inequality Is Bad for Your Health

Unequal access to health care is a major culprit behind racial disparities in health. Building high-quality cancer centers in black neighborhoods—or even just facilities with digital mammograms and specialists to read them—would reduce breast cancer deaths in Chicago. Providing high-quality medical care to everyone is an essential first step to eliminating the appalling chasm in death and disease rates based on race in this country. Yet providing equal access to health care would not be enough to close the racial divide. Nancy Baxter, a surgeon and health advocate at University of Toronto, notes that even the generous Canadian health care system many U.S. citizens envy is insufficient to equalize health in her country. "It would be naive to believe that equal treatment at the point of care could obviate economic, educational and social inequities that, in some cases, have affected our patients not just throughout their lives, but even in utero," she writes.[13] Studies in Canada have found that babies in low-income neighborhoods are more likely to be born premature and underweight, poor children have an increased risk of being treated in an emergency room for asthma attacks, and there is a higher rate of suicide in indigenous First Nations communities.

Racial inequality causes health disparities apart from blocking access to high-quality care. It makes people of color sicker in the first place—before they get to a doctor's office or a hospital emergency room. The only other possible explanation—that there is something inherently different about these groups of people that affects their health—fails, as previous chapters have shown. Neither genetics nor access to health care alone determines an individual's health because health is affected primarily by the social environment. Throughout your lifetime, your parents' income and education, the neighborhoods you live in, the schools you attend, the jobs you hold, your experience of discrimination and privilege, and the resources you and your community have ultimately govern how the genetic hand you were dealt is translated into well-being. The way society is organized drives group disparities in health. Because they reflect social inequality, a more accurate word for the racial gaps in health is health *inequities*. As British public health champion Margaret Whitehead defines it, health inequities result from the "systemic and unjust distribution of social, economic, political and environmental resources needed for health."[14] Racism is not just a matter of wounded feelings or an uneven playing field: it determines the life and health of whole populations.

It has been firmly established that the best predictor of health is an individual's position in the social hierarchy. Hundreds of studies tracking the health of people along the social ladder show that health gradually worsens as status declines.[15] In any society, people with low socioeconomic status have poorer health than people with higher socioeconomic status. The classic Whitehall Study of British Civil Servants, lasting for more than two decades, compared heart disease and mortality in employees at four civil service levels: administrators, professional and executive employees, clerical staff, and menial workers. As you might expect, the study found that the administrators at the top had far better health than the janitors at the bottom—high status bought them ten more years of life. But it also found that health got worse and mortality increased with each step down the occupational ladder.[16] The clerical staff had higher rates of heart disease and early death than their professional supervisors. These social gradients in disease occurred despite everyone's access to the British universal health care system.

The question, then, becomes whether this relationship between economic inequality and health applies equally to America's glaring racial disparities?

Race also produces huge gaps in advantage and disadvantage that parallel the relative health of people in different racial groups. A new field of scientific research shows that racial inequality, like income inequality, causes health disparities—and provides the missing factor that many scientists are erroneously substituting with genetic explanations. Understanding this impact requires changing the way we think about the relationship between race and biology. Thomas LaViest, a leading public health expert at Johns Hopkins, surveyed the use of race in epidemiological studies in the 1990s. He found that most of the articles on U.S.-based populations did use race, but the most common use was as a control variable—to filter out the impact of race. So while geneticists were homing in on the biological impact of race, epidemiologists were ignoring it. "What is needed is not simply *more* research on race," LaViest concluded, "but *better* research on race."[17]

Thinking on this issue tends to fall into two camps: either race is a social category that has nothing to do with the biological causes of disease, or race is a biological category that causes differences in disease. Both approaches fail to grasp the way in which race as a social grouping can affect health—because of different life experiences based on race, not because of race-based genetic difference. In this sense, race *is* biological.[18] This is where many people get confused. So let me be clear: race is not a biological category that naturally produces health disparities because of genetic differences. Race is a political category that has staggering biological consequences because of the impact of social inequality on people's health. Understanding race as a political category does not erase its impact on biology; instead, it redirects attention from genetic explanations to social ones. This new conceptual model disrupts the dichotomy between biological and environmental causes of health inequities by suggesting complex biological interactions between racism, socioeconomic disadvantage, and poor health. According to sociologist Troy Duster, "The task is to determine how the social meaning of race can affect biological outcomes."[19] How does racial inequality get under the skin? How is racism embodied?

Nancy Krieger, a prominent epidemiologist at Harvard's School of Public Health, is the leading architect of the science of embodiment. Krieger sparked a revolution in public health research in 1986 when she published a paper, co-authored with public health researcher Mary Bassett, titled "The Health of Black Folk: Disease, Class, and Ideology in Science."[20] Her paper repudiated the leading explanations for high premature death rates among blacks, which had to do with genes and lifestyle, and argued instead that racial discrimination and poverty were to blame. In the two decades since, Krieger has written a prodigious body of scholarship that lays the theoretical and empirical groundwork for a new field centered on the social determinants of health. "In the case of racial/ethnic inequalities in health, the scientific challenge is to understand whether—and, if so, how—these disparities arise from the literal embodiment of unjust race relations," she writes.[21] Krieger is adamant about building a *science* of socially determined health disparities that "requires rigorous thinking: logical, sociological, and biological." She likens the potential impact of explicitly naming and scientifically investigating racism as a determinant of population health to the 1962 publication of the first scientific article identifying battered child syndrome.[22]

I had a chance to meet with Krieger in 2008 when we both were speakers at a conference in Cambridge, Massachusetts, and I asked her to explain what she meant by embodiment. "Embodiment to me refers to the many, many, many different ways that we literally incorporate the world outside of us in us in the expression of our biology," she told me. "Embodiment gives you a frame to allow you to appreciate what the connections are, but recognizes society as the key driver. That's the profound difference of having an embodiment approach versus one that sees innate biology as the key driver which just happens to get expressed. My focus is on how inequity becomes embodied and harms health."[23]

Krieger took the first scientific step by partnering with physician Stephen Sidney to specifically measure research participants' exposure to racial discrimination and test its association with high blood pressure. Instead of treating race as a biological risk factor, as was typical in epidemiological research, Krieger zoomed in on *racism* as a cause of disease and developed a fledgling methodology to measure its health impact directly. Her findings, published in the *American Journal of Public Health* in 1996, were the first to show that experiencing racial discrimination raises the risk of high blood pressure.[24] By 1999, a literature review conducted by Krieger titled "Embodying Inequality" identified fifteen studies that examined the health consequences of racial discrimination.[25] Krieger's focus on systemic inequality rather than lifestyle choices garnered the ire of conservatives, who argued that it undermined traditional American values of self-reliance and blamed society for personal failings. In "Public Health Quackery," published in the Manhattan

Institute's *City Journal* in 1998, lawyer Heather MacDonald charged that "[b]ehind all their talk of racism and sexism, Kreiger and her colleagues' real prey is individual responsibility."[26]

In response to these criticisms, Krieger and her colleagues designed a follow-up investigation that tracked patients' experiences of racial discrimination. In 2004, Krieger co-authored a study of 352 births, published in the *American Journal of Public Health*, that found that women reporting high levels of racial discrimination were almost five times more likely than those reporting no racial discrimination to deliver low-weight babies and had three times the risk of preterm births. Self-reported discrimination helped to account for the racial gap in birth outcomes.[27] In the same issue, James Collins of Northwestern Medical School reported that his study of African American mothers giving birth at two Chicago hospitals similarly found that lifelong accumulated experiences of racial discrimination are an independent risk factor for very low birth weight, which accounts for most of the black–white gap in infant mortality.[28] Krieger has added geocoding to her methodological arsenal, mapping concentrations of poor health at the local, state, and national levels, to understand health disparities from a geographical perspective.[29]

Krieger still sees herself today in a battle for the soul of public health—except that now the criticisms tend to come from people advocating a genetics model rather than focusing on individual risk behaviors. Krieger recognizes that embodiment research is temporarily dwarfed by the genetics juggernaut, "but I think what's also going to happen is that some of their genomics is going to fall flat." Nevertheless, her work—and the work of other public health researchers who have similarly documented the health effects of racism—needs to pinpoint the biological mechanisms of embodiment. If racial discrimination causes health disparities, how does it get into the body?

A growing number of scientists from a variety of disciplines, including medicine, biology, psychology, anthropology, and epidemiology, are attempting to answer this question by investigating how racial inequities in income, housing, and education, along with experiences of stigma and discrimination, translate into bad health. More than one hundred studies now document the adverse effects of racial discrimination on health.[30] Three of the main biology-related pathways embodiment researchers have identified are: chronic exposure to stress, segregation in unhealthy neighborhoods, and transmission of harms from one generation to the next through the fetal environment.[31] Their effects on health stem specifically from racism that is separate from, and added to, the harmful consequences of lower socioeconomic status.

Stress Takes Its Toll

Stress is a key biological mechanism. When the human body undergoes stress, it responds by producing higher levels of cortisol, often called the stress hormone. This temporary spike in cortisol helps you deal with an occasional stressful situation like a final exam or a suspicious bump in the middle of the night. The body typically returns to normal after the stressor goes away. But trouble occurs when stress is relentless and the stress response, known as the allostatic load, stays on. People who experience repeated exposure to stress have constantly high levels of cortisol in their bloodstreams. This produces a chronically higher allostatic load, which disrupts internal systems involved in stress response, including the endocrine, metabolic, cardiovascular, and immune systems. It is as if the body is always in a heightened state of alarm. When these systems become dysregulated from repeated stress response cycles, it can cause tissue damage and inflammation, increasing the risk of diseases like hypertension, asthma, and diabetes.[32]

Embodiment scientists believe that experiencing racial discrimination on a daily basis throughout life is a form of chronic stress that pushes allostatic load to dangerous levels—suppressing the immune system, driving up blood pressure, and increasing blood sugar levels. By stress, they are not referring to discomfort from a perceived racial slight or a problem within the psyche of discrimination victims. (Though continual racial harassment does have a cumulative impact: as Troy Duster puts it, "If you get stopped by police eight more times than whites on average, if you get followed around Neiman-Marcus, if you get fewer bank loans from the Philadelphia banks, you might be more likely to develop hypertension."[33]) This is a serious disruption of bodily processes in response to enduring institutionalized racism day in and day out—for instance, the kinds of institutionalized racism that has resulted in American black men spending, on average, three years of their lives in prison.[34] The resulting bodily wear and tear increases the risk of disease and contributes to glaring health disparities. Investigators are testing this theory with new methods that use biomarkers to measure the real-time effects of experiencing stress and discrimination. Research participants wear devices that record cortisol levels and blood pressure so that readings can be taken as they document their experiences in electronic diaries. Blood samples from a finger prick can yield measures of virus antibodies that allow researchers to explore stress-induced changes in immune function.[35] In this way, investigators have been able to document the physiological impact of racial discrimination.

Tené T. Lewis, an assistant professor of epidemiology at Yale School of Public Health, focuses on C-reactive protein (CRP), found in the blood, to investigate the physical toll of discriminatory treatment. Levels of CRP increase in response to inflammation, and higher levels are associated with cardiovascular disease. Lewis and fellow researchers collected blood samples from 296 older African Americans and examined the relationship between their CRP levels and the experiences of "everyday" discrimination they reported on a nine-item questionnaire. In a 2010 article in *Brain, Behavior, and Immunity*, the scientists reported a significant association between CRP levels and degrees of discrimination. This association remained strong even after adjusting for demographic variables as well as depressive symptoms and chronic health conditions (heart disease, diabetes, hypertension) that might influence inflammation, although it was not completely independent of participants' body mass index.[36]

Another example of the effects of stress on health sheds some light on the appalling breast cancer mortality gap. Although black women are less likely than white women to have breast cancer, those who are diagnosed with it are more likely to be under age thirty-five and to die by age fifty. The disparity is not only in the care they receive; it is in the tumors themselves. Black women are more likely to have tumors that spread quickly, are less responsive to hormone therapy, and develop before menopause. Because they are not fueled by estrogen (estrogen-receptor negative or ER-negative), these tumors are not affected by drugs like Tamoxifen, which modulate the hormone. Black women also have a higher incidence of tumors that are "triple negative"—negative for three key hormone receptors—which rules out additional targeted therapies.[37]

The oncologist who knows this aggressive form of cancer best is Olufunmilayo Olopade, director of the Center for Clinical Cancer Genetics at the University of Chicago. Born in Nigeria, Olopade received her medical degree from the University of Nigeria before completing a residency at Chicago's Cook County Hospital. She went on to win a MacArthur Foundation "genius" award in 2005 for her pioneering translation of breast cancer research into effective therapies for black women. When Olopade began to notice that her black patients were more likely to have triple negative tumors, her research veered toward the molecular genetics of breast cancer in women with African ancestry. She has taken her lab work to Ghana, Nigeria, and Senegal to compare the tumors of breast cancer patients in Africa to those of African Americans.

"I started off thinking it was all genetics," says Olopade.[38] But she has recently joined a novel investigation into environmental factors that might explain the tumors. Her University of Chicago colleague Martha McClintock, a biopsychologist who studies the psychosocial origins of malignant and infectious disease, suggested an unsuspected cause for the racial differences in tumors. While still an undergraduate at Wellesley, McClintock first became known for her discovery that the menstrual cycles among her dormitory mates became synchronized. After investigating the mystery for her senior thesis, published in *Nature* in 1971, she concluded that the synchronization was caused by pheromones transmitted through social interaction. Today, McClintock is both the founder of the Institute for Mind and Biology and the co-director of the Center for Interdisciplinary Health Disparities Research (CIHDR). Her research still explores how social interaction influences biology, including breast cancer.

McClintock and a team of Institute for Mind and Biology scientists experimented on rats to test the effects of social isolation on breast tumors.[39] She randomly assigned genetically comparable, cancer-prone rats to two social conditions: half lived in groups of five female rats and half lived alone. The socially isolated rats had exaggerated responses to both everyday and acute stressors, elevating their stress hormone levels. McClintock also discovered the socially isolated rats developed breast tumors 40 percent earlier and almost four times more often than the rats raised in groups. They also experienced a 135 percent increase in the number of tumors and an 8,000 percent increase in tumor size.

Could an increase in stress hormones caused by negative social experiences be responsible for the especially aggressive cancers in young black women? Supported by a $10 million grant from NIH, McClintock and other researchers at the CIHDR are investigating this question. In 2009, McClintock's team reported in *Cancer Prevention Research* that "we have used a mouse model of human breast cancer to show for the first time that a chronically isolated social environment correlates with specifically altered mammary gland gene expression."[40] Other studies support the hypothesis that psychosocial stress plays an important role in breast cancer risk and biology, "but the evidence is far from consistent," concludes a 2010 review of several epidemiologic studies.[41] Further research and reanalysis of existing data by tumor subtypes are required to understand the relationship between stress and cancer; still more is needed to know if it contributes to the racial disparities in breast cancer tumors. Nevertheless, breast cancer studies like the CIHDR project bolster the view that stress is a chief pathway that translates racism into poor health.

Unhealthy Neighborhoods

Despite the end to official Jim Crow policies during the civil rights era, America remains a terribly segregated country. The 2000 census showed that two thirds of blacks would need to move to create an even residential distribution across the national population.[42] Segregation in many cities like Detroit, New York, and Chicago almost equals levels in South African cities during apartheid. Unlike other racial groups, blacks live in segregated neighborhoods regardless of income, despite the fact that blacks express a greater preference to live in integrated areas than any other group. The ramifications are enormous. According to sociologists Douglas Massey and Nancy Denton, residential segregation is "the key structural factor responsible for the perpetuation of black poverty in the United States."[43] Residential segregation also perpetuates health inequities based on race. Neighborhoods where racial minorities, especially African Americans, are concentrated are systematically deprived of jobs, services, decent housing, fresh air, nutritious food, and other healthy living conditions. These systemic inequities create another pathway by which racial discrimination produces worse health, even for those earning the same income.[44] A mountain of recent studies has tied neighborhood segregation to obesity, low birth weight, cardiovascular disease, tuberculosis, and mortality.[45]

One disadvantage of minority neighborhoods that is easily tied to poor health is the toxic waste that gets dumped there. Consider the experience of Dickson, Tennessee, a thriving black community that grew in the heart of a rural county that is 96 percent white. Some of the black farmers live on land purchased by their ancestors after the Civil War. In 1968, county officials converted the black community's only park into a dump where local companies began burying drums of industrial waste. Chemicals from the dump seeped into the well water that the black families used to drink, bathe, and cook. By the time they were alerted to the danger in 2000, it was too late. Virtually every family in the black community in Dickson County has someone who has died from or is battling cancer.[46]

Dangerous facilities such as landfills, electrical power stations, incinerators, and waste treatment plants are more likely to be located where people of color live. The systematic location of health hazards on the basis of race spawned the term *environmental racism* and an environmental justice movement dedicated to fighting it. In 1987, the United Church of Christ issued a landmark report documenting environmental racism, *Toxic Wastes and Race in the United States*. In celebration of the report's twentieth anniversary, the UCC commissioned an updated edition, *Toxic Wastes and Race at Twenty—1987–2007: Grassroots Struggles to Dismantle Environmental Racism in the United States*. The new report is the first to use 2000 census data, a current national database of commercial hazardous waste facilities, and geographic information systems to assess the extent of racial disparities in facility locations at the national, state, and local levels. The study found that people of color make up the majority of those living in neighborhoods within two miles of hazardous waste facilities, and toxic neighborhoods have twice the percentage of minorities as nontoxic neighborhoods. Racial disparities in hazardous waste locations are widespread throughout the country: in forty of forty-four states with these toxic sites, minorities are disproportionately represented in neighborhoods where the facilities are located. In short, people of color are concentrated in neighborhoods with the greatest number of toxic facilities, and in 2007 they were more concentrated in these dangerous areas than in 1987. What's more, race independently predicts hazardous locations apart from socioeconomic status.

Recall the studies linking children's asthma to cockroach and rodent droppings in the home and to air pollution from diesel exhaust. Do black and Puerto Rican children have more severe asthma because they are genetically more susceptible to these triggers? Or could it be because they are more likely to live in neighborhoods where these triggers are concentrated? Using 2001 hospital records and data on residential traffic density, UCLA researchers analyzed differences in asthma severity among neighborhoods in Los Angeles and San Diego with varying levels of traffic.[47] Not surprisingly, they found that living near heavy traffic increases children's asthma severity: children with asthma whose homes were in neighborhoods with high-density traffic were three times more likely to be hospitalized or visit an emergency room than those who live near low traffic density. The study also discovered that these neighborhoods varied by race: Latino children with asthma are nearly two-and-a-half times more likely than whites to live in high-traffic-density neighborhoods, while "Asian/Other" and black children were almost twice as likely.

A recent study of children with asthma in Chicago provides additional evidence that unhealthy conditions concentrated in minority neighborhoods make asthma worse. In Chicago, like other cities, children's odds of having asthma depend on where they live.[48] In some neighborhoods, the chances are fifty-fifty. In others, the rate is virtually zero. Children who live in Chicago's inner-city neighborhoods are five times more likely to die of asthma than children in the suburbs.[49] Pediatrician Ruchi Gupta of Children's Memorial Hospital in Chicago noticed for years that her asthma patients from certain parts of the city had more severe symptoms. When she overlaid data on fifty thousand children with asthma on a map of Chicago, she saw that cases clustered in a few neighborhoods on the South and West sides. "That made us think there's something else about the neighborhoods themselves that's causing these differences," Gupta says.[50] On a hunch, she overlaid a map of police reports on the asthma map. The concentration of asthma rates matched the city's high-crime neighborhoods perfectly. Although this correlation didn't prove that neighborhood crime caused severe asthma, it prompted Gupta to investigate further.

To confirm the association between asthma and neighborhood violence, Gupta led a study of 561 children with asthma in which investigators interviewed caregivers to determine their stress level and exposure to violence.[51] Neighborhood violence rates were supplied by Chicago police department records. At the May 2010 annual meeting of Pediatric Academic Societies, Gupta reported that, even after adjusting for family histories of asthma and socioeconomic status, children were nearly twice as likely to suffer moderate to severe symptoms if their caregivers reported high levels of stress or if their neighborhood had a high rate of violent crime. The risk of severe asthma was better predicted by the actual incidence of neighborhood violence. Gupta suspects that when children are routinely exposed to violence down the street, their chronic stress response increases inflammation that worsens their asthma.

Gupta's study did not rule out other plausible pathways; asthma and violence might both be symptoms of inequitable conditions in these neighborhoods. If she had overlaid the asthma data on a map of neighborhoods with high incarceration or foster care rates, she would have found the same match. It is also possible that children stay inside more to avoid the violence outdoors, exposing them more to allergens inside their homes. Nor did these asthma studies examine whether race, independent of poverty, put nonwhite children at greater risk. But other research demonstrates that residential segregation makes the neighborhoods where blacks live especially unhealthy. The greater the percentage of blacks living in a neighborhood, the higher the neighborhood death rate, regardless of neighborhood income level.[52] It is the neighborhood that makes people unhealthy, not the susceptibility of black people living there. The rate of death is higher for *all* residents, including whites, who live in predominantly black neighborhoods.[53] These neighborhoods put their residents at high risk of illness and death because of the dangerous conditions that have been systematically concentrated there—pollution, substandard housing, inadequate public services, and stress from economic deprivation and racial discrimination.

All of these studies reinforce the importance of neighborhoods as sites where multiple forms of racial injustice converge to injure the people who live there. More broadly, this research confirms that institutionalized racism inflicts concrete damage on the health of its victims. The two main pathways scientists have investigated are stress and residential segregation. But there are others. I was struck by the multiple ways that racial inequality devastates people's bodies when I read about sixty-eight-year-old Dorothy Tanksley, who lives with her two grandsons, Rahsuni, fourteen, and Marvin, eleven, in Bedford-Stuyvesant, Brooklyn. Tanksley has had custody of the boys since 1997, when child services removed them from their mother's custody. Their father, Tanksley's son, is in prison. Because of back problems, Tanksley gets around in a wheelchair and sleeps in a reclining chair in the living room. "I hurt most of the time, 24-7," she told *New York Times* reporter Kari Haskell. "'I take all kinds of things,' she said, rattling off a pharmacy list: over-the-counter pain medications and prescriptions that include insulin for her diabetes and medication for high blood pressure, glaucoma and high cholesterol. 'I just bear with it,' she said, rubbing her arthritis-swollen hands." Tanksley grew up in a "tiny, primitive" house on a former plantation in rural McBean, Georgia, where she fetched water from a pump and gathered wood for the stove. She was working as a home health aide in New York when a car accident left her disabled. Now she and her grandsons barely make ends meet on a Social Security check, food stamps, and public assistance.[54]

Tanksley's life story flashes with possible causes of her poor health that are disproportionately borne by black women—caring for grandchildren rescued from the foster care system, a son in prison, reliance on inadequate public aid. Her experiences brought to mind the words of the great civil rights champion Fannie Lou Hamer, who spent years toiling in the cotton fields of Mississippi: "I'm sick and tired of being sick and tired."[55] But one other feature caught my eye: her job as a home health aide. Long-standing racial discrimination in employment intersects with sex segregation to relegate women of color to the bottom of the occupational ladder. Most women of color, like Dorothy Tanksley, are employed in low-skilled clerical, manual, or service jobs—jobs that pay the least and are the most hazardous to employee health.[56] Home health aides and nurse assistants do some of the most hazardous work in hospitals and nursing homes, lifting incapacitated patients and heavy equipment while being exposed to viruses and infectious diseases. Domestic work in private homes can also be backbreaking and is not subject to safety regulations. On top of these physical injuries is the psychological stress from holding a job that is utterly devalued, offers little control over working conditions, and often comes with sexual harassment that plays on racist stereotypes. Embodiment research tells us that the disproportionate exposure of women of color to workplace hazards damages their health and the health of the next generation in ways yet to be fully investigated.

Programming in the Womb

Racism doesn't affect just those who experience it—it also affects their children while still in the womb. A growing body of research has demonstrated that discrimination and disadvantage make their way into the uterine environment to inflict lasting damage on the next generation.

Experiencing extreme stress during pregnancy, for example, can affect birth outcomes. A study conducted by Pierre Buekens, dean of the Tulane University School of Public Health and Tropical Medicine, found that although weathering Hurricane Katrina in 2005 didn't cause a general increase in pregnancy complications, *"high exposure"* to the disaster did increase risk. Pregnant women who experienced three or more "severe hurricane events," such as sustaining significant home damage, walking through floodwaters, or seeing someone die, had a "markedly increased risk" of delivering low-birth-weight and premature babies.[57] Similarly, women who were in their first trimester when an earthquake struck Northridge, California, in 1994 delivered their babies two weeks early on average.[58]

A study of birth outcomes in the wake of the 9/11 attacks shows that this type of impact on fetal health can result from intensified discrimination. Based on prior embodiment research, Diane Lauderdale, a health researcher at the University of Chicago, hypothesized that the intense surge of anti-Arab harassment, violence, and workplace discrimination in the weeks immediately following September 11, 2001, would elevate the risk for unhealthy babies born to women of Arab descent who were pregnant at the time.[59] Lauderdale conducted her study in California, the site of the largest Arab-origin population in the country and of rampant anti-Arab discrimination after 9/11. California reported a 345.8 percent increase in "anti-other" ethnic hate crimes in 2001, primarily because of violence against Arabs. "People are being attacked," a coordinator from the Muslim Public Affairs Council in Los Angeles told a reporter shortly after the World Trade Towers fell.[60]

The research project was also an interesting study in racial classification. Its first challenge was identifying mothers with Arab ancestry from a data source listing all California birth records for 2000, 2001, and 2002. Because OMB Directive 15 defines the racial category "white" to include people with Middle Eastern ancestry, Arabs are not classified separately as a race or ethnicity in U.S. vital statistics. Instead, Lauderdale identified Arab-origin women based on a carefully selected list of Arabic names developed with the help of a special algorithm. (Unlike the list of Spanish names used by the endocrinologist mentioned in chapter 3 to demarcate genetic difference, this list was used to identify a social grouping.) Lauderdale found that women with Arabic names—and *only* Arabic-named women—who were pregnant in September 2001 were more likely to have a low-birth-weight baby compared with similar women who gave birth a year earlier. What's more, babies who were given Arabic names had a greater risk, suggesting that pregnant women with stronger ethnic identification were affected more. The study does not prove that stress from discrimination caused the unhealthy birth outcomes, but its results are consistent with that hypothesis and suggest that stress experienced by pregnant women is an important pathway for scientists to explore further.

Other mechanisms are more subtle but just as powerful. One researcher exploring how fetal origins affect health later in life is Northwestern University anthropologist Christopher Kuzawa, whose basement laboratory in a large Victorian house took some searching to locate. Kuzawa's office sits in the Laboratory for Human Biology Research, which he co-directs with two other anthropology professors. The lab is conducting research around the world on such topics as the impact of cultural change on the immune function in Samoan adolescents, the long-term effects of prenatal environments on metabolic function in Filipino women, and the health consequences of the collapse of the Soviet Union on indigenous Siberians.

"I define race the way most anthropologists have defined it for the better part of a half century," Kuzawa began. "It is a socially defined category that has biological consequences. There is a biological reality to it. But we have no evidence that the source of that biological reality traces to genes. It overwhelmingly traces back to the environment."[61] The particular environment Kuzawa focuses on is the womb: he studies how the fetal environment leaves lasting effects on health, creating a pathway for social inequalities to be transmitted from one generation to the next.

Kuzawa became intrigued by the biological origins of disease in graduate school when he learned about the pioneering work of David Barker, a British physician and epidemiologist, showing an inverse relationship between birth weight and cardiovascular disease. In the late 1980s, Barker discovered that babies who are born small have a higher risk of cardiovascular disease when they grow up. This means there is something happening in utero to accommodate being undernourished and this process has consequences for future risk of disease. "Organisms come equipped with capacities to adjust to their realities, and this starts in utero," Kuzawa says. The fetus faced with too little nutrition not only slows its growth rate to reduce nutritional needs but alters other aspects of its developmental biology in ways that have a lingering impact on its health.[62] The fetal origins hypothesis that has grown out of Barker's initial discovery posits that poor nutrition in the womb leads to adaptive changes in growth and development that may be beneficial in the short run but produce susceptibility to coronary heart disease and the related disorders hypertension, stroke, and diabetes later in life. In a study published online in *PLoS One* in 2010, Swedish and Nigerian researchers found that babies born during the Biafra famine in Nigeria, which lasted from late 1968 to early 1970, grew up forty years later to be more susceptible to obesity, hypertension, and diabetes than Nigerians born in the three years either before or after the famine.[63]

For his dissertation, Kuzawa traveled to Cebu, the second-largest city in the Philippines, to take advantage of a study that had enrolled more than three thousand pregnant women back in 1983 to compare the effects of infant formula versus breast feeding on their babies. Those babies were teenagers when Kuzawa arrived in 1998. He was interested in tracing their health conditions to the environment they experienced while still in the womb. Like Barker, Kuzawa found that women who did not eat well during pregnancy had babies who grew up to have high cholesterol. In the two decades since Barker's discovery, hundreds of studies have replicated findings of developmental "programming" that links small birth or infant size with higher blood pressure, diabetes, and obesity in adults. Today, Kuzawa is following the same children, now in their twenties, some of whom have babies of their own. "We are moving beyond the point of wanting to conduct yet another study that shows this inverse association. We want to understand the mechanisms," Kuzawa tells me. "What I'm really excited about now is to design a study to get at some of the epigenetic mechanisms that might underlie these changes."

Epigenetics refers to changes in the genome that do not involve any change in the structure of genes themselves. While mutations alter the sequence of the DNA code, the science of epigenetics concerns which genes are expressed and at what level. Scientists call each epigenetic change a *mark*, and the total set of epigenetic marks in an organism are the *epigenome*.[64] Epigenetic markings are like volume controls for genes. The epigenome can be modified by the environment, and the changes are durable. But just as volume can be turned back down or back up, epigenetic markings are also reversible.[65] "Things like nutrition and stress early in life are going to have lingering effects—not changing the genes themselves but changing the chromosomes in a way that silences genes or amplifies genes," Kuzawa tells me. We not only inherit genes from our parents; we also inherit the epigenome. It seems, then, that genes can remember the environmental impacts of prior generations. "The lives of your grandparents—the air they breathed, the food they ate, even the things they saw—can directly affect you, decades later, despite your never experiencing these things yourself," announces *The Ghost in Your Genes*, a BBC documentary on epigenetics that aired in 2006.

A prominent 2002 study tested the effect of nutrition on female mice carrying the agouti gene, which causes mice to be obese and yellow instead of the typical brown color. Pups born to a group of mothers who were fed a regular diet turned out fat and yellow, as expected. But another group fed a special diet fortified by B-vitamins gave birth to pups who were brown and healthy.[66] Similar epigenetic mechanisms have been detected in human mothers and their children. Diethylstilbestrol (DES), an estrogenic drug given to pregnant women from the 1940s to the early 1970s to avoid miscarriage, seems to have interfered with the epigenetic programming in the girl fetuses they were carrying.[67] The women who took DES gave birth to daughters *and* granddaughters with reproductive disorders and rare forms of cancer at alarming rates. Animal tests suggest the harms of DES were transmitted from mother to daughter partly via epigenetic changes.

Yet epigenetics does not operate only through the maternal environment. In a study published in 2006 in the *European Journal of Human Genetics*, a team of scientists from England and Sweden found, after controlling for confounding factors, that fathers who started smoking before age eleven had sons, but not daughters, with greater body mass index. The authors acknowledged that discovering that transgenerational effects can be sex-specific represents only a first step toward identifying the underlying mechanism.[68]

Through epigenetics, then, the effects of racism on parents might be transmitted to their children, perpetuating inequalities across generations. These pathways may explain, for example, why African Americans have higher rates of both low birth weight and adult cardiovascular disease. Kuzawa suspects that stress and discrimination experienced by black mothers throughout their lives and during pregnancy cause them to have smaller babies who, partly because of epigenetic effects of the intrauterine environment, have an elevated risk for cardiovascular disease as adults.[69] Epigenetic influences on children's health may have fooled some scientists into seeing genetic causes for health disparities that do not exist. Epigenetics may masquerade as genetic difference, but its biological effects stem from the environment, not mutations of the genetic code. Still, the line between the genome and the epigenome can seem blurred. When scientists write that epigenetic effects of racial discrimination are durable across generations, it sounds perilously close to biological theories of race. The point of this research should not be to consign another generation to the biological fallout of past discrimination. To the contrary, its hopeful message is that epigenetic changes are caused by the environment and therefore can be environmentally interrupted so that future generations can enjoy better health.

Policy Makes a Difference

Finding social explanations for health disparities has led to an even more exciting aspect of embodiment science—further research proving that policies promoting racial justice can close racial gaps in health. Krieger was motivated to investigate long-term trends in health inequities when a series of articles published around 2005 asserted that health disparities were inevitable because the most privileged segment of society would always have greater access to the latest health information and medical innovations.[70] If the magnitude of health disparities changed historically in response to social policies, Krieger theorized, then the gaps can be reduced or even eliminated through social change. To look for trends, Krieger collected data on premature mortality and infant death among U.S. counties, ranked by income level, for the period 1960–2002, for both the total population and divided by race. She found that the black–white mortality gap shrank from 1966 to 1980, in tandem with a decline in socioeconomic disparities, and then began to widen again in the 1980s.[71] What was going on in the United States during that period that could explain this momentary reduction in health disparities? Krieger believes that the most plausible explanation lies in two major social policy determinants of health: economic priorities and civil rights. She argues that the creation of Medicaid and Medicare, the construction of community health centers, the War on Poverty, and the Civil Rights Act of 1964, all of which expanded opportunities and economic resources along with access to health services, explain the narrowing of the mortality gap. Conversely, the subsequent growth in disparities stems from conservative policies that began under the Reagan administration and cut federal support for public health, antipoverty programs, and affirmative action. "The recent trend of growing disparities in health status is not inevitable," Krieger says.[72]

A number of studies conducted by population economists have supported Krieger's thesis by linking specific civil rights programs to improvements in black health. Douglas Almond, an economist at Columbia, investigated the role of public policies in the remarkable convergence of black and white infant death rates from the mid-1960s to the early 1970s, especially in Southern states. Almond and colleagues looked at whether the end of widespread segregation of Southern hospitals during this period contributed to the sharp reduction in deaths among black infants from causes like pneumonia and diarrhea.[73] Their analysis showed that Title VI of the 1964 Civil Rights Act, which forced hospitals receiving federal funds to desegregate, saved the lives of 5,000 to 7,000 black infants from 1965 to 1975 and at least 25,000 black infants from 1965 to 2002. Another study found additional improvements for black infants from 1965 to 1975, including increased birth weight and gestation length, in response to the rollout of the largest nutrition program in the United States—the Food Stamp Program.

Berkeley economist Rucker Johnson investigated the long-run impact of court-ordered school desegregation plans on later-life health.[74] Using the Panel Study of Income Dynamics, he linked neighborhood attributes and the quality of school resources to the health trajectories of children born between 1950 and 1975, followed through 2007. Johnson applied sophisticated statistical techniques to analyze the timing and scope of the implementation of desegregation plans during the 1960s, '70s, and '80s and to disentangle the effects of neighborhoods and school quality. He found that school desegregation significantly narrowed black–white adult health disparities for those who attended integrated schools as children. How did attending an integrated school as a child create health benefits in adulthood for blacks? Johnson believes that improvements in school quality, reflected in reductions in class size and increases in per-pupil spending, enhanced black children's prospects for education and socioeconomic mobility, which in turn had far-reaching impacts on health.

What could be better news for closing the despicable racial chasm in health? Yet despite these promising findings, research dollars continue to flow into genetic explanations of racial health inequalities. To give one example, the links between asthma and unhealthy urban environments are clear and undisputed. Public health officials agreed that substandard housing, decreased access to health care, and exposure to cockroaches and rodents all increase a child's risk of asthma. As one study put it in 1988, "asthma should not be thought of as an irremediable genetic problem of some population subgroups, but rather as a consequence of exposure to a modern urban environment."[75] Yet the genetic origins of this problem continue to be researched today, with no sign of abatement in the numbers of black and brown children suffering from the disease.

The search for race-specific genes linked to asthma focuses more on children's genetic susceptibility to pollutants already known to trigger the disease than on cleaning them up. Why? To build a stronger case for environmental interventions? At an NIH conference on "Understanding and Reducing Health Disparities," which I attended in October 2006, a researcher commented that adding genetics to studies examining the impact of toxins on inner-city children with asthma might help to bring attention to the issue. But why shift focus from the deplorable housing conditions that many children live in to their genetic makeup?

A chief reason why genetic explanations are emphasized over social ones is that genetic causes can be treated with a pharmaceutical product. The automatic response to disease-causing genes is to develop a drug to target them. You are just as likely to find a gene discovery reported in the business section of a newspaper as its science section. When Reuters announced in April 2008 that researchers had identified two common genetic mutations that increase the risk of osteoporosis, the story quickly moved on to discuss the implications for drug companies. It quoted the researchers at King's College London, stating, "Eventually, a panel of genetic markers could be used in addition to environmental risk factors to identify individuals who are most at risk for osteoporotic fractures." The rest of the article focuses entirely on gene-related drugs and makes no mention of environmental interventions that could help people avoid fractures, despite acknowledging the medicines' potential side effects. The article concluded with a nod to personalized medicine: "These two genes are important targets for treatments, and drugs are already under development, the researchers said."[76] Discovering a genetic risk opens a fresh avenue for profit. Dealing with the environmental risks we already know exist and are killing people costs money. I now turn to the keen interest genomic researchers have in producing pharmaceutical products and the role that race plays in bringing them to market. Race-specific drugs are just one example of increasingly popular biotechnologies that treat race as a genetic category that can be packaged in marketable commodities.

PART III

The New Racial Technology

7

Pharmacoethnicity

The Human Genome Project opened not only a new horizon for applying genetics to health, but also a new horizon for profit. From the outset, the gene-mapping initiative was marketed to Congress as a means of catapulting the United States into the forefront of a technological revolution while turning genetic knowledge into a gold mine for big business. Genomic research is conducted with a constant eye toward market applications. Molecular anthropologist Jonathan Marks argues that this close synergy between the production of genetic knowledge and the production of capital has radically transformed the basic mission of science. Unlike researchers a generation ago, contemporary geneticists use "technology, professional expertise, and the authority of the scientific voice" to pursue monetary rather than classically scientific goals, argues Marks. The resulting financial conflicts, he writes, "make it even more difficult to gauge the truth value of any claims in human genetics."[1]

Upon the prospect of genetic discoveries has been built a burgeoning gene-based biotechnology enterprise, including biomedical firms, genetic and ancestry testing services, and manufacturers of supplies and equipment used in gene sequencing. The potential to make huge sums of money in the biotech business has drawn countless molecular biologists from the ivory tower into the marketplace. Scientific researchers start with a discovery in the lab and parlay it into a commercial enterprise with the guidance of market-savvy business partners, the backing of venture capitalists, and the protection of federal regulations. Many of the genetic scientists involved in the Human

Genome Project had a commercial interest in its results—as founders, directors, officers, and stockholders in biotechnology companies.[2] "No prominent molecular biologist of my acquaintance is without a financial stake in the biotechnology business," writes Richard Lewontin.[3] Most tantalizing is the chance to tap into the pharmaceutical industry—the third most profitable business in United States, with annual prescription sales hitting $300 billion in 2009.[4]

When Craig Venter left the NIH in 1992 to accept a lucrative offer from venture capitalist Wallace Steinberg to set up a private molecular biology institute, TIGR (The Institute of Genomic Research), *Nature* reported that Venter's move signaled "both an acknowledgement that the genome project is ready for scale-up and that industry recognizes the long-term commercial potential."[5] Within two years of launching the genome project, genomic science became entangled with big business; property rights to genes were the key to reaping astronomical profits from scientific discoveries. DNA had become a marketable commodity.

The market itself was not a surprise: scientists had promised from the Human Genome Project's inception that understanding human genetic variation would transform medicine. The information gleaned from the human gene map, they said, would enable doctors to predict, diagnose, and treat illnesses according to each patient's unique genetic code, an approach that came to be known as personalized medicine. "I thought it held the promise of giving mothers a little gene card to bring home from the hospital with their newborns which would tell them what their genetic profile is," Bill Clinton stated in 2003.[6] But how would the market for gene-tailored pharmaceuticals be structured? Race soon became the linchpin for turning the vision of tomorrow's personalized medicine into today's profit-making drugs.

The Unfulfilled Promise

Today, most drugs are developed for the whole population based on a "one size fits all" approach. But there is wide variation in the safety and efficacy of drugs among the individuals who take them. A particular medication may cure one person and kill another. There are more than 2 million cases of adverse drug reactions each year in the United States, resulting in 100,000 deaths.[7] Even more patients simply are not helped by the therapies prescribed for them.

Now imagine if, instead of undergoing this risky trial and error, each patient could be tested *before* receiving a prescription to identify precisely which medication would work best. Variations in genes that code for drug-metabolizing enzymes, drug transporters, and drug targets help to determine how an individual patient will respond to a medication.[8] A central component of personalized medicine is using genetic information and technology to prescribe therapies that are safer and more effective than conventional medicines. Researchers in the field of pharmacogenomics, which studies how genes affect people's response to medications, are trying to identify the genes that affect drug response. By dividing patients into groups using a genetic test, this information can be used to "tailor" medical treatment to each individual's personal genotype.[9]

Finding drug-related genetic markers offers pharmaceutical companies an avenue for translating genomic discoveries into drug sales.[10] It should be more profitable to sell medications that work better for individual patients because doctors are more likely to prescribe these tailored drugs, and patients are more likely to comply with prescriptions. With pre-prescription genotyping, companies would spend less on lawsuits resulting from adverse drug reactions. They could also use genetic data on efficacy to direct marketing to patient groups most likely to benefit and show that their product is better than a competing drug. Instead of the blockbuster model that targets "a relatively small share of a large pie," explains Lilly CEO Sidney Taurel, "tailored therapy could expect to claim a relatively large share of a smaller pie."[11] Personalized medicine can increase profits further by cutting the costs of developing new drugs. Pharmaceutical companies can design drugs to fit a genetically identified population. Instead of conducting large, expensive, and random clinical trials to test the efficacy of a new medication, drug makers can narrow the pool of research subjects to those whose genes indicate they are more likely to benefit and less likely to suffer adverse reactions.[12]

Genetic diagnostics also promises to rescue existing drugs that are unmarketable. By screening patients, drug makers can find markets for "orphan drugs" that had to be abandoned because they helped too few people or harmed too many from side effects. Take the case of the expensive cancer drug Erbitux, developed by ImClone Systems. At the steep price of $10,000 a month, it was approved for colon cancer only after other therapies have failed. A 2008 study found that Erbitux did not help patients with a particular genetic mutation in their tumor but was beneficial for the two thirds of patients without the mutation. Based on these findings, European regulators approved the drug as an initial treatment, but only for the patients without the mutation.[13]

Dozens of small- and medium-size firms are already making commercial use of pharmacogenomic research. Most focus on services and products supporting drug development, such as genotyping services or diagnostic tests that aid in prescribing therapies, rather than developing new drugs.[14] Two highly effective cancer drugs—Genentech's Herceptin for breast cancer and Novartis's Gleevec for chronic myeloid leukemia—are commonly seen as the poster children of personalized medicine because they were developed to target molecular alterations in cancer cells and doctors use diagnostic tests to predict which patients will benefit from them. Herceptin treats the 30 percent of patients with breast cancer in which the HER2 protein is overexpressed, which can be detected with the HercepTest. Herceptin sales grossed $5.2 billion in 2009. Gleevec binds to the abnormal gene for an enzyme called tyrosine kinase and, without harming healthy cells, blocks the enzyme's ability to cause cancerous growth of white blood cells. Meanwhile, the global giant Bayer has entered into two partnerships with Curagen worth $1.3 billion to integrate Curagen's genetic databases into its drug discovery and development programs.[15]

In its publication *The Case for Personalized Medicine*, the Personalized Medicine Coalition (an industry trade group) points to an expanding infrastructure of laws, policy, education, and clinical practice that is preparing the health care industry for the "tectonic shift" about to take place in medicine. Medical institutions throughout the country have dedicated centers to the study of pharmacogenomics and are launching genomics-based education programs to train the next generation of health care providers; the FDA is integrating genetic testing into drug labels; nearly every major pharmaceutical development project is incorporating information about genetic variation into its evaluation of safety and effectiveness; and the federal government is supporting personalized medicine in funding, policy, and legislation.[16]

The infrastructure and funding are there. Now all that is needed is for researchers to identify the genes that cause common illnesses like cancer, heart disease, and diabetes. Yet after a decade of intense and expensive digging for genetic drug targets, scientists have come up virtually empty-handed. It turns out they were banking on a flawed hypothesis about the relationship between genes and disease. They assumed that these common diseases were caused by common mutations. Identifying these mutations has turned out to be more expensive than anyone had imagined. In 2002, a multinational consortium of two hundred scientists backed by $120 million in private and public funds launched the International Haplotype Map Project (affectionately known as the HapMap Project) to track patterns of human genetic variation across the globe. Haplotypes are large blocks of adjacent SNPs that are passed down intact from one generation to the next. The idea was that the HapMap Project's findings would allow researchers to associate common SNPs with specific diseases using computerized genome-wide association studies. It soon became clear, however, that the cost of a single study can exceed $10 million and that it would require hundreds of such studies to arrive at any replicable findings.[17] Calling them a "wild goose chase," Neil Holtzman, a genetic epidemiologist at Johns Hopkins, argued that the NIH should stop funding these gene-hunting expeditions. "I think this is a waste of resources," he said in a 2004 interview.[18]

Billions of dollars later, Holtzman proved to be right. In the July 30, 2007, *New York Times*, Nicholas Wade heralded "a new, advanced gene-hunting method called Whole Genome Association, which has racked up a string of successes with major diseases in the last few months."[19] A year later, it was evident that such scanning of the genome for disease-associated DNA was not working as expected. "There is absolutely no question that for the whole hope of personalized medicine, the news has been as bleak as could be," stated molecular biologist David Goldstein, director of Duke's Center for Human Genome Variation, in September 2008.[20] Despite statistically linking hundreds of common variants to various diseases, scientists discovered that they either account for only a tiny fraction of the genetic risk or tell us nothing useful about disease risk at all. Instead, it now appears that most common diseases are tied to a host of rare genetic variants that evade detection by genome-wide association studies. What's more, the impact of isolated disease-linked genes depends on their interaction with an individual's other approximately 24,000 genes in ways we do not understand.[21]

The prospects looked no better at the turn of a new decade. DeCode Genetics filed for bankruptcy in 2009 and abandoned its plans to develop genetically tailored drugs.[22] A study conducted by a medical team led by Nina P. Paynter of Brigham and Women's Hospital in Boston to test the predictive ability of gene variants statistically associated with heart disease found that the genes failed to identify the actual incidence of disease.[23] The researchers created a genetic risk score from 101 genetic variants that had been statistically linked to heart disease in a slew of genome-wide association studies published between 2005 and 2009 to see if the score accurately predicted heart disease in 19,313 initially healthy white women who had been followed for twelve years. Dr. Paynter reported in 2010 in *JAMA* that a high genetic risk score was not associated with cases of heart disease among the women studied and that the "old fashioned method of taking a family history" worked better at identifying women at risk.[24] (On October 22, 2010, the American Society of Human Genetics issued a press release declaring family health history "the gold standard" for assessing personal disease risk.)[25] "The hunt for the genetic roots of common diseases has hit a blank wall," lamented Wade in January 2010.[26]

Three months later, Wade announced that a dramatic shift in genomic research had revitalized hope for personalized medicine. Instead of scanning thousands of genomes to find common links to disease, researchers were beginning to zoom in on the complete genome of unhealthy individuals. On March 10, 2010, under the headline "Disease Cure Is Pinpointed with Genome," Wade reported, "Two research teams have independently decoded the entire genome of patients to find the exact genetic cause of their diseases." One of the patients is a prominent medical geneticist, Dr. James Lupski, who has a debilitating nerve disease. The cost to sequence his genome was only $50,000—a huge reduction compared to the $500 million price tag on the first genome map. The sequencing revealed the culprit as mutated copies of a gene called SH3TC2 that Dr. Lupski had inherited from both his parents. "We are finally about to turn the corner," said Goldstein, the molecular biologist at Duke who had expressed such disappointment two years before. "I suspect that in the next few years human genetics will finally begin to systematically deliver clinically meaningful results."[27]

Given the track record, we should take these predictions of imminent success with a grain of salt. While the focus has shifted from widespread common mutations to whole individual genomes, the underlying unproven assumptions—that gene scanning will unveil the causes of common diseases and that these causes can be treated with genetically tailored medicine—remain the same. So far, these hypotheses have generated no genetic cures, despite billions of dollars having been spent in their pursuit.

"I think what we are finding is, understanding the genetic cause of disease doesn't really tell you anything about how to fix it," Mildred Cho, a

Stanford bioethicist, pointed out to me. She noted that geneticists have long known about the genetic mutations that cause cystic fibrosis, a chronic, progressive disease that causes thick mucus to build up in the respiratory and digestive systems. It used to be common for children with cystic fibrosis to die from lung infections. "There have been great strides made in cystic fibrosis, but none of it comes from the understanding it's caused by a particular gene or even what that gene does. We have doubled the life expectancy of individuals with cystic fibrosis, but it really has to do with management of the [consequent] infectious disease," Cho says.[28] When I later heard a radio program about a girl suffering from cystic fibrosis whose parents could no longer afford her infection-fighting medications, it seemed clear to me that the public money invested in gene hunting would be better spent using already-proven therapies to treat sick people who aren't getting the care they need.

Still, the genetic engine of personalized medicine pushes forward full steam ahead. It continues to attract huge financial allocations from both the federal research and private pharmaceutical industries despite the abysmal return on their ten-year investment. While the biotech, pharmaceutical, and health industries wait for results, what direction will drug tailoring take?

From Personalized Medicine to Pharmacoethnicity

A 1999 *Science* article with the practical title "Pharmacogenomics: Translating Functional Genomics into Rational Therapies" provides a clue. The authors predicted the explosion of industry interest in using genomic strategies to discover new drug targets. They acknowledged that developing medications for every member of a population was a "pharmacological long shot" and recommended instead developing drugs "targeted for specific, but genetically identifiable, subgroups of the population."[29] They hinted at how these subgroups might be defined. Noting that "all pharmacogenetic polymorphisms studied to date differ in frequency among ethnic and racial groups," they concluded that this "marked racial and ethnic diversity" in drug-metabolizing enzymes "dictates that race be considered in studies aimed at discovering whether specific genotypes or phenotypes are associated with disease risk or drug toxicity."[30]

Within ten years, the prediction that race would be used in pharmacogenomics had materialized. A special issue of the journal *Clinical Pharmacology & Therapeutics* published in September 2008 was devoted to ethnic and racial differences in drug response. Its cover displayed a map of the United States with photos of people of all racial backgrounds above the provocative banner PHARMACOETHNICITY.

Despite the lack of genetic data—or perhaps because of it—race has become a tantalizing fix to bridge the gap between the promise and the disappointment of personalized medicine. Pharmacogenomics homes in on genetic differences among people. According to the NIH Pharmacogenetics Research Network, "it is important to understand the 0.1 percent difference because it can help explain why one person is more susceptible to a disease or responds differently to a drug or an environmental factor than another person." That 0.1 percent of difference could be catalogued in a variety of ways, starting by identifying the genes themselves that have an impact on drug response. Many scientists are grouping their findings by race as a crucial first step to producing tailored treatment because, they argue, race can serve as a proxy for individual genetic difference. Until science is able to match therapies to each individual's unique genome, race stands in as a convenient surrogate.

Writing in *Slate* in 2008, William Saletan nicely summed up the paradigm that sees race as a temporary biological stand-in for individual genotype in the march toward personalized medicine:

> Race is the stone age of genetics. Biologically, race is real. It's an extension of extended family. But it's also transitional and, in the long view, crude. Any theory of heredity that starts with observed racial patterns has to end with genetic differences that cross racial lines. . . . In the stone age of genetics, we've often had to settle for racial medicine.[31]

You would think that since scientists know so little about the genes associated with common diseases, they would have little to say about how these undiscovered genes vary by race. Yet many have already concluded that differences in genetic predisposition to common diseases must be racially allocated. Likewise, racial disparities in disease rates are attributed to genetic difference without any evidence of the underlying genotypes or how they function. It is increasingly clear that, unlike rare single-gene disorders like Tay-Sachs, common chronic diseases are influenced by multiple genes that are found in virtually all populations at varying rates. Imagine how complicated it would be to calculate a "genetic susceptibility score" for any particular racial group. This measurement would require "summing the frequencies of these susceptibility alleles in all genomic regions, while taking into account the environmental factors that are either difficult to measure or wholly unknown," write Richard Cooper, Jay Kaufman, and Ryk Ward.[32] The scientists who leap to inferences about racial predisposition to disease are not even at step one of this analysis.

An article in the *Washington Post* gave a simple account of racial differences in drug response: "Race influences which people are genetically predisposed to lack various enzymes needed to break down medications. Without those enzymes, the medication can have either a heightened or lessened effect."[33] Pharmacogenomic scientists have a more sophisticated explanation. They know, of course, that the genes that influence disease susceptibility and drug response are not neatly grouped by race. Whatever variants elevate the chance of having diabetes or benefiting from a particular drug for it are not found exclusively in one race or another. They are found in all human populations at varying degrees of frequency. It's a matter of statistical probability.

"I think where we are headed is that we will do genetic testing directly on Dorothy and be able to say, 'Dorothy, regardless of your racial background, these are the drugs that should work for you,'" Esteban Burchard predicted when we spoke at his lab at University of California at San Francisco in 2008.

"What about in between that point and where we are now?" I asked.

"Right now, *race is a proxy*," he emphasized.

I wondered how Burchard could apply this race-is-a-proxy approach to patients who racially identify based on social definitions. "I call myself black or African American, but I'm a mixture," I told him.

"You're right, but you will always be treated as a black woman—just like Obama [will always be treated like a black man]."

"I absolutely agree with you, but that doesn't tell me if I have the mutation for asthma which you said you found in Africans."

"There is a *higher likelihood* that you will have that genetic factor."

In other words, race, an admittedly social category, can be used as a proxy for genes that affect drug response because it permits scientists to make a "probabilistic guess" at an individual's genes.[34]

The problem with this picture is that prescribing medication on the "race is a proxy for genetic difference" theory takes the statistical correlation too far. It is one thing to show that gene variants that affect drug response differ in frequency across racial groups. It is quite another to use these racial differences to predict which drug an *individual* should use. "Self-identified race is a surrogate for ancestral geographic origin, which is a surrogate for variation across the genome, which is a surrogate for variation in disease-relevant alleles, which is a surrogate for individual disease risk," points out Francis Collins.[35] Accuracy gets lost in the translation from race to an individual's risk of disease or drug response. Allelic frequency gives a rough estimate of the likelihood members of a racial group will carry the mutation. But an individual group member may not carry the mutation at all, regardless of the group frequency. Statistical roulette is too imprecise for prescribing drugs.

Besides, there is little evidence that race really does predict the genetic differences that influence drug response.[36] Duke molecular biologist David Goldstein and his colleague Sarah Tate surveyed twenty-nine medicines that were claimed in scientific or medical journals to differ in safety or efficacy across racial or ethnic groups. In a widely cited article published in *Nature Genetics* in 2004, Goldstein and Tate stated at the outset that "these claims are universally controversial and there is no consensus on how important race or ethnicity is in determining drug response."[37] They determined that in only one case of the twenty-nine studied was there evidence of a genetic basis for racial difference: beta blockers were shown to be more effective in Europeans than in African Americans for treating hypertension. For a majority of drugs or types of drugs (fifteen out of twenty-two), Goldstein and Tate found that evidence of a physiological basis for racial differences was thin or completely absent. Although differences in drug response are likely "influenced to some degree by genetics," they wrote, "in most cases it is difficult to separate genetic from environmental factors."

Nevertheless, scientists often point to variations in genes that regulate drug-metabolizing enzymes (DME), proteins that break down drugs in the body, as an uncontroversial example of a medically meaningful racial distinction. Races that are more likely to have gene variants that help to metabolize drugs will respond better to certain therapies, they argue. That seems like a simple case for race-based pharmacogenomics: drugs should be prescribed to the racial groups most likely to have the enzymes that will metabolize them. For instance, the authors of an article on variable drug response noted that 5–10 percent of Europeans, but only about 1 percent of Japanese, have "loss-of-function" variants at the CYP2D6 locus that affect the metabolism of many drugs.[38]

Before reaching the conclusion that these drugs are less effective in Europeans than in the Japanese, consider three complications. First, the authors went on to note that the metabolizing alleles vary in frequency just as much *within* Europe, from about 10 percent in Northern Spain to 1–2 percent in Sweden. So dividing patients by race is not the most accurate way of representing genetic differences in metabolism. In fact, prescribing the drugs at issue according to race would be positively inaccurate. It would group Swedes with Spaniards when they should be grouped with the Japanese. A 2002 comprehensive review article reported similar overlap in racial differences in the frequency of an allele involved in metabolizing the blood-thinning drug Warfarin: the frequency variation ranged from 0.9 percent to 20.4 percent in Caucasians and 0.0 percent to 8.6 percent in Africans.[39]

This is exactly what Craig Venter and fellow investigators at his research institute discovered when they examined the recently sequenced genomes of two of the most famous geneticists in the world—Venter himself and James Watson. Although both men are white, Venter has two fully functional CYP2D6 alleles, making him a high metabolizer, while Watson has a version that reduces enzyme activity.[40] The trait Watson carries is usually classified as Asian; it is estimated to be found in half the population of China but in only 3 percent of whites.

A second problem is the mismatch between racial categories and the genetic groupings that might be useful in pharmacogenomics. In addition to measuring differences in DME allele frequencies across eight ethnic groups (Bantu, Ashkenazi, Ethiopian, Norwegian, Armenian, Chinese, Papua New Guinean, and Afro-Caribbean), the researchers measured these differences across four genetic clusters inferred by the Structure program. In other words, they compared results using race as a proxy for genetic differences with results using genetic clusters created by the data itself. They discovered that commonly used ethnic labels did not match the genetic clusters and were not reliable at predicting variation in the DME genes. One glaring lack of correspondence was the fact that 62 percent of Ethiopians, who would socially be labeled as black and grouped with the Bantu and Afro-Caribbeans, fell in the same genetic cluster as Ashkenazi Jews, Norwegians, and Armenians. A gene variant involved in metabolizing codeine and antidepressants "is found in 9%, 17%, and 34% of the Ethiopian, Tanzanian, and Zimbabwean populations, respectively."[41] The prevalence of an allele that predicts severe reactions to the HIV-drug abacavir is 13.6 percent among the Masai in Kenya, but only 3.3 percent among the Kenyan Luhya, and 0 percent among the Yoruba in Nigeria.[42] Grouping all these people together on the basis of race for purposes of drug tailoring would be disastrous.

Finally, the percentage of people with certain poor-metabolizer alleles in any group is small. In 2006, a team of pharmacologists at University of Alabama reported a racial difference in dihydropyrimidine dehydrogenase deficiency (DPD), which causes poor metabolism of a widely prescribed cancer drug, fluorouracil.[43] Despite its established place in chemotherapy, fluorouracil is toxic for approximately 30 percent of patients, and about half of the patients with severe toxicity are DPD deficient. The researchers figured that the increased toxicity seen in African American cancer patients might be traced to reduced DPD enzyme activity. Using a breath test, they evaluated a sample of healthy African American and white volunteers and discovered that DPD deficiency was three times higher in African Americans—8 percent versus 2.8 percent.

Now what? It wouldn't make sense to prescribe fluorouracil to white but not black patients. Most patients in both groups are not DPD deficient and probably will tolerate the drug. Conversely, concern about DPD-related toxicity should not target African Americans alone since some whites also have the enzyme deficit. Instead, the researchers proposed a far more sensible solution that did not hinge on race. Noting that "DPD deficiency was observed in several ethnicities," they recommended that oncologists use the breath test to screen all their patients, regardless of race, to identify those at risk for developing toxicity to the drug.[44]

Some direct-to-consumer genetic testing companies have taken the next step toward personalized medicine by analyzing DNA submitted by customers to predict their drug response, a practice known as pharmacodiagnostics. Since the diagnosis is based on whether or not an individual carries specific genes, his or her race should be irrelevant to the analysis. Yet companies have found it useful to include race in their marketing strategies. DNA Direct helps customers predict drug response based on whether they carry alleles that make them "Poor Metabolizers" or "Ultra Metabolizers."[45] The Web site offers a table titled "Drug Metabolism Variation by Ethnicity," listing common drug metabolism gene variants and their frequencies in Caucasians,

African Americans, and Asians. It advises that it is "useful to understand . . . what the rates of CYP450 gene types are in your ethnic group." Then the Web site switches to a personalized message. It warns customers not to determine individual drug response by race, noting "it would be a mistake to assume that every member of a group has the same genetic profile and will react to a drug the same way." Of course, this provides a reason for customers to pay the fee to have their own individual genotypes analyzed rather than rely on the racial table. Why include the information about race at all, if ultimately customers should be tested individually? Even when a genetic test is available, racial difference helps to sell the importance of our molecular makeup to drug response.

AutoGenomics, a company based in Carlsbad, California, is developing an even more direct racial pitch to market its test for gene variants that predict warfarin metabolism. Researchers have identified specific alleles in the CYP2C9 gene and VKORC1 gene that account for 30 to 50 percent of the difference in individuals' responses to warfarin.[46] When, in August 2007, the FDA updated the label for Coumadin, warfarin's trade name, to explain how genetics may affect users' responses to the drug, companies began to apply for approval of diagnostic kits that test for the relevant alleles.[47] AutoGenomics obtained the green light to market its Warfarin assay, which screens for common variants of CYP2C9 and VKORC1. But the company also devised an extended Warfarin XP assay, not yet approved by the FDA, that tests for rarer forms of the genes. While awaiting FDA approval of its basic warfarin gene test, AutoGenomics promoted the Warfarin XP assay on its Web site by highlighting its benefit to particular racial groups. A 2008 PowerPoint presentation, for example, titled "Infiniti Warfarin XP: Because Ethnic Diversity Matters When Dosing with Warfarin," discusses racial variation in the frequency of the particular alleles covered by its test, noting that it is the only product that includes these variants.[48] Apparently, AutoGenomics' marketing plan for its Warfarin XP assay is to associate the rare alleles with specific races to persuade people who belong to those races to buy its test. This creates an additional racial hook: patients who might have been satisfied with a test for the more common alleles may think they need to use Warfarin XP because of their race.

A more basic problem is that there are many factors besides genes that affect drug metabolism. Whether a drug will be safe and effective for a particular individual depends on a long list of physical and lifestyle traits, which may affect gene-related metabolism and may in fact be more important than genes: weight, sex, age, diet, smoking, drinking, the cause and length of the illness, other coexisting health conditions, interactions with other medications, and the type of care the patient has already received. After World War II, for example, malnutrition caused some Germans to die when they were injected with a local anesthetic that was ordinarily safe.[49] "There are many, many other things that are much more sensible to attend to than race," epidemiologist Jay Kaufman told me. "There isn't any special heart failure drug that's only for the obese population. And yet there is much stronger evidence that obesity modifies the dosages of drugs. There's a tradition of elevating race above all these other things."[50] Even if race were a good proxy for genetic determinants, it isn't necessarily the chief factor in predicting how a patient will respond to medication. In the seemingly simple case of drug-metabolizing enzymes, race is both a faulty test for the telltale gene and a diversion from nongenetic determinants of drug response that may be more useful.

Equally as intriguing as the concept of personalized medicine is the proposal to develop the first drugs based on race. Think of the paradox: a classification system constructed centuries ago to enslave people became the portal for the most cutting-edge biomedical advance of the twenty-first century. Predicting drug response based on a patient's race rather than on genetic traits, says Lawrence Lesco of the FDA's Center for Drug Evaluation Research, is "like telling time with a sundial instead of looking at a Rolex watch."[51] Still, he believes it is acceptable to rely on the sundial until the precision timepiece is available. But thinking of genotype as a more accurate version of race is a misconception that only perpetuates the myth that race is a biological category. There is no better example of the psychic grip that race has on science. How can a drug be personalized—crafted or chosen to respond specifically to my unique genotype—if it is formulated, marketed, and prescribed to me according to my race? The idea that the racial category I fit into can substitute for my genetic makeup reflects the continued power of race to define people.

Racial Markets

What's in it for a pharmaceutical company to develop or market a new drug to members of a particular race? Why not the entire national—or even international—market?

First, specifying a race-based application for a pharmaceutical invention may be a way of gaining or extending patent exclusivity. In the case of BiDil, the first drug to receive race-specific approval from the FDA, turning the therapy into a product for blacks allowed the producer, NitroMed, to parlay a patent due to expire in less than ten years into twenty years of monopoly over the black heart-failure market. A decade earlier, the pharmaceutical industry had already considered emphasizing racial distinctions as a strategy for resisting pressure to cut the runaway costs of prescription drugs and to protect their marketability.[52] A 1993 report on "ethnic and racial differences in response to medicines" by a trade group called the National Pharmaceutical Council addressed "cost containment tactics," such as the mandatory use of generics or approved drugs listed on a formulary, instituted by managed care programs. The report proposed contesting the presumption underlying these measures that "related medicines having the same general effect are almost identical in their actions and are therefore interchangeable," with evidence that races respond differently to drugs. Restricting the range of drugs physicians can choose from, the industry could argue, can "lead to inappropriate or suboptimal therapy for minority groups."[53]

Second, rather than limiting market size, race actually creates a new market for personalized medicine. Because pharmacogenomics teases out genetic differences in drug response, it naturally narrows the market for each drug developed. Personalized medicine is individualized; it is supposed to target therapies to each person's unique genome. But as the President's Council of Advisors on Science and Technology explained, personalized medicine "does not literally mean the creation of drugs or medical devices that are unique to a patient but rather the ability to classify individuals into subpopulations that differ in their susceptibility to a particular disease or their response to specific treatment."[54] It may not be profitable to sell products to genetically differentiated population groups if each group is too small. As one industry insider said, "XXXX is the biggest drug in the world. Why would we want to do genetic studies on it? We wouldn't play with it."[55]

Herein lies the Catch-22 of personalized medicine. As genomic research helps pharmaceutical companies zoom in on the patients who will truly benefit from their drugs, it simultaneously reduces the pool of consumers who will buy them. The challenge is to determine where to draw the line between the subgroups so that they are personalized enough to define drug targets but large enough to guarantee a profit. What's more, even if enough people have the relevant gene to make a drug profitable, there has to be an economically viable way of identifying who they are. Scanning the genomes of every patient or testing them for a particular gene variant adds a cost to personalized medicine that some industry experts see as prohibitive. Even if this technological transformation takes place, the medical profession would have to catch up with it. Doctors would have to be trained to translate genetic information into their practice, counseling patients about genetic testing and determining what the results mean for treatment. "Some practitioners tell me they hope they retire before pharmacogenomics hits," says Gary Merchant, director of the Center for the Study of Law, Science, and Technology at Arizona State University in Tempe.[56]

Race solves both marketing problems inherent in personalized medicine. It provides a large, identifiable group of consumers for drugs developed using genetic research. While it would be expensive for drug makers to segment the pharmacogenomic market into relatively small ancestral groups with shared allele frequencies—creating special drugs for people of Yoruba, Ethiopian, Korean, Japanese, Swedish, Spanish, or Mexican descent—it does make market sense to sell their products to vastly larger patient pools of blacks, Asians, whites, or Hispanics. Instead of developing small batches of designer drugs, companies can develop drugs for entire racial groups.

Marketing drugs to racial minorities represents a potential for market growth because of their increasing numbers and higher risk of disease. In 2004, the announcement for the 5th Annual Multicultural Pharmaceutical Marketing and PR Conference, noted: "Major U.S. Drug manufacturers are making it a high priority to cultivate relationships with ethnic consumers, physician groups, community networks and other key stakeholder groups to uncover new market growth. Disproportionately high incidence of diabetes, obesity, heart disease, cancer, HIV/AIDS, asthma and other health conditions among these segments require many strategic and tactical moves in pharmaceutical marketing and PR."[57]

The brochure for the conference the following year similarly touted the profit potential in racial markets: "The unprecedented growth in ethnic populations across various regions in the United States opens doors to a wide array of new market opportunities for healthcare and pharmaceutical companies." These new markets, moreover, could help overcome the barrier to industry profit. "With the onslaught of generics, pricing battles and DTC competition," the brochure stated, "reaching out effectively to America's emerging majority is a clear road to brand building and market growth."[58] The meeting of the Pharmaceutical Marketing Research Group held in September 2007 included a session on how to target new biomedical products to specific racial groups with the catchy title "When the Ivory Tower Goes to the Ebony Hood."[59]

Businesses already have established marketing infrastructures based on race that can be tapped for new race-specific drugs.[60] Hair products, sports shoes, music, media, cigarettes, and alcohol are all marketed by race. A 2007 meta-analysis of peer-reviewed studies that directly compared tobacco advertisements in African American and white markets found that there were almost three times more tobacco billboards per person in black communities, and the odds of a given billboard hawking tobacco were 70 percent higher in their neighborhoods.[61] A study of alcohol and cigarette billboards in Chicago similarly revealed "market segmentation aimed at saturating poor and minority neighborhoods with messages to buy and use dangerous products."[62] The authors note that advertisers select themes to target specific market segments, including racial and ethnic groups. Because many African Americans prefer menthol cigarettes, for example, billboards featuring glamorous black models smoking Newports, Kools, and Salems proliferate in their communities. The same techniques that have sold special brands of cigarettes, beer, and other products on the basis of race are already being put in the service of therapeutic distribution.

With race-specific drugs and diagnostics, pharmaceutical companies have at their disposal a marketing tool that is even more persuasive than the typical gimmicks. The message about race and genetics the products themselves convey has a special appeal. Marketing a drug or genetic test for a particular race seems to say that it was scientifically developed to target the genetic makeup of patients who belong to that race. Drug makers can exploit this false belief that race represents individual genetic difference; when patients hear about a drug indicated specifically for their race, they may assume that it is "just right for me." The misconception of race underlying pharmacoethnicity is itself a persuasive marketing trick.

The commercial exploitation of race to sell personalized medicine is not merely hypothetical. On June 23, 2005, the federal agency charged with regulating pharmaceuticals made this startling announcement: "The Food and Drug Administration (FDA) approved BiDil (bye-DILL), a drug for the treatment of heart failure in self-identified black patients, representing a step toward the promise of personalized medicine." With these words, the federal government put its seal of approval on an emerging biotechnology that can turn the science of racial genetics into marketable products. For the first time, the FDA approved a pharmaceutical for use solely by patients belonging to a specific race. It also positioned racial marketing of drugs as an essential phase in the development of drugs tailored to individuals' personal genotypes. The next chapter will take a closer look at this controversial product.

Scientists and entrepreneurs see race as an avenue for quickly translating the embryonic science of personalized medicine into marketable products. At the same time, race-specific medicine provides moral legitimacy as well as profitability to the new racial science. By claiming that health disparities are caused by race-based genetic differences, scientists have gone beyond validating race as a biological concept. They made the biological understanding of race appealing in the post–civil rights era. This medical benefit makes the genomic clustering studies relevant to the average person, who is far more concerned about which drug to take than prehistoric human migrations. Telling the public that clusters resembling races can advance medicine gives people an extremely positive frame for interpreting the research findings. "The new medical interest in race and genetics has left many sociologists and anthropologists beating a different drum in their assertions that race is a cultural idea, not a biological one," Nicholas Wade wrote in his 2002 front-page *New York Times* article "Gene Study Identifies 5 Main Human Populations."[63]

Eschewing the damaging ideological aims of past racial scientists, the new generation's goals are morally unassailable—to diagnose and cure disease at the genetic level. This medical mission permits a dramatic retelling of the sordid history of race and genetics. In this version, the very revelations of the last century about the dangerous myth of race as biology have been recast as blinders to the "truth" about race as biology. Consider the opening paragraph of a 2005 *Christian Science Monitor* article, "A Place for Race in Medicine?":

> Ever since the fall of the Nazis, the world has tried to keep the biology of racial disparity under wraps. It has been acceptable to link racial differences to social and cultural factors. One race might underperform another because of upbringing or poverty. But suggesting biology as the cause for those differences—like "The Bell Curve" did a decade ago

> when it looked at academic achievement—was strictly taboo. Now, a new and unexpected force—medicine—is pulling back the covers. By taking a close look at minute differences in people's genetic codes, researchers and drug companies are beginning to create racially based drugs and treatments.[64]

Race-based medicine gives people a morally acceptable reason to hold on to their belief in intrinsic racial difference. They can now talk openly about natural distinctions between races—even their biological inferiority and superiority, at least when it comes to disease—without appearing racist. This would be a case of public enlightenment—"pulling back the covers"—if the science supporting racial therapies were sound. But to the contrary, the purported benefits of racial medicine provide an excuse to overlook the scientific flaws in research claiming to show race-based genetic difference. These technologies are not just products of racial science. They are driving racial science.

8

Color-Coded Pills

BiDil, the first race-specific drug approved by the FDA, was not what it seemed. It did not contain new ingredients. Nor was it designed only for black people. It was not even developed to target any particular genetic profile. Jay Cohn, the University of Minnesota cardiologist who patented BiDil, combined into a single pill two generic drugs that had been prescribed to patients regardless of race for over a decade. In fact, he originally intended to market it to patients of any race who could benefit from it. But after Cohn licensed the rights to NitroMed, a private biotech start-up in Lexington, Massachusetts, BiDil was repackaged as a drug specifically for African Americans.[1]

BiDil's race-specific label forecasts the role race will increasingly play in translating genomic research into profitable products. The reason why BiDil was marketed according to race has far more to do with its commercial appeal than its medical benefits. Ultimately, the plan to profit by marketing the drug to self-identified African Americans failed. But understanding why NitroMed targeted this market in the first place, and why a group of scientists, doctors, activists, and politicians vehemently supported this plan, sheds light on the ways that contemporary science is reinforcing, rather than eliminating, antiquated ideas about race.

BiDil's Conversion

Heart failure is a debilitating illness that affects millions of Americans and kills tens of thousands every year. It occurs when the heart muscle becomes so weak from heart attack, high blood pressure, or infection that it no longer pumps enough blood. People suffering from it are so tired and short of breath that they are usually unable to work and may struggle to climb stairs or walk across a room. Half die within five years of their diagnosis.

In the 1970s, Jay Cohn had a hunch that vasodilators—drugs that relax the blood vessels—might help to treat heart failure.[2] He discovered that giving hypertensive patients sodium nitroprusside, a powerful vasodilator that is administered intravenously, improved their heart function. Cohn combined two generic drugs, hydralazine and isosorbide dinitrate (H-I), to develop a therapy with the same properties as sodium nitroprusside that could be taken orally. The duo seems to work by helping the body make nitric oxide, a gas that widens the arteries and lets blood flow through them more easily. The result is lower blood pressure and less strain on the heart. BiDil combines these two ingredients in one pill.

At the time, there was virtually no effective therapy for the deadly illness. Working with the Veterans Administration, Cohn and other cardiologists conducted the Vasodilator Heart Failure Trial (V-HeFT I) to see if drugs that widen blood vessels would reduce mortality. He tested the H-I combination against a placebo and prazosin, a drug for high blood pressure, on 642 white and black men. Lasting five years, V-HeFT I showed that the H-I combination worked to stave off death from heart failure. In the late 1980s, the researchers conducted a second trial, V-HeFT II, to compare the efficacy of the H-I combination against an angiotensin-converting enzyme (ACE) inhibitor called enalapril. The second study found that enalapril lowered the overall death rate even more than H-I.

The results of the V-HeFT trials were spectacular news for the treatment of heart failure. They soon revolutionized cardiac care, as ACE inhibitors became the preferred therapy for most heart failure patients. Unfortunately, the prospects for H-I were not as rosy. Cohn wanted to study the effects of combining H-I with enalapril in the next phase of the Veterans Administration research. But because the ingredients in H-I were generics, there was no commercial reason for a pharmaceutical company to invest in such a large-scale clinical trial. Instead, a calcium blocker was substituted for H-I in V-HeFT III. Cohn then took matters into his own hands: he would patent and market the H-I combination as a new fixed-dose pill.

Like a growing number of academic scientists in the 1980s, Cohn partnered with a biotech firm to bring his newly patented drug to market. The patent gave Cohn a monopoly on the combined H-I pill without preventing sales of the individual generic ingredients. As Marcia Angell, the former editor-in-chief of the *New England Journal of Medicine* explains, "exclusivity is the lifeblood of the industry because it means that no other company may sell the same drug for a set period."[3] The patent Cohn filed in 1989 made no mention of race. He licensed the intellectual property rights to a small pharmaceutical company in North Carolina called Medco to develop, manufacture, and market BiDil—a single pill to treat heart failure, regardless of race. In 1996, Medco submitted a New Drug Application for BiDil to the FDA. Like Cohn's patent application, this application for marketing approval did not mention race. Its evidence of the drug's efficacy consisted of a retrospective analysis of the V-HeFT trials conducted in the 1980s.

The FDA Advisory Committee that reviewed Cohn's application was not convinced. The committee found that the reanalysis of old data, which had not been collected specifically to test a new drug for FDA approval, failed to meet its criteria for statistical significance. Another problem was that BiDil would only be prescribed as an adjunct to standard ACE inhibitor and beta blocker therapies, which had not been combined with H-I in the V-HeFT trials. (V-HeFT II compared the ingredients in BiDil *against* the ACE inhibitor enalapril.) The FDA therefore denied Medco's request in 1997.

Having lost the bid for FDA approval, Medco let the rights to BiDil revert to Cohn. With his original patent due to expire in less than ten years, Cohn desperately needed a strategy for salvaging his pharmaceutical venture. It was only after the FDA rejection that Cohn turned BiDil—the exact same drug that he had patented without regard to race—into a therapy for African Americans.

Cohn and another lead researcher on the V-HeFT trials, Peter Carson, went back to the original data to parse it for some way to satisfy the FDA. What if they broke down the statistics by race? In 1999, Cohn and Carson published a paper in the *Journal of Cardiac Failure* reporting that their retrospective racial analysis of the V-HeFT data showed "the H-I combination appears to be particularly effective in prolonging survival in black patients." In contrast, they reported that the ACE inhibitor enalapril worked particularly well in white patients. These findings, they argued, suggested that "therapy for heart failure might appropriately be racially tailored."[4]

But did they? There were several serious problems with going back to the old V-HeFT data and dividing it into racial groups who received the H-I combination—what experts call a post hoc subgroup analysis.[5] Divvying up a randomized, controlled trial inevitably involves a loss of statistical power; the tiny subgroups make it hard to know if any statistically significant differences were real or by chance. Besides, observed differences in response between subgroups are far more likely to be a matter of degree than evidence that a drug works for one group and not for another.[6] The numbers of black patients in the original studies were relatively small: in V-HeFT I, only 49 black and 132 white patients received the H-I combination. V-HeFT II included 109 black and 282 white participants.[7]

In addition to the numerical imbalance between the black and white patients, there was a socioeconomic and medical imbalance. To tell if race made a difference to drug response, the researchers would have to control for other factors that might explain the difference between the subgroups. One such factor affecting drug response might be that the black participants in V-HeFT I were more likely to have a history of hypertension and less likely to have a history of coronary artery disease than the white patients. It was also difficult to apply the V-HeFT results to patients almost twenty years later because the treatment for heart failure had changed so dramatically in the meantime. Those older trials were conducted to test the effectiveness of ACE inhibitors, which are now widely used as the first-line therapy for heart failure.

Despite these weaknesses, BiDil was on its way to becoming a race-specific drug. In September 1999, the same month his article was published, Cohn relicensed the intellectual property rights to NitroMed, a Boston-area firm specializing in nitric oxide products. Cohn and Carson also submitted a new patent for BiDil—this time claiming a race-specific method. The critical difference between the new patent and the original one Cohn filed in 1989, set to expire in 2007, is this key language: the "present invention provides methods for treating and preventing mortality associated with heart failure *in an African American patient*" (emphasis added). The new patent, issued in 2000, lasts until 2020, buying Cohn thirteen more years of intellectual property control over the drug. As legal scholar Jonathan Kahn astutely notes, "patent law did not spur the invention of a new drug, but rather the reinvention of an existing therapy as race-specific."[8]

In 2001, Cohn and NitroMed approached the FDA to test the idea of selling BiDil as a drug specifically for African Americans. Cohn was now in a far better position to do so: there was growing concern that African

Americans died younger and faster from heart failure and benefitted less from ACE inhibitors, and the retrospective analysis of V-HeFT suggested BiDil could fill the gap. The FDA responded with a letter suggesting that it might approve BiDil if NitroMed could prove in a clinical trial that the drug worked effectively for black patients. With that promising signal from the FDA, NitroMed began to raise venture capital to pay for a clinical trial testing BiDil on African Americans. NitroMed CEO Michael Loberg predicted revenues of $120 million the first year blacks with heart failure started taking the pill.[9] Within a year, NitroMed had raised $31.4 million in private financing. It now had the funds needed to conduct the clinical trial proposed by the FDA.

The African-American Heart Failure Trial, or A-HeFT, was launched in 2001. NitroMed partnered with the Association of Black Cardiologists to cosponsor the study. It enrolled 1,050 self-identified African American men and women suffering from advanced heart failure at 170 sites around the country. Half the clinical subjects received the BiDil formulation in addition to standard heart failure medicine, while a control group received the standard therapies alone.

When asked at a 2006 conference at MIT why the clinical trial did not include more diverse groups, a NitroMed representative replied that the money raised for research was insufficient to enroll a larger number of subjects. FDA officials agreed that it would have been too costly for NitroMed to test its claim of racial efficacy in a well-designed study involving various racial groups.[10] In fact, NitroMed had a financial *disincentive* for finding that BiDil worked regardless of race—its patent (and market monopoly) applied only to its use by African American patients.

On November 6, 2003, in the midst of the A-HeFT trial, NitroMed went public. Managed by Deutsche Bank Securities and J.P. Morgan, the initial public offering was for 6 million common shares at a target price of $11 per share for a total market value of $66 million.[11] The prospectus filed with the SEC listed as "the BiDil Market Opportunity" African Americans' greater risk for heart failure, the lack of effectiveness of other heart failure medicines, and the belief that "African Americans as a group may produce less nitric oxide in their arteries." "As BiDil is intended to treat these specific characteristics," the company stated, "we believe that BiDil, if approved, will be broadly prescribed for African-American heart failure patients." Then investors waited with bated breath to see if the study would confirm the drug's market viability.

BiDil worked. In fact, it worked so spectacularly that NitroMed's Data Safety Monitoring Board stopped the trial ahead of schedule, in July 2004, so all patients in the study could take it. BiDil increased survival of patients by an astonishing 43 percent. Hospitalizations were reduced by 33 percent. Patients taking BiDil reported feeling better and having an improved quality of life.

Loberg hastily issued a news release and held a conference call to update investors and financial analysts. "We will now move up our commercial preparedness and seek to have launch elements in place in the first quarter of 2005, one year ahead of our prior schedule," he announced. He then presented the company's glowing financial forecast: "There are approximately 750,000 diagnosed African Americans with heart failure and that number is projected to go upwards to 900,000 by the end of the decade. We have looked at that market potential and just in aggregate terms, the market opportunity, if you were just to dollarize the patient base, at current heart failure therapy prices, you would get a market opportunity approaching $1 billion." Jennifer Chao, a senior analyst for biotechnology at Deutsche Bank, predicted "relatively rapid market penetration" during BiDil's first year of sales.[12]

The FDA Hearing

The FDA's Cardiovascular and Renal Drugs Advisory Committee met all day on June 16, 2005, to assess the evidence that BiDil improves outcomes for heart failure patients.[13] In the public portion of its meeting, the committee listened to thirteen speakers who debated the pros and cons of FDA approval. The point of contention was not whether BiDil should be available at all. It was whether BiDil should be labeled as a drug for blacks.

Black cardiologists, activists, and members of Congress testified that approving BiDil would help the agency make amends for America's racist history of medical maltreatment and demonstrate its concern for black people's health.[14] The very first voice heard was Donna Christensen, the Democratic representative of the U.S. Virgin Islands, who testified on behalf of the Black Congressional Caucus. She told committee members that BiDil presented "an unprecedented opportunity to significantly reduce" a leading cause of death among blacks. She urged them to authorize the drug as a remedy for medical wrongs against African Americans "for whom treatment has been denied and deferred for 400 years."[15]

But concerns about the "for blacks only" indication persisted throughout the day. One worry was the lack of scientific evidence to back up the race-specific indication: NitroMed's application was based on only one clinical trial (the FDA usually required at least two) that tested only African American patients, which meant that there was no control to test NitroMed's race-specific claim. In his remarks, Jonathan Kahn argued that there was reason to approve BiDil for general use by heart failure patients, without regard to race, but there was no justification for a racial label. "Most drugs on the market today were approved by the FDA based on trials conducted almost exclusively in white patients," he noted, "but those drugs are not designated as white drugs, and rightly so."[16]

The next person to take the microphone was Charles L. Curry, chief of the Division of Cardiovascular Diseases at Howard University College of Medicine and president of the board of trustees of the International Society of Hypertension in Blacks. He agreed that BiDil should be approved without regard to race, noting that American cardiologists "jumped on the statin drugs" once the Scandinavian Simvastatin Survival Study showed they were effective. "Would you restrict the results of the Scandinavian trial to Scandinavian people?" he asked. "I don't think so."[17] Dr. Curry's colleague Charles Rotimi, from Howard University's National Human Genome Center, echoed this position. Rotimi warned that upholding an unproven biological explanation for health disparities would steer biomedical research in a dangerous direction. "It would be tragic not to approve [BiDil]," Rotimi said, "and it would be even more tragic just to approve it for African Americans."[18]

The issue crystallized in a debate between two committee members. Vivian Ota Wang, also from Howard's National Human Genome Center, challenged the use of race in the A-HeFT trial as a proxy for an underlying biological trait that explained how BiDil worked. "There is a presumption here that somehow this self-identified social identifier is somewhat equivalent or representative of a biological process, and I am not sure it really is," Ota Wang pointed out.[19] "We need to really carefully look at the issue of self-identified racial categories, because if the assumption is that these population differences are biological, the self-identified population is a social and political construct."[20] The committee chair, Cleveland Clinic cardiologist Steven Nissen, dismissed Ota Wang's concerns. "We are using self-identified race as a surrogate for genomic-based medicine and I don't think that is unreasonable," he said. "I wish we had the genetic markers . . . to decide who is going to respond to what drug. But in the absence of that, we have to use the best available evidence, and that evidence was used in this trial and it worked."[21] Nissen dispatched the worry that this rationale for approving a race-specific drug would reinforce a genetic definition of race by asserting, "Drugs aren't racist; people are."[22]

Ota Wang and Nissen also differed on the standards the FDA should use to evaluate BiDil. Nissen took the position that the remarkable improvement in health experienced by the A-HeFT patients on BiDil should outweigh concerns about the statistical strength of trial data. "I have to approve a drug when I think there's evidence you can reduce mortality by 43 percent," he said. "As a clinician, I find the evidence more than adequate to vote for approval."[23] Comparing heart failure among African Americans to an orphan disease, he argued that "you make some adjustments sometimes because you want to encourage trials in special populations and diseases which are of public health importance which we have few therapies for."[24] In other words, to Nissen, heart failure suffered by African Americans was a special type of illness that warranted exceptions to the rules, as in the case of rare medical conditions.

Ota Wang objected to "the notion that for some types of research, for some types of communities or populations we can actually lower the bar in terms of scientific integrity that we are using to evaluate the research."[25] In response, Nissen reiterated the orphan disease analogy. "So, if you are developing a drug for a disease and there are not many people that have it, you get some points for doing that," he said. "I am arguing that it is not unreasonable public policy to make some adjustment for that."[26] Statistical weaknesses in the data that would ordinarily pose problems for FDA approval were overlooked because BiDil was a drug for a minority group.

Nissen may have been right about the importance of the clinical findings. The overwhelming evidence from the A-HeFT trial that BiDil was beneficial for many patients was a compelling reason to make it widely available on the market. What the A-HeFT trial did not do was give the FDA grounds to base its decision on race. NitroMed offered no evidence of the biological process that explains why BiDil supposedly works differently in blacks. Nissen was satisfied with the "biological plausibility" of NitroMed's claim based on the untested assumption that race is a surrogate for genetic difference.[27] Also disturbing is the way the racial indication became an excuse for "lowering the bar" the agency usually applies in evaluating new drugs. As chair, Nissen was able to steer the discussion away from the points Ota Wang raised, abandoning any serious scrutiny of NitroMed's racial claims.

In the end, the nine committee members voted unanimously to approve BiDil for the treatment of heart failure. Ota Wang and John Teerlink, a cardiologist at the San Francisco VA Medical Center, dissented from the recommendation that the drug be approved for self-identified African American patients alone.[28] As is typical, the FDA followed the advisory committee's recommendation and granted NitroMed's request to market BiDil as a drug for blacks. With no scientific confirmation of its racial claim, NitroMed was nevertheless able to persuade the FDA to back its strategy to sell its product according to race.

The Mysterious Mechanism

As should now be clear, there is no scientific proof that BiDil works differently in black people. A-HeFT, the clinical trial that tested BiDil, enrolled only self-identified African Americans. Because there was no comparison group, the A-HeFT researchers never showed that BiDil functions only or even better in blacks than in patients of other races. Rather, NitroMed argued that, because BiDil was tested only on blacks, the FDA should label it as a drug for blacks only. As Jane Kramer, NitroMed's vice president of corporate affairs, would later explain, "That doesn't mean that it works in all African Americans and it doesn't mean that it doesn't work in other patients. It just means that we know it clearly works in African Americans."[29]

By the logic used to approve BiDil, drugs tested on Americans should never be marketed overseas, or drugs tested only on whites should not be made available to anyone else. That logic had never resulted in a racial indication before. In the past, the FDA has generalized clinical trials involving white patients to approve drugs for everyone because it is assumed that white bodies function like all human bodies. By approving BiDil only for use in black patients, the FDA emphasized the supposedly distinctive—and, it is implied, substandard—quality of black bodies.[30] The FDA treated white heart failure patients as the norm and blacks as a special case that had to be given a specialized therapy that Nissen compared to an orphan drug and that could not be assumed to work for other people. The message is: black people cannot represent all of humanity as well as white people can.

But how or why does BiDil work differently in black patients than in patients of other races? Why do black patients with advanced heart disease need a race-specific therapy? The most common answer is that race is a proxy for an unknown physiological mechanism that explains how BiDil works. The reason for higher mortality rates among black heart failure patients, the theory goes, lies in some unspecified biological difference, either in the reason for getting heart disease or the reason for responding differently to medications for it. Because the researchers had not discovered this mechanism nor shown how it causes heart failure or differential drug response, they used race as a surrogate in the meantime. Eventually, this mysterious mechanism was attributed to genes.

Cohn and his colleague Peter Carson began the background section of their patent application for a race-specific drug by claiming innate differences between blacks and whites suffering from heart failure: "Heart failure in black patients has been associated with a poorer prognosis than in white patients. In diseases such as hypertension, blacks exhibit pathophysiologic differences and respond differently to some therapies than whites." A later section notes, "Black and white patients exhibit differences in etiology, neurohormonal stimulation and pharmacologic response in heart failure." Cohn and Carson elaborate that the "uniquely favorable" effect of BiDil on black patients "suggests the possibility that blacks, particularly with a hypertensive history, may have a greater deficiency of nitric oxide generation that is restored" by the drug's combined ingredients.[31]

In a March 2001 press release, NitroMed similarly attributed higher black death rates from heart failure to an unspecified "pathophysiology found primarily in black patients."[32] A second press release reiterated, "Observed racial disparities in mortality and therapeutic response rates in black heart failure patients may be due in part to ethnic differences in the underlying pathophysiology of heart failure."[33] The 2004 report of the clinical trial testing BiDil in the *New England Journal of Medicine* referred to "the hypothesis that the balance of mechanisms leading to the progression of heart failure may vary with geographic origin."[34] Even in his FDA testimony, Jay Cohn could not identify the precise biological mechanism that explained the differential response.[35]

The FDA explained its decision to approve BiDil specifically for black patients in similar terms but suggested that this unknown mechanism might be genetic. Robert Temple, FDA associate director of medical policy, elaborated the race-as-proxy theory in a January 2007 article in *Annals of Internal Medicine*. "We hope that further research elucidates the genetic or other factors that predict the usefulness of hydralazine hydrochloride–isosorbide dinitrate," he wrote. "Until then, we are pleased that one defined group has access to a dramatically life-prolonging therapy."[36] In other words, a racially defined group could serve as a temporary substitute for the yet-undiscovered genetic or other biological factor that identifies who will benefit from BiDil. Nissen, the FDA advisory committee chair, put it more bluntly: "We're using self-identified race as a surrogate for genetic markers."[37]

Writing in the *Journal of the National Medical Association*, South Carolina surgeon Robert Sade defended the use of race as a proxy in the BiDil case as a preliminary yet beneficial "step forward toward treating heart failure." Of course, using more specific markers, based on genes associated with diseases, would be preferable, he acknowledges. "Yet, until we know what the specific markers are and can identity them fairly easily, we must use more approximate markers." In the case of the BiDil clinical trial, writes Sade, "self-identification by race was undoubtedly not the most precise marker to use, but the investigators did not have a better one."[38]

The familiar defense that, despite being a "crude marker," a "blunt tool," an "imprecise proxy," "a makeshift solution," or "an imperfect placeholder," race is the best that science can do at the moment is not a justification. The reason the BiDil investigators did not have a better marker is that they did not look for one. They stopped at race. NitroMed sought FDA approval based on pure speculation that race was a good enough proxy for some underlying genetic or pathophysiological difference without conducting any investigation to test whether or not this difference existed. The researchers who reported BiDil's effectiveness for African American heart patients recognized this flaw and simply promised to correct it at some future time. "A future strategy would be to identify genotypic and phenotypic characteristics that would transcend racial or ethnic categories to identify a population with heart failure in which there is an increased likelihood of a favorable response to such therapy," they wrote.[39]

Sade argues that NitroMed, as a for-profit enterprise, had no obligation to spend any additional funds on "uncovering underlying biological factors," explaining, "Commercial companies do not and should not be expected to expend their limited resources on research and development that will not, in their view, lead to commercially viable products."[40] This sounds like an excuse for pharmaceutical companies to market race-specific drugs because it is profitable, even if they have conducted insufficient research to justify their use of race as a diagnostic category.

In fact, there were nonracial explanations for BiDil's success that should have been investigated. At the American Heart Association Scientific Sessions held in 2006, Dr. William Abraham from Ohio State University noted that the black patients enrolled in the A-HeFT trial stood out because they were more likely to suffer from hypertension, obesity, and diabetes and less likely to have ischemic heart disease (resulting from hardening of the arteries) than white participants in similar studies. African Americans tend to have nonischemic heart failure, which results from chronic strain to the heart due to complications from hypertension, diabetes, and kidney disease. "So is this a color-of-the-skin debate, or is it really some other phenotypic marker that predicts response?" he asked.[41] In the A-HeFT trial, there was a striking difference in the prevalence of diabetes between the black patients who received BiDil compared to those who received the placebo. Forty-five percent of the patients on BiDil had diabetes, while only 37 percent of those on placebo did.[42] Instead of pointing to race, A-HeFT researchers might have focused on this statistically significant difference in a medical factor that could affect the trial results.

Since the A-HeFT study never tested genes, there was no reason to associate BiDil with pharmacogenomics. Yet in its 2005 announcement of BiDil's approval, the FDA claimed that the drug represented "a step toward the promise of personalized medicine." The press also treated the first race-specific drug as the initial foray into gene-tailored therapies. The *Times* of London began an article on BiDil thus: "It sounds like science fiction: genetically based medicine offering drugs that work best for you because they perfectly suit your racial profile."[43] That is because it *is* science fiction. Similarly, the *New York Times Magazine* described BiDil as "on the leading edge of the emerging field of race-based pharmacogenomics."[44] In *FDA Week*, Michael Warner, a former regulatory affairs specialist for the Biotechnology Industry Organization, an industry trade group, linked BiDil to personalized medicine with the false assertion that "BiDil is the first time, the highest profile time, the model of 'let's identify a target population and let's develop a drug for that population' has been pursued."[45] With this sleight of hand, BiDil was flaunted as evidence that using race in genomic research can improve health, even though the drug's developers had conducted no genomic research. It became the symbol for the new racial science even though it offers absolutely no proof of its validity. BiDil is not a pharmacogenomic drug.

As Common as Viagra

As soon as it received FDA approval, NitroMed began promoting BiDil to black patients. The company had already tested the waters for physicians' willingness to prescribe by race by funding a national Web survey of six hundred physicians from June 20 to June 22, 2005, in response to the FDA advisory panel's recommendation to approve BiDil. On June 23, the day of the FDA's final decision, the health care marketing research firm HCD released the results. The survey found that 81 percent of physicians believed that "race should be used as a biological basis for determining ailments or diseases" and agreed the BiDil should be approved "for use among African Americans."[46]

The beauty of this marketing strategy is that actual sales need not be limited to a particular racial group, regardless of the marketing pitch. Jay Cohn repeatedly stated in the media that he prescribes the combined ingredients in BiDil to his white patients and that, as he told the *Los Angeles Times*, "everybody should be using it."[47] Although racial labeling restricts marketing of medications, it is not an absolute bar to off-label prescriptions. Many cardiologists know that the ingredients in BiDil have long been listed by the American Heart Association and the American College of Cardiology as a therapy for heart failure without regard to race. The *Harvard Heart Letter*, distributed by Harvard Medical School, told its readers with heart failure that is not controlled by standard therapies to ask their doctors about BiDil or the generic combination "no matter what your skin color."[48]

NitroMed, meanwhile, had already been investing in goodwill among leaders of the black community. On the one hand, some prominent African Americans, such as the sociologist Troy Duster and the geneticist Charles

Rotimi, criticized drugs labeled specifically for blacks as a scientifically flawed misuse of biomedical research that threatened to reinforce dangerous biological understandings of race. On the other hand, several prominent black organizations supported racial therapeutics precisely to redress past discrimination in medical research and health care and fulfill long-standing demands for science to attend to the needs of African Americans. Members of the Black Congressional Caucus, the Association of Black Cardiologists, the National Medical Association, the National Association for the Advancement of Colored People (NAACP), and the National Minority Health Month Foundation urged the FDA to authorize BiDil and promoted its use by black patients. Many fervently believed that access to BiDil would improve the health of their constituents.

To some African Americans, BiDil represented the first tangible recognition of their particular health needs and the first sign of progress in addressing them. NitroMed reached out to the Association of Black Cardiologists to collaborate in designing medical research aimed at prolonging the lives of black heart failure patients. The BiDil clinical trial, focused on improving black people's health, represented a long-fought-for departure from traditional studies designed to benefit white patients only. The A-HeFT trial included the largest number of black heart failure patients ever studied and the largest number of black women ever to participate in *any* clinical trial.[49] Imagine the excitement felt by the black cardiologists who led the study when their black patients, so long overlooked, began responding to the drug they were testing. "*This time*," said one, "African-Americans will be the first to reap the benefit."[50]

Rallying around BiDil as a cure for black heart failure was reminiscent of the black protest movement's embrace of sickle cell disease as a political symbol. "The pain and suffering of sickle cell anemia victims became nationally politicized, used by many black Americans—from college sororities to women's auxiliaries to the Black Panther Party—to raise consciousness about their long-ignored social condition and their civil right to health equality," writes historian Keith Wailoo.[51] Wailoo notes that the disease was transformed into a "cultural commodity" that represented African identity and helped to express patient-centered activism. It is important to keep in mind, however, that BiDil was a *market* commodity that furthered commercial interests far beyond medical justice.

The organizations supporting BiDil were interested in getting the new therapy to the hundreds of thousands of blacks suffering from heart failure. They were not focused on the drug having a racial label. In fact, most acknowledged the dangers lurking behind "racial profiling" in medicine. This wariness was evident in a joint press conference of key physician, research, and civil rights groups concerned with minority health held on June 15, 2005, the day before the FDA advisory committee meeting on BiDil. Convened by the National Minority Health Month Foundation, the group included representatives from the Alliance of Minority Medical Associations, Association of Black Cardiologists, Genetic Alliance, NAACP, and National Medical Association. The point of the gathering was to endorse BiDil's approval. But several people at the gathering expressed worry that BiDil had been characterized as a drug for blacks. "The assertion that this is a race drug is misguided," warned Randall Maxey, president of the Alliance of Minority Medical Associations. Gary Puckrein, executive director of the National Minority Health Month Foundation, noted that BiDil could "benefit every person who suffers from heart failure, regardless of social race." Ngozi Robinson, director of Health Disparities Initiatives at Genetic Alliance, Inc., proposed another phase of research that would "determine at the cellular level who would benefit from BiDil," so that doctors could test patients for "molecular sensitivity to BiDil" instead of prescribing it according to race. Dr. Gail Christopher similarly called for a larger study with a more heterogeneous patient population. "It would be 'bad science' to label or market a drug as a 'Black drug,'" she concluded.[52]

NitroMed was able to count on the support of the black organizations that argued for BiDil's approval. A sales force of 195 representatives focused on the market of African Americans with cardiovascular disease. NitroMed hired Vigilante, a black-run agency specializing in urban ethnic communities, to advise its marketing campaign. NitroMed ran radio spots and print ads featuring black models in black newspapers in Detroit, Houston, and Washington, D.C., cities with large African American populations, and placed an ad in the national black magazine *Jet*.[53] It also funded projects that linked health advocacy outreach in black communities to spreading the word about the pill. The University of Pittsburgh's School of Pharmacy received a grant to begin "Helpful Hands for Healthy Hearts," a patient-focused educational program to increase awareness of cardiovascular disease in African American communities in the Pittsburgh area.[54] Faculty and students would visit churches and community centers to provide information about hypertension and chronic heart failure—and about BiDil.

The NAACP, the nation's oldest and largest civil rights organization, was BiDil's staunchest supporter. Shortly after the FDA approved the drug, NAACP chairman Julian Bond and interim president Dennis Hayes signed an emergency resolution pledging that the organization would encourage black patients to use it. The statement cast BiDil as the start of an era of "targeted therapies" that would tackle health disparities. Finding that "Blacks and other racial and ethnic minorities experience worse patient health outcomes for a variety of diseases and conditions than their white counterparts," the document declared that "BiDil has the potential to save thousands of lives a year in a population that has been disproportionately impacted by congestive heart failure."[55]

In December 2005, the NAACP announced a strategic alliance with NitroMed "to implement measures to narrow health disparities that exist between African Americans and Caucasians."[56] As part of the partnership, NitroMed promised the NAACP a three-year, $1.5 million grant to build an infrastructure for the organization's advocacy initiatives promoting equal access to quality health care. The NAACP placed its support of BiDil in the context of ongoing efforts to improve black health. "We have been on a health care march for some time," NAACP president Bruce Gordon told the *Baltimore Afro-American*.[57] In return for NitroMed's $1.5 million grant, the NAACP vigorously promoted BiDil in black communities as part of a grassroots marketing strategy that relied on ordinary African Americans to spread the word.[58] Michael Loberg described one of the partnership's chief aims as "together with the NAACP, . . . doing our part to remove all barriers to access to BiDil."[59]

Juan Cofield, president of the NAACP New England branch, was one of NitroMed's most ardent foot soldiers. "I would like to see the name BiDil [become] as common in our community as Viagra is in the general public," he told delegates at the annual meeting of the New England branch in October 2006, when BiDil had begun to lose its luster. Cofield invited NitroMed's senior vice president of sales, Gerald Bruce, to give an hour-long pitch to the audience. "What you can do to help is just building awareness in the community," Bruce said to rally the delegates. "Talk to your family members, your constituents, your friends, your neighbors. Tell them you heard about a lifesaving medication that makes patients feel better and be able to function better. Let's, as Juan said, make BiDil as commonly known in the African American community as Viagra is in the broader community."[60]

The image of BiDil as a white knight rescuing dying black patients proved to be a powerful defense against criticisms. "Should a drug be withheld simply because it may play into the fear of a racist agenda?" asks a guest editorial in the black newspaper the *Chicago Defender*.[61] Gary Puckrein says that "concern about the medical and scientific validity of the concept of race . . . is valid, but, under present circumstances, impractical."[62] Dr. Winston Price, president of the National Medical Association, urged that BiDil be rushed to market, warning that "any day of delay represents an unacceptable missed opportunity to save lives."[63] When I stated at an April 2006 MIT conference on race-based therapeutics that there was no consensus among African Americans on their benefits, Juan Cofield stood up in the audience and emotionally objected. "There *is* a consensus supporting BiDil," he shouted at me. "The NAACP supports it, the Association of Black Cardiologists supports it, and the Black Congressional Caucus supports it." He accused me of jeopardizing the lives of black people. According to this view, the urgent crisis of African American heart disease must take precedence over political objections to the use of race as biomedical category. Indeed, these objections are seen as a form of racial discrimination or betrayal on grounds that they block black heart patients' access to the medicines they need. But other critics and I never argued that BiDil should be withheld from the market. To the contrary, we argued that, if it were to be marketed, it should be made more widely available, without regard to race.

NitroMed also made payments to other black organizations that promoted BiDil. Gary Puckrein's organization received an undisclosed amount of an "unrestricted educational grant to undertake epidemiological research on chronic heart failure patients." NitroMed paid the Association of Black Cardiologists $200,000 for its part in organizing the clinical trial, but the organization's members had far more to gain in the perks that drug companies offer medical professionals.[64] Of course, this is no different from the way the pharmaceutical industry typically promotes its products. Marcia Angell reports that 94 percent of physicians take money or gifts from drug companies.[65] Now BiDil gave black biomedical researchers and cardiologists an edge; they had special access to black research subjects and patients and would be instrumental in persuading them to take race-specific pills.

My quarrel with the commercial aspects of BiDil's development is not that the people involved made money. Congress has ensured that profit is the central incentive for the pharmaceutical industry to research and market medications. Jay Cohn once objected to my criticism of BiDil by asking, "How else do drugs get developed in this country except on the basis of commercial potential?"[66] Yet it is one thing for biomedical researchers and pharmaceutical companies to profit from scientific innovation; it is quite another for the profit motive to distort the underlying scientific research. Commercial interests induce pharmaceutical companies to exaggerate or invent the therapeutic importance of race. NitroMed did not make money from a drug that was developed to treat heart failure in black patients. It made money by converting a drug for heart failure into a drug for African Americans based on unsubstantiated claims about racial difference.

BiDil's Demise

In the end, NitroMed's racial marketing strategy failed. Sales started out sluggish, reaching only $4.5 million by the end of 2005.[67] The first sign of trouble was the 2006 resignation of its top executives, Michael Loberg and Lawrence Bloch, amid reports that the company lost $108 million in the first year after its IPO.[68] "We lost a lot of momentum after launch when the reimbursement coverage was so poor," says Kenneth Bate, the new chief operating officer.[69] By 2007, two years after BiDil had hit the market, NitroMed's stock price had plunged from $23 to $3 per share. Only 1 percent of the 750,000 black heart failure patients had prescriptions for it. In January 2008, NitroMed announced that it was firing most of its staff, suspending its marketing campaign, and looking for a buyer.[70]

There is as much speculation about why BiDil failed as there is about why BiDil worked. The company misread the public's willingness or ability to pay a high premium for a race-specific medication whose generic equivalent was available at a fraction of the cost. NitroMed priced BiDil at $1.80 per pill and recommended a dose of six pills daily, totaling $10.80 per day or $3,942 per year. At that price, BiDil cost patients four to seven times the estimated price for the combined generic equivalents.[71] Insurance companies refused to spend the extra $3,000 a year that BiDil cost compared to taking the two generic tablets.

Some states excluded BiDil from their Medicaid formulary, the list of reimbursable drugs, and the federal Centers for Medicare and Medicaid Services (CMS) did not require Medicare Part D plans to cover it. In January 2005, in an effort to expand drug coverage for the elderly, Medicaid patients who were also eligible for Medicare were required to switch to Part D plans. Tens of thousands of African American heart failure patients who were eligible for BiDil under Medicaid found themselves on Part D plans that did not pay for it.[72]

The New England area chapter of the NAACP charged that the Medicare decision smacked of racism. In a letter to six leading prescription drug insurers, Cofield wrote, "Anything short of immediate and full, affordable access to BiDil for African-American heart failure patients enrolled in your Part D plan is inadequate."[73] The Washington Legal Foundation, a Washington, D.C., public interest law and policy center, petitioned CMS to reverse its reimbursement policy, arguing the federal agency was putting cost-cutting ahead of black lives. "CMS's unwillingness to support coverage for the one drug approved solely for African-Americans, an unwillingness that undoubtedly is contributing to the woeful undertreatment of black heart failure patients, suggests insensitivity to the need for the federal government to lead the way in promoting racial equality in healthcare delivery," said the organization's chief counsel, Richard Samp.[74]

Telling black patients they should use cheap generics instead of the drug the FDA had specifically approved for them unfortunately seemed like yet another instance of state disinterest in black health. This was a long-standing issue; in the past, some blacks had questioned advice that old-fashioned diuretics were more effective for their hypertension than the advanced medications prescribed to whites. Ironically, these protests reveal a danger in racial labeling itself: defining a drug as black means it can easily become a target for discriminatory treatment by private insurers and government agencies.

But racial discrimination against blacks taking BiDil was not the only concern. The pharmaceutical industry has a financial stake in the government's reimbursement of expensive brand-name drugs. The National Pharmaceutical Council championed racial segmentation as a way to contest cost-cutting measures by health insurance providers. Consider, for example, a pernicious use of the Medicare reimbursement controversy by the same Washington Legal Foundation that petitioned CMS. On June 16, 2008, the organization ran a large paid essay by its chairman Daniel J. Popeo in the *New York Times* titled "Depriving Americans of Treatment." In it, Popeo claimed that taxpayer money going to hospital care for undocumented immigrants was bankrupting U.S. health care and draining funds for treating sick Americans. The diatribe was both an argument for tougher border enforcement by the federal government and an argument against universal health care. But what caught my eye was the connection Popeo made between illegal immigration and BiDil: "Disgracefully, the U.S. Department of Health and Human Services has denied full coverage for BiDil, a medication approved for use by African American heart failure patients. . . . How can it be that government must pay to care for illegal aliens, but deprive our citizens of life-saving drugs?" Once again, the "life-saving drug" label became a powerful weapon in a political battle, this time perversely pitting the health of Latino immigrants against that of African Americans.

While private and state insurance decisions helped doom BiDil, an episode of the popular Fox medical drama *House MD* that aired three months after the FDA approved BiDil suggests another reason.[75] Early in the episode, Dr. Eric Foreman, played by African American actor Omar Epps, treats an older African American man who is having trouble climbing stairs. Noticing that his blood pressure is high, Dr. Foreman tells the patient, "I have something new that should help you out. Combines a nitrate with a blood pressure pill. It's targeted to African Americans." Foreman and the patient argue over whether to believe studies indicating blacks tend to have nitric oxide deficiency.

"I've had white people lying to me for sixty years," the patient says.

"You really want to screw whitey? Be one of the few black men to live long enough to collect Social Security. Take the medicine," Foreman replies, handing him a prescription.

The next day, the same man returns without having filled the prescription. This time he is seen by Dr. House.

"I'm not buying into no racist drug, okay?" the patient tells House.

"It's racist because it helps black people more than white people? Well, on behalf of my peeps, let me say, thanks for dying on principle for us," House shoots back.

"Look. My heart's red, your heart's red. And it don't make no sense to give us different drugs."

House writes a new prescription. "I'll give you the same medicine we give Republicans."

Black suspicion was also the theme of a cartoon poking fun at BiDil in the

K Chronicles by Keith Knight. Four black figures stare warily at a pill bottle sitting in the middle of a table. "I ain't gonna try it. . . . You try it!!" one says. "No, way!! I ain't gonna try it!!" says another. To the side, the caption reads: "BiDil, the first government approved drug specifically made for a particular race, is supposed to help black folks combat heart failure. . . ." The next frame depicts reminders of U.S. medical experimentation and exploitation of African Americans: "Tuskegee Syphilis Experiment (1932–1972); Eugenic Sterilization (1907–1970s); Radiation Experiments (1932–1988). "Now, given the United States government's less than stellar track record concerning the health of black citizens. . . . It's really no surprise that folks may be a bit reluctant to try the drug." The cartoon ends with, "I can't wait to hear the apology the Senate will give fifty years from now for the infamous BiDil 'mistake.'"[76]

A 2003 study published in *Genetics in Medicine* shows that this skepticism about race-specific drugs is not fictional; it is widespread in the black community.[77] Participants in an anonymous survey and two focus groups that oversampled for minority groups reported that they would be highly suspicious of race-labeled drugs. Nearly half said they would be very suspicious of their safety, and 40 percent said they would be very suspicious of their efficacy. In fact, 13 percent of African Americans said they would choose a drug labeled for whites over one designated for blacks. At a conference on BiDil, an elderly black woman in the audience stood up and said, "If I were sick and somebody told me that they had a drug just for black people to help me, I'd say to them: give me what the white people are taking."[78]

BiDil's projected customers—middle-aged and older African Americans—grew up in the shadow of the Tuskegee Syphilis Study and segregated medical care. Dr. Theodore Addai of Meharry Medical College, a historically black institution in Nashville, recounted the suspicions that arose when he tried to recruit black patients for the BiDil clinical trial: "We had to try to persuade them that this was not another Tuskegee."[79] Older African Americans are intimately familiar not only with the history of medical experimentation on black people but also with colored-only institutions that were separate and unequal. In the United States, a BLACK ONLY sign means inferior or even hazardous. Racial profiling tends to work against black interests in everything from criminal justice to employment, education, and medicine. This explains why NitroMed's marketing campaign invested in partnerships with black doctors, politicians, and organizations who could credibly persuade blacks that a race-specific drug was good for them. But even these black professionals could not overcome a mistrust—and common sense—forged over centuries of medical abuse.

None of these reasons offers a complete explanation. Black mistrust and the Medicaid and Medicare decisions not to cover BiDil cannot fully account for why only a tiny percentage of black heart failure patients are taking even the generics. NitroMed established NitroMed Cares, a patient assistance program that offered the drug at no cost to low-income patients with no health coverage. And some blacks are perfectly comfortable with a medication tailored for them. It is possible BiDil flopped for more mundane reasons unrelated to the controversy over race. Some physicians and patients balked at adding yet another six pills daily to an already complicated regimen of diuretics, ACE inhibitors, and beta blockers. NitroMed relied heavily on a grassroots marketing strategy that did not include a single television spot. Perhaps most damaging, by 2008 the entire biotechnology sector was collapsing as cash-strapped small firms canceled drug trials, filed for bankruptcy, and sold out to larger companies.[80] It's possible that BiDil was an early victim of the financial downturn, not protest against its race-specific label.

The Future of Race-Based Therapies

BiDil's demise does not spell a halt to race-based pharmacogenomics. To the contrary, NitroMed's success at using race to gain FDA and patent approval, as well as support from influential political players, signaled the potential profitability of race-specific therapies. Jonathan Kahn argues that BiDil's racial labeling gave the pharmaceutical companies "a new model of how to exploit race in the marketplace by literally capitalizing on the racial identity of minority populations," providing "a cheaper, more efficient way to gain FDA approval for drugs" as well as to distinguish their patents from the competition.[81] Supporting his view is evidence of the growing use of race as a genetic category to gain patent protection. Using the U.S. Patent Office database, Kahn reviewed gene-related patent applications filed between 1976 and 2005 that employed racial or ethnic categories. He discovered a fivefold increase in racial patents during that period. Race was not mentioned in any application filed between 1976 and 1997. Then the number of gene-related patents and patent applications using race or ethnicity skyrocketed. Twelve such patents were issued in the period 1976–2005, and sixty-five such patent applications were filed in the period 2001–2005. This trend of treating racial identity as a component of a genetic commodity further solidifies the contemporary view that race is biological. "The patent process takes race as a social category and recodes it as 'natural,'" Kahn writes.[82]

Using racial categories to patent an invention and carve out a racially defined market for it is nothing new. There is a long history, dating back to the 1800s, of patenting all sorts of products that involve race—chemicals to straighten kinky hair, creams to lighten dark skin, toys that celebrate or mock people of color.[83] A patent from 1940 for an arcade game featured a figure of a "negro stealing a chicken" as a target. "As soon as the target is initially moved, with the negro moving toward the hen house, a successful hit will cause him to reverse his direction of movement and leave the hen house," the inventor explained.[84] The civil rights movement ushered in patents for dolls, games, and teaching materials that celebrated diversity, ethnic holidays, and civil rights leaders. A more recent patent, filed in 2006, for a device that quickly removes natural or synthetic braids that have been attached to human hair, claims that African Americans will benefit from the device because they "genetically have hair that resists the formation of longer lengths" and so tend to use hair attachments.[85]

Race and ethnic heritage are used by patent applicants in a variety of ways, some more harmful than others. The problem with recent gene-related patents is that they treat race as a biological category. I was curious to see what these racial gene-related patents looked like and whether they were still being filed. In 2009, a research assistant helped me comb the patent database for recent applications to patent biotechnological inventions that incorporated race. An application filed in April 2009 seeks to patent a method of predicting response to antipsychotic drug therapy by testing patients for the presence of a polymorphism in dopamine D2 receptor genes.[86] This broad claim is followed by a narrower one: "The method of claim 1, wherein the subject is an African American subject." The inventor hopes that if the broad claim is struck down, the more specific racial claim will survive. The application elaborates how the genetic screen works by describing a study conducted on thirty-one African American patients diagnosed with schizophrenia who received clozapine treatment. Although the genetic test is supposed to identify the alleles that predict whether someone of any race would benefit from the antipsychotic drug therapy, the inventor seems to tack on a racial claim as an added hook for securing a patent.

Other inventions relied even more directly on race. One June 2009 application is for a genotyping technique that identifies "a male Caucasian human subject at increased risk of experiencing hypersensitivity reaction to a therapeutic regime of abacavir," an anti-HIV medicine.[87] Another filed in June 2008 seeks to patent a genetic test for "analyzing a biological sample from an African American woman for presence of a polymorphism or mutation associated with breast cancer," specified later as the BRCA1 and BRCA2 genes. The background section explains that "African-American women under 50 years old have the highest rate of new cases of breast cancer in the nation and tend to present at an earlier age with larger tumors and more advanced-state disease." Given this higher burden of breast cancer and the high cost of conventional testing for BRCA1/2, the inventor argues, this efficient screening panel would "particularly" benefit African American women.[88] All of these patent applications involve a method for testing people for a specific gene variant that can predict disease or drug response. This is the sort of test that is supposed to allow personalized medicine to move beyond race as a temporary proxy for genetic difference. Yet each application makes race central to its bid for market protection.

The growing number of biomedical studies and patents that rely on race suggest that biotech, pharmaceutical, and medical device companies are poised to launch a new generation of racial products. A 2007 report by the Pharmaceutical Research & Manufacturers of America (PhaRMA) highlighted almost seven hundred medicines in development for diseases that disproportionately affect African Americans. Although the *Memphis Business Journal* described them in a headline as new medicines "for the black population," the drugs are not tailored specifically for black patients but treat diseases—such as prostate cancer, hypertension, asthma, HIV, and diabetes—that afflict blacks at higher rates.[89] "African Americans, and all Americans, should have timely access to the medicines their physicians believe would work best for them," PhaRMA president and CEO Billy Tauzin wrote in the report's introduction. "These 691 medicines in development by America's research companies may one day help to bridge the health gap and increase the likelihood that all Americans will share in the benefits of medical progress."[90] Tauzin's statement echoes the claim central to BiDil's marketing strategy that developing pharmaceuticals specially for African Americans can help to eliminate health inequities.

The package inserts for a number of medications reveal that race currently plays a role in pharmaceutical marketing and prescriptions. Many companies differentiate their drugs by claiming they work especially well in a specific race. Between 1995 and 1998—before the personalized medicine revolution had begun—the labels of 8 percent of new drugs included a statement about racial differences in effectiveness.[91] In 2001, the FDA approved Travatan, an eye drop manufactured by Alcon for lowering ocular pressure in glaucoma patients. Clinical studies testing Travatan, as well as a randomized trial comparing the efficacy of three glaucoma drugs, found that the ingredient in Travatan worked better in black patients than in others. Alcon latched on to this finding to advertise its product as "the first glaucoma drug to demonstrate greater effectiveness in black patients."[92] Some physicians saw Travatan as the perfect match for the striking racial differences in glaucoma prevalence: African Americans are more likely to suffer from glaucoma, experience the disease at an early age, and go blind from it than people of other races. "We think there's something different genetically about their optic nerve that may make them more susceptible to changes in pressure," says Philadelphia surgeon Marlene Moster.[93]

The studies of Travatan lacked any hypothesis to explain why the black patients responded better on average. Socioeconomic or lifestyle factors may have distinguished the black patients. Or the difference in efficacy may have resulted from something even simpler: the black patients in the study were more likely to have dark eyes. Darker iris color is associated with greater intraocular pressure, which increases the risk of glaucoma.[94] After noting the higher pressure reduction among black clinical subjects, the Travatan insert discloses, "It is not known at this time whether this difference is attributed to race or to heavily pigmented irides."

Race also features in the marketing of statins, which millions of Americans take to lower their cholesterol. Crestor, a cholesterol-lowering drug manufactured by the pharmaceutical giant AstraZeneca, broke into the multibillion dollar statin market in 2003. The year before, the statin Lipitor earned Pfizer $8 billion, making it the bestselling drug in the world. AstraZeneca counted on Crestor to capture 20 percent of this global market. In August 2003, the FDA granted market approval with a racial qualification. The FDA allowed Crestor to be labeled with a different initial dosage for Asians because blood levels of the drug rose twice as high in research subjects of Asian origin compared with a Caucasian control group, increasing the risk of muscle damage.[95] The prescribing information AstraZeneca provides doctors includes a special section, "Dosage in Asian Patients," advising that Crestor therapy start with 5 mg daily for such patients rather than the usual starting dose of 10–20 mg. Another section, "Race," explains that studies found no clinically relevant differences among Caucasian, Hispanic, and black or Afro-Caribbean groups.

At first glance, it might seem that racial dosing is a careful way to prevent harmful side effects. Perhaps this is an instance where using race really does "enhance diagnostic and therapeutic precision," as Jay Cohn has argued.[96] But if the usual starting dosage of Crestor can be toxic for some patients, is race the most accurate way to determine the right amount? If there is a genotype that increases Crestor blood levels, there must be some patients who identify as Caucasian, Hispanic, or black who carry the allele and would therefore be harmed by the high recommended dosage. They should start on the 5 mg pill along with Asians. And then there is the perpetual problem of patients who identify as one race but have mixed ancestry. Is it safe for Tiger Woods, whose father is African American (mixed with Chinese and Native American) and whose mother is Asian (part Chinese, part Thai), to take the higher dose for blacks? Based on his looks, Woods, who once called himself "Cablinasian," representing his Caucasian, black, Indian, and Asian heritage, might easily be prescribed the dosage for blacks even though most of his ancestors came from Asia.

But the problem with racial dosing runs deeper than assigning someone to the wrong race. It is simply bad medicine to prescribe potentially toxic drugs on the basis of a social category. As a patient, I personally would insist on the lower dosage to make sure the drug is safe for me, regardless of race. If there is a genetic trigger that makes Crestor dangerous for some patients, doctors should test for the risky genotype. If this test is unavailable, everyone should start with the safest dosage. As one health advocate writes, "You don't want the strongest statin. You want the mildest statin that works for you."[97]

Instead of prescribing by race, we should ask why Crestor is usually prescribed at a dose known to have catastrophic side effects. In the first year of Crestor's launch, AstraZeneca invested an estimated $1 billion in a promotional campaign to convince doctors to prescribe the drug.[98] At the time, Crestor was competing for market share against four other cholesterol-lowering drugs—Mevacor, Pravachol, Zocor, and the blockbuster Lipitor. AstraZeneca had been plagued by evidence that high doses of rosuvastatin, the active ingredient in Crestor, can cause rhabdomyolysis, the rapid breakdown of muscle tissue that can lead to acute kidney failure. Bayer withdrew cerivastatin, known as Baycol or Lipobay, in 2001, after it was associated with fifty-two deaths from kidney failure. AstraZeneca withdrew an 80 mg version of Crestor because it was too risky.

But many patient advocates expressed concern that Crestor's safety remained unproven. Public Citizen, the consumer watchdog group founded by Ralph Nader, tagged Crestor with its "Do Not Use" warning and urged the FDA to pull it from the market.[99] In a blistering editorial, the respected British medical journal *The Lancet* accused AstraZeneca of driving its "marketing machine too hard and too fast" without regard to patient safety. "Since there are no reliable data about efficacy and safety—and AstraZeneca is facing unusually acute commercial pressure to force rosuvastatin into the market—doctors should pause before prescribing this drug," wrote *Lancet* editor Dr. Richard Horton. Horton called on AstraZeneca chief executive Tom McKillop to "desist from this unprincipled campaign."[100] (Since then, statin therapy has also been associated with an increased risk of developing diabetes.)[101] Instead, the company resolved its safety problem by marketing a standard dose for most patients, with an exception for Asians. Using a crude and ambiguous social marker cannot possibly identify the patients at risk for kidney failure. Race-based dosing may make prescribing and marketing drugs simpler, but it does not make drugs safer.

Nevertheless, AstraZeneca embarked on a full-scale strategy to test, and apparently market, its drugs by race. When a 2004 clinical trial testing its lung cancer drug Iressa, approved by the FDA in 2003 under its accelerated approval regulations, failed to show any benefit, the company claimed that it worked in the subgroup of Asians who were enrolled in the study. Unlike its decision on BiDil, the FDA rejected the evidence altogether and refused to approve it for any new patients at all.[102] AstraZeneca also launched a galaxy of race-specific trials to test Crestor: ARIES (African American Rosuvastatin Investigation of Efficacy and Safety), IRIS (Investigation of Rosuvastatin In South-Asian Subjects), and STARSHIP (Study Assessing RosuvaStatin in the Hispanic Population).

Instead of limiting Crestor prescriptions, as health advocates urged, the FDA recently expanded the market for the drug. In February 2010, the agency approved the pills as a preventive medicine for millions of people who show no sign of heart or cholesterol problems.[103] The revised label permits doctors to prescribe Crestor to healthy men aged fifty or older and women aged sixty or older who have one other risk factor, such as smoking or high blood pressure, along with elevated inflammation in the body. This adds 6.5 million potential Crestor users to the 80 million people who already have high cholesterol levels. Prescriptions for statins already doubled in less than a decade, from roughly 100 million in 2001 to 200 million in 2009. Critics questioned whether the benefits of taking statins outweighed the risk of muscle damage and diabetes when people prescribed the drug are in good health. "It just turns a lot of healthy people into patients and commits them to a lifetime of medication," says cardiologist Steven W. Seiden.[104]

In other words, the FDA has sanctioned marketing a powerful medication to healthy people because they are at risk of becoming unhealthy in the future. Could this be a framework for selling drugs on the basis of genetic risk? Prescribing drugs based on risk alone takes personalized medicine to another level that exponentially expands drug makers' profit potential. We are not only talking about drugs that are tailored to fit the genotypes of sick people; we are talking about drugs marketed to people predicted to become sick based on a genetic test. If race is considered a proxy for genetic risk, risk-based prescriptions offer another avenue to market pharmaceuticals to specific racial groups.

Besides race-specific prescription drugs waiting in the wings, there is already an expanding assortment of race-based therapies on the market in the emerging field of nutrigenomics. Nutritional genomics is a science that studies how genes interact with dietary chemicals to affect people's health and, conversely, how genes metabolize nutrients from the food we eat.[105] As in the case of pharmacogenomics, some companies are using race as a quick and dirty way to apply the link between genes and nutrition to sell products in supermarkets, drug stores, and the Internet. Many nutritional supplements, which do not require FDA approval, are sold according to race. The labels on these products are based on the premise that absorption of nutrients and their impact on people's health varies by racial identity.

GenSpec, a company in Florida, sells a line of multivitamins formulated specially for African Americans, Caucasians, and Hispanics. The Web site claims these vitamins are "the first genetically specific nutritional supplements" on the market. But closer inspection of the vitamins' ingredients reveals that the difference hinges on the amount of vitamin D in each race-specific formulation. GenSpec's racial vitamins rest on the premise that races are divided by a genetic color-coding: African Americans are darkest,

Caucasians lightest, and Hispanics are somewhere in between. Races with dark skin are predicted to have a vitamin D deficiency because the melanin affects the amount of vitamin D their bodies absorb naturally from sunlight. The company also boasts weight-loss pills and metabolic boosters "specifically formulated to work in harmony with the Caucasian physiology and basal metabolic rates." The same claim is made for the African American and the Hispanic physiologies and metabolisms.

The over-the-counter pill Citracal provides vitamin D_3 and calcium to ward off osteoporosis. A message on the side of the box says: "Regular exercise and a healthy diet with enough calcium helps teen and young adult white and Asian women maintain good bone health and may reduce their high risk of osteoporosis later in life." This advice suggests either that only white and Asian women are at risk for osteoporosis or that only these women will benefit from a healthy diet with enough calcium. It was once commonly believed that black women have a lower risk of osteoporosis because their bones are denser. But recent studies suggest these differences may be explained by factors apart from race: after controlling for bone size, body weight, height, diet, and exercise, racial differences disappear.[106] Other studies show that African American women are more likely to suffer from vitamin D deficiency because they are more likely to have dark skin and a diet poor in calcium—and therefore are in greater need of the advice on the Citracal box. The Citracal Web site clarifies that whites and Asians are not the only "ethnicities at risk," noting that "African Americans are also still at risk when they go through menopause, just at a slightly lower percentage rate." This confusing racial message does more harm than good.

Back to Genetics

In 2008, the Department of Health & Human Services Advisory Committee on Genetics, Health and Society addressed the controversy over race-specific medicine in a report, "Realizing the Promise of Pharmacogenomics." Calling race-based prescribing information "suboptimal and medically impractical," the report recommended that the FDA encourage biomedical investigators to conduct gene-based studies instead of relying on race as a proxy for genetic difference. The latter, the report cautioned, "can result in imprecise prescription guidelines and reinforce a public view of biologically defined races."[107]

The NitroMed research team began searching for the gene that explained BiDil in 2005—after it was approved as a race-specific drug. The Genetic Risk Assessment in Heart Failure Trial, or GRAHF, compared the frequency of aldosterone synthase (CYP11B2) alleles in 354 patients who participated in A-HeFT to the frequency of this same allele in white participants in the Genetic Risk Assessment of Cardiac Events, or GRACE, study conducted at the University of Pittsburgh.[108] Activation of aldosterone appears to hasten the progression of heart failure. The GRAHF study found that a specific variant of CYP11B2 influenced clinical outcomes in the African American patients and that it was more common in African American patients than in the white patients who participated in GRACE. The authors observed that these findings "suggest that the genetic variation in aldosterone production may contribute" to differences in heart failure in blacks and whites. "In determining optimal heart failure treatment for an individual, race is likely a surrogate marker for differences in genetic background," they concluded.[109]

By comparing genotypes in black and white patients (instead of patients who were helped by BiDil versus those who weren't), the researchers seemed stuck on finding a biological mechanism based on race. NitroMed's vice president of corporate affairs stated in 2007 that the company might eventually use the genetic data to develop a diagnostic test for BiDil, though it is not clear what financial incentive it had to invest in genetic screening since it was approved to use race instead.

Finding an underlying genetic reason why BiDil works in black patients does no more to address disparities in cardiovascular disease than does labeling BiDil as a drug for blacks. Neither avenue for developing a pharmaceutical remedy for heart failure will eliminate the social conditions and barriers to medical care that create health inequities in the first place. By turning to a genetic explanation, the researchers foreclosed a potentially more fruitful investigation of the environmental factors that separate white and black health—and that could improve prevention of heart disease for everyone. In fact, conducting genetic research on the heels of BiDil's approval as a drug for blacks only reinforces the erroneous view that race is a genetic category that causes disparities in disease. It seemed the researchers were looking not for gene variants that predicted who would benefit from the drug, but for gene variants in blacks that explained why it benefited them in particular. In other words, NitroMed sought a post hoc genetic justification for marketing its product based on race.

All the research on BiDil—both the race-based clinical trial and gene-based analysis—was distorted by the myth that race is a biological category, by the unwillingness to come to grips with social inequality, and by the commercial interest in finding new markets for pharmaceuticals. Genetics has its place in developing personalized medicine, but genes are not the place to look for racial health disparities. Likewise, race has its place in understanding the impact of racism on people's health, but it is not where we should look for genetic differences. Health inequities—which are caused by social injustice, not genetic dissimilarity—cannot be fixed by color-coded pills. Understanding race as a political system that produces health disparities would lead instead to research into the social factors that drive black hypertension, diabetes, and kidney disease and to policies aimed at eliminating them.

Will We Forget About Race?

In February 2009, I gave a lecture at the University of Minnesota titled "What's Wrong with Race-Based Medicine?" Before my lecture, I was interviewed on Minnesota Public Radio's *Midmorning Show.* Halfway through, Duke molecular biologist David Goldstein joined the program to give a scientist's take on the issue of race and personalized medicine. Goldstein argued that the media are to blame for exaggerating the importance of race to pharmacogenomic research. He recalled the wild misrepresentations of his study finding little evidence backing claims of racial differences in drug response. But Goldstein predicted that the tide is turning.

"It's perfectly clear that we will not be using race or ethnicity in any important way going forward," he said. "In many cases, the differences are not genetic. In others, there will be a genetic component to differences. But whenever that is the case, what you will look at is the underlying genetic difference and you will forget about race. . . . However we roll out personalized medicine in the future, we will not be relying on race and ethnicity." Goldstein's tone was confident. The editors of *Scientific American* recently agreed that the rapid sequencing of whole genomes "will likely make the segmentation of drug therapy by race a mercifully short chapter in the evolution of personalized medicine."[110]

"This should have been the vision of researchers all along," I interjected, "and yet race came into play. Many scientists say race is the first step to personalized medicine. Please let me know what happened to stop the market and political influences on science."

Goldstein replied, "I think what's happening, frankly, is that science is overwhelming the tendency of media to emphasize race and ethnic differences. Everything that we're finding causes us to zero attention on those individual genetic differences. So the way I view it is the reality of genetic science is winning."

This prediction from a lab scientist was in stark contrast with the views of Jay Cohn, who was invited to comment immediately after my lecture a couple hours later. Of course, I expected Cohn to defend his invention, BiDil, but I was surprised at the pivotal importance he ascribed to race in every aspect of medicine. Far from seeing race as irrelevant going forward, Cohn envisioned race becoming the focal point of clinical trials, drug labeling, and doctors' diagnoses and prescriptions.

"The FDA's position has always been that they will approve a drug for use in the population in which it was studied," Cohn said. "We should do a study in Native Americans, we should do a study in Asians, we should do a study in whites to demonstrate how each of those populations, which can be identified imperfectly," responds to a new drug. "By discriminating in practice, you see a patient and if you know that patient to be a Native American and you know that there's a trial in Native Americans that shows specific benefits of a therapy, you would be inclined to use that therapy, even though you don't know if it might also work in other people. You're only dealing with that person sitting in front of you who is a Native American."[111]

Cohn puts little stock in individualized genetic tailoring; he wants to develop and prescribe drugs according to race, not genetics. "Maybe someday we will be able to replace crude [racial] identification with a more precise genetic identifier. We would all love that to be the case. But I'm not sure it would solve our problem as easily as you might wish it would," he said to me during a panel discussion after my lecture. To Cohn, it will be harder than Goldstein foresees to stop classifying people by race in clinical trials, FDA approval, and medical practice. "Right now, we have only skin color to identify populations," Cohn said on another occasion. "You'd have to blindfold yourself to say we're not going to pay attention to obvious differences."[112]

An article by FDA scientists explaining the agency's decision on BiDil suggests that Cohn is right about the future of drug development. The scientists noted that regulations governing new drug applications require analyses of safety and effectiveness broken down by "demographic subsets of the patient population," and labeling regulations require similar "demographic subset information" in package inserts.[113] They argued that the congressional mandate to NIH requiring inclusion of minorities in clinical trials as well as a racial breakdown of findings shows that the FDA is hardly alone in its focus on race. "Given the long history of urgent interest in searching for racial and other demographic differences, which surely accepted the possibility that such differences might be discovered, it seems surprising that there would be so much discomfort when one was found," they observe. This logic could also leave out minorities from clinical trials of drugs believed to work better in whites. Schering-Plough excluded African Americans from a 2006 study testing a hepatitis C therapy.[114] The FDA held out BiDil as a model for how pharmaceuticals should be developed in the future, not as an exception to the rule.

Even Goldstein curbed his prediction about personalized medicine with a caveat. He noted that the focus on individual-level genetic difference required sequencing individual genomes. "We have to be quite careful because it's so difficult to interpret that information. There's so much change in the scientific literature, there's so much information inherent in our genomes that, in order to make maximal use of your own genetic information right now, you would need a highly paid consultant to advise you about your genome," he said. This raises the potential for another type of health inequity in personalized medicine: between those who can afford all the mapping and consulting and those who cannot. "So we have a lot of concern about how fairly information about the genome will be made available to the broadest communities," Goldstein added. This disparity, which is directly tied to social inequality, means race will continue to play a role, despite Goldstein's scientific ideal of forgetting about it.

It is hard to predict whether or not BiDil will serve as a template for spinning racial science into pharmaceutical gold. The scientific and commercial contexts are in flux. Not only has genomic science changed direction, but the business model of biotech startups making millions from tiny market segments has fizzled. Even if never repeated, the BiDil story still reveals how securely the biological meaning of race is fastened to medicine and how easily it can be exploited by the pharmaceutical industry to sell its products. Let's not forget that ten years ago, when the mapping of the human genome revealed no genetic basis for race, some predicted that the science of human diversity would unseat race as the focus of genetic research. Yet it was precisely then that scientists intensified their efforts to redefine race in genetic terms, find genetic explanations for health disparities, and treat race as an essential ingredient of personalized medicine. If scientists are now promising to diagnose individual genomes as a gene-based racial science emerges, future genetic technologies are likely to continue to make use of race despite having no scientific need for it.

9

Race and the New Biocitizen

"I'd rather spend my money on my genome than a Bentley or an airplane."[1] That's millionaire Dan Stoicescu, explaining why, in January 2008, he became the second person in the world to buy the complete map of his personal genetic profile. Sequencing your genome used to be quite pricey. Stoicescu went to Knome, a Cambridge, Massachusetts, firm founded by Harvard geneticist George Church with the motto "Know Thyself." Knome charged private clients $350,000 to read their entire genetic code and provide a face-to-face, customized analysis of what it revealed about them. That hefty price tag included being flown to Boston to spend a day at Knome discussing the results.[2] Then the cost of personalized gene maps began to plummet. By the end of 2009, a start-up biotech firm called Complete Genomics began taking orders for whole genome sequences at the astonishingly low price of $5,000. Scientists predict that in the near future they will be able to decode everyone's genome at an even more reasonable cost, allowing doctors to make more accurate predictions about our health and prescribe medications designed to match each person's individual genotype. Yet it remains to be seen whether personalized genomics will empower all individuals to take control of their health or whether it will be marketed and consumed in accordance with existing social inequalities, including race.

Genetic Testing Comes Home

For as little as $400, you can have a portion of your DNA scanned for a personalized analysis of your risk for specific, supposedly genetic, diseases. A burgeoning industry of close to a hundred biotech companies offers this more limited form of direct-to-consumer (DTC) genotyping. Using the slogan "Genetics just got personal," the Silicon Valley start-up 23andMe is an online service co-founded in 2006 by Anne Wojcicki, wife of Google co-founder Sergey Brin, and backed by the Google fortune. Its Web site allows you to choose to "Fill in your family tree" for $399, "Take charge of your health" for $429, or do both for an economical $499. Customers receive an at-home kit to collect a saliva sample, which they mail back to a lab for genetic screening. In a few weeks, they can log on to the Web site for an interactive report. Using the latest genomic association research, 23andMe gives customers quantitative estimates of their risk for a variety of diseases, like breast cancer, Parkinson's disease, diabetes, and age-related macular degeneration. It also informs them about gene-related traits ranging from earwax type to resistance to HIV/AIDS, baldness, and sprinting ability. The predictions require constant updating because current findings are often overturned by later studies. In November 2010, the company launched a monthly fee plan so customers can access updated genetic information after the one-year contract expires. Other companies provide more targeted testing. For less than $200, CaffeineGen diagnoses the gene for caffeine metabolism. GeneLink uses genetic profile information to recommend a targeted regimen of its vitamins and skin care products. The industry's message is clear: genetic testing empowers you to take control of your life.

Can a DTC genetic reading really enable us to take charge of our health? Of course genes matter to health; we are not born as blank slates. But nor are we programmed by our DNA to have predictable bodies or abilities. As I discussed in chapter 6, we inherit not only genes but the complicated cellular machinery that reads the information in our DNA while perpetually interacting with the internal and external environments. "Genes do not simply *act*: they must be *activated* (or inactivated)," notes MIT science philosopher Evelyn Fox Keller.[3] It is the complex regulatory system in our cells, not genes themselves, that determines the final transcript of instructions. Because cellular dynamics are so messy and unpredictable, some molecular biologists are questioning whether there are even genes for proteins, let alone traits.

The very term *gene* may now be an impediment to our understanding of human biology. Gene sequences look less and less like a blueprint; instead, they function more like resources that cellular mechanisms use in different ways in a multitude of molecular processes that are, in turn, influenced by the multifaceted environments in which we live.[4]

The media's penchant for announcing a new "gene for X," accompanied now by the proliferation of personalized genotyping services, are fortifying the popular misconception that genes operate on their own.[5] Telling consumers that their DNA can predict their chance of getting diabetes, doing well on a test, or winning a marathon simply falsifies what DNA does. Having a gene that a computer has linked to a disease or a trait may not even be relevant because it could be overridden by other mutations that were not tested or whose impact remains unknown. And the consequences of those genetic interactions depend on a host of nongenetic factors that genotyping does not begin to capture. "If you could take the integrated influences of a bunch of genetic variants and a number of environmental inputs, and you put that into a formula, then you have the potential to make a good prediction," says Leslie Biesecker, a senior investigator in the National Human Genome Research Institute.[6] The analyses performed by DTC genetic testing services do not deliver the complexity needed for an accurate prediction.

Yet personal genome services are diagnosing disease risk without the same oversight that applies to drug companies and health care providers. There is no check on the accuracy or validity of the information dispensed. Without proper genetic counseling, consumers can easily misunderstand the test results, and no provision is made for appropriate follow-up care. "Direct-to-consumer genetic testing is a buyer-beware market," warns Dr. Jennifer House, president of the March of Dimes. "Consumers need to be very, very cautious."[7] In 2008, the California Department of Public Health issued cease and desist orders to 23andMe, Navigenics, and ten other genetic-testing companies, demanding that they comply with state and federal regulations. The companies continued operating while they obtained licenses under the California regulations that apply to traditional clinical laboratories. Companies doing business in New York received similar letters from the New York State Department of Health, and some stopped offering their tests for sale in the state. Health and Human Services Secretary Kathleen Sebelius is studying the need for federal oversight, as are Congress and the FDA. The urgency heightened in July 2010 when the General Accountability Office reported that federal investigators posing as fictitious customers purchased ten kits each from four genetic testing companies and received results that were "misleading and of little or no practical use." One donor was told by the various companies that he had below-average, average, and above-average risk for prostate cancer and hypertension.[8]

Companies like 23andMe market their services as tools not only for gathering personal information, but also for creating social connections. In September 2008, the front page of the *New York Times* "Styles" section featured a colorful story about a celebrity "spit party," hosted by media moguls Barry Diller, Rupert Murdoch, and Harvey Weinstein, where the glitterati spit into test tubes so their DNA could be analyzed by 23andMe. In the lobby of Diller's IAC building in Chelsea, Anne Wojcicki explained that her company helps people use their genomes as a platform for social networking: "If you want to have a community around psoriasis, we'd like to be able to allow you to form a psoriasis-specific community."[9] The company's Web site prominently displayed "sharing and community" as a chief service, noting, "Seeing your own genetics is just the beginning of the 23andMe experience. Our features also give you the ability to share and compare yourself to family, friends and people around the world."

Customers can join other organizations based on shared genetic identity, like FORCE (Facing Our Risk of Cancer Empowered), an advocacy group centered on hereditary breast and ovarian cancer. FORCE helps women to determine whether they are at high risk for hereditary breast and ovarian cancer; gives them support for managing their genetic risk; raises awareness of hereditary breast and ovarian cancer; represents the concerns and interests of its high-risk constituency "to the cancer advocacy community, the scientific and medical community, the legislative community, and the general public"; and promotes research and access to information, resources, and clinical trials specific to hereditary cancer.[10] The FORCE Web site provides a forum for people with hereditary cancer to socialize through chat rooms, online groups, and message boards. Similarly, Facebook has a "BRCA 1 and 2 Genetic Ovarian and Breast Cancer Gene" group.

Technologies like in-home genetic testing kits have helped us cross a threshold into a Brave New World of relationships—with our own bodies and with each other. What some scholars are calling "biological citizenship" is grounded in the unprecedented authority wielded by individuals over their well-being at the molecular level. According to British sociologist Nikolas

Rose, "our very biological life itself has entered the domain of decision and choice."[11] Biological citizenship entails both individual control over personal welfare and a "biosociality" that links people together around their common genetic traits.[12] Genetic information enables people not only to manage their own health, but also to unite with others around their common health conditions, as revealed by DNA testing. Some scholars and activists celebrate biocitizenship because it enhances human agency, as patients "become active and responsible consumers of medical services and products ranging from pharmaceuticals to reproductive technologies and genetic tests," and as they are empowered to form alliances with physicians, scientists, and clinicians to advocate for their interests.[13]

Sharing one's DNA is fast becoming a civic duty as well as a social activity. Researchers hunting for associations between genes and disease need large amounts of "bioinformation" consisting of individuals' medical histories and genotypes. Universities across the country have begun amassing storehouses of genetic information from patients that can be matched to their medical records. My own university, Northwestern, launched the NUgene project in 2002 to collect and store DNA samples from patients of university-affiliated hospitals and clinics. By 2010, almost ten thousand patients had enrolled. Its goal is to enroll a hundred thousand people. Patients donate their DNA purely as a gift to genetic research. In fact, because the project de-identifies the samples, researchers can't inform patients of anything discovered about their individual disease risk. Under the banner "Your Genes, Everyone's Future," NUgene offers patients "no compensation for their participation except for the opportunity to contribute to the improvement of health care for generations to come."[14]

In a university environment, it can be difficult to tell whether these attempts to collect genetic information are voluntary or coercive. In May 2010, the University of California at Berkeley announced that it would ask 5,500 incoming students to submit DNA samples to be tested for gene variants that affect people's responses to folate, lactose, and alcohol as part of a personal genomics exercise. "Bring Your Genes to Cal" became Berkeley's annual "On the Same Page" program, which typically encourages incoming freshmen to read the same book over the summer and discuss it in fall seminars.[15] "This year, we'll all be on the same page exploring the theme of Personalized Medicine—the set of emerging technologies that promises to transform our ability to predict, diagnose, and treat human disease—with

Professor Jasper Rine as our guide," explained the university Web site. It told students their DNA would contribute to "understanding the impact of the variation in each of our genomes . . . the defining challenge for human biology for this century."[16]

The students were given the option not to participate, but their young age and the seemingly obligatory nature of a request by the university administration to be part of an "On the Same Page" project made the decision to participate far from voluntary. According to Berkeley anthropology professor Nancy Scheper-Hughes, the consent form read "like a promotional leaflet on the miracles of medical genomics."[17] Jasper Rine, the genetics professor heading the initiative, co-founded several California biotech firms, including a genetics testing company, and thus had a commercial interest in conscripting students to become biocitizens who believed it to be a civic duty to surrender their DNA for the sake of personalized medicine. The university promised it would maintain student anonymity in analyzing the aggregate data, but students could access their individual test results online, using their ID numbers and passwords. After months of heated criticism, the California Department of Health instructed Berkeley to cease the genetic testing component of the program. In response, Mark Schlissel, dean of biological sciences, stated that the program would now "focus prominently on the politics of genetic testing and whether individuals, rather than physicians and public agencies, ultimately control their own genetic information."[18]

Nor do all of these projects place a high value on privacy. In October 2008, George Church launched the Personal Genome Project (PGP), which will build the only public DNA database that links genes to diseases, physical traits, and abilities.[19] In exchange for having their genomes inventoried, PGP participants agree to make it all public, along with personal information about their health, ancestries, and habits. Ten people, called the PGP-10, initially volunteered. Among them was Steven Pinker, the Harvard psychologist who wrote *The Blank Slate: The Modern Denial of Human Nature* and whose grandmother suffered from Alzheimer's. Church appealed to 99,990 more of his fellow citizens to join him in donating their genetic material as part of the new civic responsibility to aid scientists in their mission to advance personalized medicine. "We're all at risk for everything to some extent," he said, "and so we need to have a rich set of data and we need to be sharing that data until we get a much deeper understanding of what all the risk factors are, environmental and genetic."[20] Church also sees the importance of giving each individual more personal genetic information as "part of an experiment that's unfolding about how much individualized self-knowledge will change us."[21]

The United States military is also planning to conscript soldiers to be biocitizens. The Department of Defense (DoD) asked JASON, a group of scientist advisers that works through the nonprofit Mitre Corporation in McLean, Virginia, to "consider the impact of anticipated advances in genome sequencing technology over the next decade" and to assess how the armed forces could take advantage of genomic information about their personnel. In December 2010, JASON released a report, *The $100 Genome: Implications for the DoD*, which presents recommendations for the collection, storage, analysis, and use of personal genomic data from members of the military. Predicting that "the $100 genome is nearly upon us," the report advocates that the DoD health system establish procedures to collect not only DNA samples from all military personnel but also their complete DNA sequence data. Drawing from more than 2 million people on active duty and in the reserves, the DoD genomic database would far surpass any university collection. By entering genomic information in the medical database, JASON says, the military can group personnel data into phenotypes and correlate those phenotypes with genetic profiles. According to the report, "Many phenotypes of relevance to the DoD are likely to have a strong genetic component, for which better understanding may lead to improved military capabilities." JASON's proposals suggest that the military may begin deploying personnel according to their genotypes that are thought to predict traits important to military performance.

A major objective of genetic testing operations, including DTC services, is to compile the genetic information into giant databases that can be used to develop future personalized pharmaceutical products. As the for-profit company 23andMe discloses on its Web site, "Because we believe 23andMe's mission extends to the advancement of science, we intend to give you the opportunity to participate in research that could improve understanding of how genetics influences our lives."[22] To be more specific, customers can decline to participate in surveys, but they cannot opt out of having their genetic data shared anonymously.[23] Privacy and consumer watchdogs, including the ACLU, the Council for Responsible Genetics, and Patient Privacy Rights, have expressed concerns about the lack of privacy protections for the DNA contained in cheek swabs and spit samples sent to DTC genetic testing companies. "Your DNA might be sold, shared or used without your consent for testing and research," warns bioethicist Lori Andrews. "Some companies are just a front end for biotech companies that use it for research."[24]

The drive to amass individuals' genetic information has also steered social policy. In April 2008, Congress passed a law that bans discrimination by employers and health insurers on the basis of genetic tests—the Genetic Information Nondiscrimination Act (GINA). Its Senate sponsor, Republican Olympia Snowe, quoted Senator Ted Kennedy in calling it "the first civil rights act of the 21st century."[25] Although the law provides needed protections against genetic discrimination, its driving force was the crucial importance of genetic data for the launch of personalized medicine. A true reflection of biocitizenship, GINA was backed by an alliance of drug companies, scientists, and patient groups that believe the legislation will foster the development of gene-tailored pharmaceuticals. GINA was meant to increase the number of people who are willing to participate in genetic testing by relieving anxieties surrounding the disclosure of genetic information.[26] Betty McCollum, a Democratic representative from Minnesota, explained from the House floor, "Patients recognize that few laws exist to prevent health insurers or employers from using their predictive genetic information to deny them coverage or jobs. As a result, fear of such discrimination could cause individuals to refuse potentially life-saving testing or [to] participate in genetic research."[27] Francis Collins, then director of the National Human Genome Research Institute, had told Congress in 2007 that a third of people with family histories of disease refused to participate in federal health research.[28] As Edward Abrahams, the executive director of the Personalized Medicine Coalition put it, "This bill removes a significant barrier to the advancement of personalized medicine"—people's fear that their genetic information would be used against them.[29]

Biocitizenship is usually discussed without regard to race. In fact, the new genetic ties are supposed to transcend race. "We envision a new type of community where people will come together around specific genotypes, and these artificial barriers of country and race will start to break down," says Wojcicki about the social connections fostered by 23andMe.[30] This mirrors the hope for a postracial society that many people believe will result from a better understanding of personal genomics. Recall David Goldstein's prediction that individual genome scanning would allow scientists working on personalized medicine to "forget about race." It is believed that if every American sent in a cheek swab, the test results would give them a new genetic identity to replace their racial identity.

But race has been made fundamental to the new genetic citizenship. Race is treated as a key—even essential—classification in the genetic research and testing that informs biocitizenship. Race is at the cutting edge of technologies that empower biocitizens, and it determines access to them. Race is integral to the public discourse about genetics that promotes biocitizenship. So far, genetic technologies are reinforcing race, not transcending it.

Who Has Access?

Will genetic testing empower everyone to take charge of their health, regardless of race? A study conducted by Katrina Armstrong and her colleagues at the University of Pennsylvania School of Medicine of genetic testing referrals of at-risk women suggests that the answer is no.[31] The racial disparities documented in most areas of health care may in fact be greatest for new technologies. Dr. Armstrong studied 408 women who had a family history of breast or ovarian cancer and who had seen a doctor in the university health system in order to determine the factors that predicted which women used counseling about mutations that increase the risk of cancer. While research and media have focused on the decisions women make after they receive genetic counseling, Armstrong was interested in finding out who was referred for genetic testing in the first place.

Mutations of the BRCA1 and BRCA2 genes have been associated with an increased risk of breast and ovarian cancer. About one in four hundred women have the mutation and, although the risk is highest for women who are Ashkenazi Jews, it has been found in women who don't identify as Jewish. Myriad Genetics, a Utah-based biotech firm, was first to isolate the mutations and patented the DNA sequences themselves, giving the company a monopoly over the diagnostic test for the patented genes. With no competition, Myriad set the price for the test at more than $3,000, and any woman who wanted to find out if she had the mutations had to use Myriad's test. (A March 2010 federal court decision invalidating Myriad's patents is on appeal.)[32]

Armstrong discovered a huge racial gap in genetic counseling: white women were almost five times more likely to undergo BRCA1/2 counseling than African American women. This inverse relationship between African Americans and genetic counseling remained even after adjusting for the probability of carrying the BRCA1/2 mutation (based on the number of relatives diagnosed with cancer and Ashkenazi Jewish heritage), socioeconomic status, cancer risk perception and worry, attitudes about genetic testing, and doctor visits and discussions about BRCA1/2 testing. In other words, the racial disparity among women receiving genetic counseling was not explained by their risk of cancer or their concern about it. Race mattered even when Armstrong looked only at women who were not of Ashkenazi descent. Race mattered even among women with the same income, education, and type of health insurance. And race mattered more with this genetic technology than with more traditional technologies such as cardiac catheterization.

This is not to say BRCA1/2 testing is advisable for all women with a family history of cancer. Although some women who tested positive for the mutations have reduced their risk of cancer by having their ovaries or breasts removed, there are downsides to being tested: testing is expensive and the information gleaned from it is imprecise; a positive test result can't tell whether or not a woman will actually develop cancer or when. It is a wrenching decision for women who test positive to weigh the possible benefits of prophylactic surgery or chemoprevention against the uncertain odds of getting cancer. It is therefore hard to pinpoint the optimal number of women with a family history of cancer who should get tested. But whoever these women are should not be determined by race. It is just as useful clinically for black at-risk women to be tested as white at-risk women.

Like the genetic trait for sickle cell, the BRCA1 and 2 genes are commonly perceived as race-specific traits.[33] While the gene for sickle cell is seen as black, predisposing black people to sickle cell disease, the BRCA mutations are seen as Jewish, predisposing Jewish women to breast or ovarian cancer. Alluding to evidence attributing cancers in non-Jewish women to BRCA1/2, Columbia research scholar Sherry Brandt-Rauf argues, "researchers compete over which ethnic group 'owns' a mutation rather than consider the possibility that the mutation is shared among people who have lives in close proximity or that ethnic identity may be a less than reliable proxy for genetic risk."[34] Flowing from this blindness is the view that people who are ethnically Jewish do not carry the sickle cell trait and black women do not carry the BRCA alleles. These beliefs are wrong, as is the conclusion that these genes demonstrate the biological nature of race.[35]

Building Better Babies

"Unable to have a baby of her own, Amy Kehoe became her own general contractor to manufacture one," writes journalist Stephanie Saul in a front-page *New York Times* article titled "Building a Baby." The story continues:

> Working mostly over the Internet, Ms. Kehoe handpicked the egg donor, a premed student at the University of Michigan. From the Web site of California Cryobank, she chose the anonymous sperm donor, an athletic man with a 4.0 high school grade-point average. On another Web site, surromomsonline.com, Ms. Kehoe found a gestational carrier who would deliver her baby. Finally, she hired the fertility clinic, IVF Michigan, which put together her creation last December. "We paid for the egg, sperm, the in vitro fertilization," Ms. Kehoe said as she showed off baby pictures at her home near Grand Rapids, Mich. "They wouldn't be here if it weren't for us."[36]

Genetic science is empowering individuals not only to manage their own genetic risk, but also to eliminate genetic risk in their children. For decades, prenatal testing has allowed women to avoid bearing children with genetic disorders by having selective abortions. Advances in reproduction-assisting technologies that create embryos in a laboratory have converged with advances in genetic testing to produce increasingly sophisticated methods to select for preferred genetic traits before pregnancy.[37] In vitro fertilization (IVF)—fertilization of the egg in a petri dish followed by transfer to the uterus—has become a staple of fertility clinics. With preimplantation genetic diagnosis (PGD), clinicians biopsy a single cell from an early embryo, diagnose it for the risk of developing hundreds of genetic conditions, and select for implantation only those embryos at low risk for these conditions. As Reprogenetics, LLC, a New Jersey–based genetics laboratory that specializes in PGD, puts it, this technique allows patients to choose "those embryos classified by genetic diagnosis as normal."[38] These technologies—sometimes called reprogenetics—allow couples to have children who not only are genetically related to them but who are genetically advantaged.

Parents also try to ensure high-quality genes for their children by picking sperm and egg donors who possess favored qualities. In 1980, California millionaire Robert Graham opened the Repository for Germinal Choice, a "genius sperm bank" that selected donors based on scientific achievement and boasted having sperm from several Nobel Prize winners, including William Shockley, also known for his racist views on intelligence.[39] Another California sperm bank offers celebrity lookalike sperm from donors who resemble actor Ben Affleck or soccer star David Beckham, for example. An ad in the University of Chicago newspaper offered $35,000 for an egg with these specifications: "You must be very healthy, very intelligent and very attractive, and most of all, very happy. Liberal political views and athletic ability are pluses."[40] The prospective parents apparently believe not only that genes determine children's features, but that all their donors' features, including personality and political preferences, are inherited. Other fertility clinics use PGD to choose embryos by gender. By hiring a surrogate to be implanted with carefully pieced-together embryos, biocitizen parents can take control of the whole reproductive process.

In the 1980s, Margaret Atwood and other feminist writers imagined dystopias in which childbearing by white women was valued and privileged while the wombs of minority women were devalued and exploited. In her dystopian novel *The Handmaid's Tale*, published in 1985, Atwood described the repressive Republic of Gilead, where handmaids were forced to serve as breeders for elite men and their infertile wives in order to perpetuate the white race. Black women, along with handmaids who failed to bear children, were exiled to toxic colonies. That same year, in *The Mother Machine*, Gena Corea predicted that white women would hire surrogates of color in reproductive brothels to be implanted with their eggs and gestate their babies at low cost.[41]

Two decades later, the opposing relationships of white women and women of color to reproduction-assisting technologies remain. At a time when wealthy white women have access to technologies designed to produce genetically screened babies, an assortment of laws and policies discourage women of color from having babies at all.[42] As NYU anthropologist Rayna Rapp stated at a Radcliffe conference, "Reproductive Health in the Twenty-First Century," "Some women struggle for basic reproductive technologies, like a clinic where sterile conditions might be available to perform C-sections, while others turn to cutting-edge genetic techniques."[43] The multi-billion-dollar apparatus devoted to technologically facilitating affluent couples' procreative decisions stands in glaring contrast to the appalling numbers of infant and maternal deaths among blacks that have remained twice and quadruple the rates for whites for decades.[44]

In my 1997 book, *Killing the Black Body: Race, Reproduction, and the Meaning of Liberty*, I placed white, affluent women who had access to high-tech reproduction and women of color who were targets of population control policies at opposite ends of a reproductive hierarchy. At the time, pictures showing the success of reproduction-assisting technologies were always of white babies, usually with blond hair and blue eyes, as if to highlight their racial purity. By contrast, black babies figured in media coverage of these technologies only in stories about their devaluation precisely because of their race. One instance was a highly publicized lawsuit brought by a white woman against a fertility clinic she claimed had mistakenly inseminated her with a black man's sperm, resulting in the birth of a mixed-race child.[45] The woman, who was the child's biological mother, demanded monetary damages for her injury, which she explained was due to the unbearable racial taunting the child suffered because the mother and daughter looked so different from each other. Because they belonged to different races, people apparently did not think they resembled each other, even though they were genetically related. The genetic trait (or taint) of race seemed to have weakened the genetic tie between them.[46] Two reporters covering the story speculated that "if the suit goes to trial, a jury could be faced with the difficult task of deciding damages involved in raising an interracial child."[47] In this particular case, the story was complicated by the fact that the mother had wanted a child with sperm from her husband, who later died of cancer. But the perceived harm of receiving the wrong sperm was intensified by the clinic's failure to deliver a *white* baby.

The recent expansion of both reproductive genetic screening and race-based biomedicine signals a dramatic change in the racial politics of reproductive technologies. Companies that market race-based biotechnologies now promise to extend the benefits of genetic research to people of color, and reproductive technologies are no exception. Today, the high-tech fertility business, including genetic screening services, no longer appeals to an exclusively white clientele. Some clinics that offer high-tech reproductive services, including PGD, explicitly market to clients of color. In contrast to the 1990s media coverage, images on fertility clinic Web sites routinely show people of color alongside claims advertising clinic services and their benefits.

Countless advertisements on Craigslist explicitly solicit egg donors by race, and many seek minority women. For example, a posting by Beverly Hills Egg Donation notes, "All Ethnicities Welcome!"[48] Williams Donor Services' listing states, "Ethnic Diverse Egg Donors Needed," and includes a photo of an Asian, a white, and a black woman.[49] Happy Beginnings, LLC, advertises, "Egg Donors Wanted All Ethnic Backgrounds," specifying, "We have a very high demand for Jewish, East Indian, Middle Eastern, Asian, Italian, and blonde donors."[50] Meanwhile, some agencies specialize in donors with a particular racial background. As its name implies, Asian Egg Donation, LLC, based in East Brunswick, New Jersey, "specialize[s] in providing an Asian Egg Donor pool by recruiting highly qualified donors between the ages of 20 to 29 with various Asian cultures and ethnic backgrounds."[51]

As with other forms of personalized medicine, race is considered an essential way of grouping reprogenetic commodities. California Cryobank, the world's largest sperm bank, organizes its Donor Catalog according to race, with separate sections for "Caucasian Donors," "Black/African American Donors," "Asian Donors," and "Other Ancestry Donors." At the top of each page is a banner in bold print identifying the race of the donors listed on that page. The company's Web site features a "Quick Search" drop-down menu that filters available donors according to hair color, eye color, and ethnic origin. Part of Cryobank's quality-assurance procedure is color-coding the samples. The caps that seal the vials of semen come in different colors—white for Caucasian donors, black for African Americans, yellow for Asians, and red for unique or mixed ancestry.[52] No doubt this reflects an effort to avoid racial mix-ups, which can generate costly legal battles.

Some fertility clinic Web sites not only market their reprogenetic services to people of color, but they also offer race-based genetic testing as part of those services. Pacific Fertility Center's Web site includes the statement "Genetic screening is also recommended, based on ethnic background."[53] Reproductive Genetics Institute, in Chicago, similarly includes race in the factors it takes into account in its genetic testing: "Screening Results and Accuracy: By combining the results of the ultrasound and blood test along with the age, race and weight of the mother, a number can be generated by computer which represents the risk of the pregnancy being affected by Down syndrome or another chromosome problem."[54] Granted, there are some rare genetic problems, such as Tay-Sachs, that are so highly concentrated in an ethnic group that it is arguably defensible to segregate testing for these conditions, but genetic mutations are not grouped by race. Race-based testing reinforces the myth that races are genetically distinct from one another and that our genetic risks are determined by our race.

Still, white donors are in greater demand than any others and they supply most of the sperm and eggs available for sale. Of the 312 donors listed in California Cryobank's April–May–June 2009 Donor Catalog, only seven were black.[55] The economic valuation of eggs is more intimate than setting the price of sperm. Cryobank charges $215 for all of its sperm vials. "Women are paid to produce eggs for a particular recipient who has agreed to a specific price for that donor's reproductive material," explains Yale sociologist Rene Almeling, based on her investigation of egg agencies.[56] The price of eggs is determined by a racial supply-and-demand system. Although most customers are white and want eggs from a white donor, egg agencies find it hard to recruit black donors for their small African American clientele. The surprising result is that black women are often paid a few thousand dollars more for their eggs than the fee typically earned by white women.[57] Nevertheless, tall, blond, college-educated donors still fetch the highest premium. In 2007, a couple advertised in Ivy League newspapers that they were willing to pay $50,000 for eggs from a donor who was five-foot-ten or taller and had scored at least 1400 on the SATs.[58]

Customer satisfaction hinges on racial results. A recent racial mix-up mirrored the one from the 1990s I described earlier—except the plaintiff was Latina. A Dominican woman, Nancy Andrews, and her white husband sued the Park Avenue fertility clinic that provided IVF for them because their daughter Jessica came out too dark. DNA tests confirmed that the baby, born in October 2004, was not conceived with Mr. Andrews's sperm. Because Jessica has "characteristics more typical of African or African-American descent," the couple claimed she was a different race from the rest of the family and would therefore "be the object of scorn and ridicule by other children."[59]

A family photo that circulated online indicates that Mrs. Andrews has some African ancestry, as do most Dominicans. In allowing the malpractice claim to go forward, the Manhattan judge noted that Mrs. Andrews "has a complexion, skin coloration and facial characteristics" typical of the Dominican Republic. Another baby born to Mrs. Andrews might have had darker skin than hers even if the clinic *had* used her husband's sperm. But the parents did not see it this way: "While we love Baby Jessica as our own, we are reminded of this terrible mistake each and every time we look at her . . . each and every time we appear in public," the Andrewses wrote in court papers. These parents' tragic words reflect once again the power of race to override the bonds of genetics and kinship.

Some parents want to design their children's physical traits more directly, rather than rely on the features of egg and sperm donors. Fertility Institutes, a Los Angeles–based chain of fertility clinics run by IVF pioneer Dr. Jeffrey Steinberg, announced in December 2009 that it planned to offer "cosmetic" screening of embryos.[60] The company's Web site promised that its custom genetic service would allow parents to make "a pre-selected choice of gender, eye color, hair color and complexion." Critics accused Steinberg of unethically crossing a crucial line between testing for medical and nonmedical traits, as well as promising unrealizable results.[61] Amid the controversy, Steinberg discontinued his cosmetic services. The company now focuses on its PGD sex selection program, which the Web site boasts is 100 percent accurate. "Gender selection is a commodity for purchase," says Steinberg. "If you don't like it, don't buy it."[62]

Obliged to Choose

As genetic screening increasingly enables individuals to manage their own health by reducing genetic risk, we may see its wider incorporation into the health care system. Using reprogenetics to select the traits of children may become more of a general duty than a privileged choice. Widespread prenatal testing has already assigned pregnant women primary responsibility for making the "right" genetic decisions. It is increasingly routine for pregnant women to get prenatal diagnoses for certain genetic conditions such as Down syndrome or dwarfism.[63] They are typically expected to opt for abortion to select against any disabling traits identified by genetic testing. Many obstetricians provide these tests without much explanation or deliberation because they consider such screenings to be a normal part of treating their pregnant patients. The director of reproductive genetics at a large Detroit hospital reported that at least half of the women referred there with an abnormal amniocentesis result were "uncertain about why they even had the test."[64] A genetic counselor told writer Lynda Beck Fenwick, "Patients will come in and say, 'I am having the amniocentesis because my doctor told me to,' but really in their hearts they are not sure that's right for them."[65]

I faced this expectation in 2000 when I was pregnant with my fourth child at age forty-four. My obstetrician recommended that I participate in a clinical trial by Northwestern University Medical School researchers investigating the potential for a blood test and ultrasound to detect in the first trimester of pregnancy the risk of Down syndrome and trisomy 18, caused by a chromosomal defect that leads to serious medical complications and developmental delays.[66] "It's a way to get a free ultrasound," he told me. Although the researchers had an ethical obligation to disclose any significant risks entailed in the procedure, neither they nor my obstetrician ever discussed with me the implications of results predicting a high risk of Down syndrome or trisomy 18: that I would be asked to decide whether or not to terminate my pregnancy. No one mentioned why I was getting genetic screening in the first place. Instead, everyone assumed that screening could only make me better off and that there was no need for serious deliberation about whether or not to be screened.

Although genetic counseling should be nondirective, many counselors show disapproval when patients decide against selective abortion. One genetic counselor asked a woman who decided to bear a child predicted to have Down syndrome, "What are you going to say to people when they ask you how you could bring a child like this into the world?"[67] Dr. Brian Skotko, a leading expert on Down syndrome, discovered similar pressures in his survey of close to a thousand mothers who received postnatal diagnoses of Down syndrome for their children. Many of the mothers were chastised by health care professionals for not undergoing prenatal testing:

> "Right after [my child] was born, the doctor flat out told my husband that this could have been prevented or discontinued at an earlier stage of the pregnancy," wrote one mother who had a child with DS in 2000. A mother who had a child in 1993 recalled, "I had a resident in the recovery room when I learned that my daughter had DS. When I started to cry, I overheard him say, 'What did she expect? She refused prenatal testing.'" . . . Another mother reported, from her experience in 1997, "The attending neonatologist, rather than extending some form of compassion, lambasted us for our ignorance in not doing prior testing and for bringing this burden to society—noting the economical, educational, and social hardships he would bring." Regarding a postnatal visit, a mother who had a child in 1992 wrote, "[My doctor] stressed 'next time' the need for amniocentesis so that I could 'choose to terminate.'"[68]

Partly because of pressure like this, many pregnant women now view genetic testing as a requirement of responsible mothering.[69]

Current tort case law creates incentives in favor of genetic testing by imposing legal duties on obstetricians to offer it.[70] While there are virtually no legal consequences for doctors who encourage genetic tests, doctors who fail to use them may be liable for damages in "wrongful birth" lawsuits. For example, the Supreme Court of Ohio recently held that parents of an unhealthy child born following a doctor's negligent failure to diagnose a fetal defect or disease may bring suit under traditional medical malpractice principles for the costs arising from the pregnancy and birth of the child.[71] Some legal scholars have argued that tort law should compensate for "procreative injury" caused by reproduction-assisting technologies.[72]

Poor women, especially women of color, currently face financial and other barriers to receiving both high-tech infertility services and genetic testing for themselves and their children. But some of these barriers are lifting. Researchers are developing a test using the pregnant woman's blood to identify fetal genetic traits. Relatively cheap genetic testing is already available to consumers. In January 2010, a company called Counsyl joined the ranks of Silicon Valley start-ups marketing direct-to-consumer genetic tests on the Internet. The procedure mirrors the 23andMe model: customers receive a spit kit in the mail, they send the saliva sample to Counsyl's lab for analysis, and, for only $349, they view the results online a few weeks later. Unlike genotyping services that tell customers about their own disease risk, Counsyl informs customers whether they are at risk of passing inherited diseases to their children. The company says it can identify genes that put children at risk for more than a hundred diseases, including cystic fibrosis, Tay-Sachs, sickle cell disease, and spinal muscular atrophy—relatively rare diseases that are caused by the mutation of a single gene. "MDs recommend genetic testing before pregnancy," Counsyl tells prospective parents on its Web site. "One of our goals is to make this like the home pregnancy test," says Ramji Srinivasan, head of Counsyl.[73]

Although Counsyl is a for-profit company, its founders, a group of friends who met at Stanford, see themselves "as social entrepreneurs on a campaign." Counsyl's Web site reads like a manifesto against genetic disease, and consumers are enlisted to join the cause of purging the population of genetic risk by eliminating it from their children. The company's values are listed prominently on the home page: "We believe that genetic testing is a human right, not a luxury. We believe children deserve healthy lives, free from genetic disease. And we believe in universal access, especially for those most in need." Harvard professor Henry Louis Gates Jr., who received stock options for his role as adviser to the company, praised the service as a "genuine breakthrough for minority health," presumably because it makes genetic testing more affordable.[74]

Because prenatal genetic screening is now considered an essential part of personalized medicine, it is becoming integrated into social welfare systems and private insurance schemes. In March 2009, a California program began offering genetic screening through blood tests to every pregnant woman in the state for $162. Women shown to be at high risk can get follow-up services at state-approved Prenatal Diagnostic Centers. Run by the California Department of Public Health, the Genetic Disease Screening Program, which also includes newborn screening, is the largest screening program in the world. Its mission is "to serve the people of California by reducing the emotional and financial burden of disability and death caused by genetic and congenital disorders."[75] Such government programs will make genetic testing of fetuses and embryos increasingly available across racial and socioeconomic lines.[76]

What are the implications when the government encourages all of its citizens, including low-income women of color, to use genetic screening to select out certain disfavored traits? State genetic screening programs provide low-income women and women of color the reproductive options more privileged women already have. They also seem to reduce health disparities by equalizing the prevention of disabling genes. But the expansion of a coercive genetic testing regime could also reduce public support for general health care for everyone. Unlike IVF, whose primary purpose is to *increase* fertility, PGD functions to help women *avoid* starting a pregnancy that entails unwanted genes.[77] The aim of IVF is to produce the birth of a live baby; the aim of PGD and fetal diagnosis is to prevent the birth of certain children. Although government welfare systems have disdained facilitating childbearing by poor women of color by declining to fund fertility treatments, they may treat prenatal genetic testing quite differently. The very same thinking that promotes laws and policies that pressure these women to have fewer children could promote laws and policies that pressure them to have genetically screened children.

Making citizens responsible for managing their health at the genetic level reflects the shift of responsibility for public welfare from the state to the private realms of market and family. As Canadian legal scholar Roxanne Mykitiuk observes, genetic testing serves as a form of privatization that makes the individual the site of governance through the self-regulation of genetic risk.[78] Today, state genetic testing programs do not force citizens to participate. Instead the government and corporate sectors rely on the sense of obligation individuals feel to control their own health at the genetic level. Turning people into "gene carriers" concentrates responsibility on them to manage their own genetic predispositions, shifting the spotlight away from state responsibility for ensuring healthy living conditions.[79] The landmark health care reform package passed by Congress in March 2010 slowed down the push toward privatization but did not stop it. It gives the federal government greater power to regulate the health insurance industry without fundamentally changing how health care is provided. President Obama's proposal of a very limited public option for uninsured Americans raised such an outcry that congressional Democrats were forced to drop it.

Genetic selection procedures have become social responsibilities reinforced not only by cultural expectations but also by legal penalties and incentives. Will making the wrong genetic choices disqualify biocitizens from claiming public support? In her book exploring the public consequences of private decisions about reproductive technologies, Lynda Beck Fenwick suggests readers ask themselves, "Are you willing to pay higher taxes to cover costs of government benefits for babies born with genetic defects, even when the parents knew of the high likelihood or certainty such defects would occur?"[80] This question suggests that the main objective of a state-supported reprogenetics program would not be to give individuals more reproductive choices but to escape public responsibility for disability-related needs.[81]

Now management of children's genetic risk has moved beyond childhood diseases. Parents are increasingly using PGD to detect genetic susceptibility to developing cancer as an adult.[82] A recent article in *JAMA* encouraged families affected by hereditary cancer syndromes, including breast, ovarian, and colon cancer, to use PGD to screen out embryos with the risky genes.[83] "Our genetic counselors now try to bring up the potential of this technology in circumstances where we think it may be empowering to young couples," says Dr. Kenneth Offit, chief of clinical genetics at Memorial Sloan-Kettering Cancer Center and a co-author of the article. In the future, the government may rely on the expectation that all pregnant women will undergo genetic testing to justify not only its refusal to support the care of disabled children, but also its denial of broader claims for the public provision of health care. Without a right to basic health care, more widespread use of genetic technologies could come at the expense of public health.

Extending access to reprogenetics to women of color does not make this future any less disturbing. States across the country have met budget shortfalls precipitated by the economic recession by slashing already underfunded services to the most vulnerable people who are sick and disabled. In November 2010, the Center on Budget and Policy Priorities reported that "cuts enacted in at least 46 states plus the District of Columbia since 2008 have occurred in all major areas of state services," including health care and services to the disabled, with even more severe reductions slated for fiscal year 2011. Washington, for example, plans to decrease by nearly 25 percent the amount of cash benefits paid to physically or mentally incapacitated adults through its Disability Lifeline program; the program's medical services for poor adults with disabilities will be completely eliminated.[84] The depletion of public resources for general health care and for supporting people with disabilities hits poor minority communities the hardest. Chilling evidence of the racial disparities that already plague care for people with disabilities is the gap in life expectancy for those born with Down syndrome. While the median age of death for whites with Down syndrome increased from two to fifty in the years between 1968 and 1997, the median age of death for blacks with Down syndrome reached only twenty-five years in 1997 from zero in 1968.[85] The expectation of genetic self-regulation may fall especially harshly on black and Latina women, who are stereotypically defined as hyperfertile and lacking the capacity for self-control.[86] In an ironic twist, it may be poor women of color, not affluent white women, who are most compelled to use reprogenetic technologies. "Having a baby with Down syndrome is becoming a luxury for the rich," sociologist Barbara Katz Rothman told me.[87]

Genetic technologies by themselves cannot undo or transcend social inequalities. The most advanced and expensive ones will be reserved for the wealthiest people and fall outside the reach of most women of color. The market will privilege a tiny elite among people of color who can afford high-tech reproductive innovations while relegating the vast majority to the most intense reproductive surveillance. The application of market logic to child-bearing is likely to expand the hiring of poor and working-class women of color for their reproductive labor. Paying these women to gestate fetuses or to produce eggs for genetic research could intensify even as the very same women are encouraged to use genetic technologies to screen their own children, limiting their reproductive freedom.

"Reproductive tourism" is a growth industry. More and more Europeans and Americans are traveling to India and other developing countries to hire desperately poor, often illiterate women as gestational surrogates.[88] Surrogacy clinics in Anand, in the western Indian state of Gujarat, recruit women from neighboring villages to be implanted with fertilized eggs of couples from around the world and carry the fetus to term.[89] The clinics run hostels where the surrogates bunk together while they are pregnant. Their diet, medicine, and daily activities are kept under constant surveillance to ensure healthy babies for the couples who hired them. The wombs of poor, dark-skinned women have become marketable commodities for more privileged people to rent. By outsourcing surrogacy to India, Americans pay a fraction of the U.S. fee—$4,000 or $5,000, rather than $80,000 to $120,000. They also avoid U.S. regulations and the legal hassles that arise if the surrogate wants to keep or visit the baby. In allowing wealthy couples to take greater control of their reproductive lives, genetic technologies do not erase the deep social chasm that separates them from the Indian surrogates they hire. The celebration of biological citizenship forgets that some people's autonomy over "life itself" comes at the cost of devaluing the lives of others who must sell new forms of personal "biocapital" to survive.[90]

Contrary to the illusion of biocitizens liberated by a personalized genomics that transcends racial divisions, the reality includes racially segregated markets for eggs and sperm, gender selection to weed out girls, renting wombs of impoverished surrogates, lawsuits branding dark-skinned children as genetic "mistakes," and a growing obligation to use high-tech testing and interventions to ensure the genetic fitness of children. Though genetic technologies have thrust "our very biological life itself" into "the domain of decision and choice," as Nikolas Rose writes, this does not mean they give us greater freedom, justice, or equality. By extending individual management of health to the molecular level, the state and big business exercise *greater* ability to monitor and influence our lives. While we are expected to choose products and services that promise to reduce genetic risk, we cannot expect guaranteed health care for everyone who needs it.

Biocitizens or Bioconsumers?

When Steven Pinker received his genetic diagnosis from 23andMe, he found out just how confusing and inaccurate dabbling in recreational genetics can be. Although the company informed him of a lower-than-average chance of getting prostate cancer before age eighty, his own research turned up studies that linked several of his genes to an elevated risk of the disease. "Assessing risks from genomic data is not like using a pregnancy-test kit with its bright blue line," Pinker writes. "It's more like writing a term paper on a topic with a huge and chaotic research literature." Of course, most people buying in-home DNA test kits are not as equipped as a Harvard expert on genetics to delve into the research required to understand the bright-line data they receive online. Pinker observes that the "horoscopelike fascination of learning about genes that predict your traits" does not have a sound scientific basis because "individual genes are just not very informative." But instead of calling for an end to the personal genomics craze, Pinker continues to embrace it: he has faith that "the science will improve as efforts like the Personal Genome Project amass huge samples, the price of sequencing sinks and biologists come to a better understanding of what genes do and why they vary. . . . And besides, personal genomics is just too much fun."[91]

Pinker ignores the societal implications of marketing to the public the false belief that genes on their own can determine health and personality traits. Giving people more genetic data at a lower cost will only reinforce this misunderstanding. Pinker misreads the danger posed by genetic determinism when he writes that "many of the dystopian fears raised by personal genomics are simply out of touch with the complex and probabilistic nature of genes. Forget about the hyperparents who want to implant math genes in their unborn children, the 'Gattaca' corporations that scan people's DNA to assign them to castes, the employers or suitors who hack into your genome to find out what kind of worker or spouse you'd make. Let them try; they'd be wasting their time."[92] But the problem is not that hyperparents, corporations, and employers really can do these things with genes; the problem is that they *think* they can—and the private sector and government have a long history of using the myth of genetic determinism to gain public support for discriminatory policies.

When I gave a talk at MIT in 2008 about the new biocitizen enabled by genetic testing, Harvard epidemiologist Nancy Krieger called out from the audience: "That's not biocitizenship; that's bioconsumerism!" Purchasing genetic testing kits that are marketed as status symbols and joining a chat room based on the results does not entail the kind of civic engagement that is the mark of true citizenship. Although patients and customers are encouraged to participate in university and commercial genomic research ventures in language that suggests active citizenship, their actual involvement is typically limited to donating DNA samples.[93] In the case of some DTC services, their involvement requires no action at all: it is the default unless you opt out. Patients and customers who donate their DNA are not making truly informed decisions to participate fully in the research enterprise—to share in its governance, monitor its progress, and distribute its benefits equally. There is no democratic process to make sure the research truly serves the public interest and improves the lives of people throughout our society. More fundamentally, its focus is as atomistic as can be, right down to each individual's unique genetic code. Individuals may merge where their genotypes intersect, but not as whole, self-determining people who join together as a collective based on shared values to change society for the better. The central objective of dominating one's individual life at the molecular level is the opposite of working in solidarity to eliminate unjust social structures like race.

Yet I think it is correct to recognize the emerging relationship among individuals, the market, and the state as biocitizenship. It is a new form of citizenship that threatens to replace active, collective engagement to create a better society with providing information to the biotech industry and consuming its goods and services. We do not have to accept this consumer-citizen role being marketed to us. The dawning era of biocitizenship can be an opportunity for people dedicated to social justice to intervene collectively in biopolitics—not just to gain greater access to products of biomedical research, but also to change the relationship between biotechnology and power, especially biotechnology and race, to create a more humane world.

10

Tracing Racial Roots

Eric London, a teacher at the Laboratory School of Finance and Technology in the Bronx, engaged his middle school students in an unusual science project. He collected cheek swabs from his eleven black and Latino students and sent the samples to DNAPrint Genomics, a Florida-based company that uses DNA to trace the ancestry of its customers. Using its AncestrybyDNA technology, the company compared the student samples to its proprietary database of genetic profiles from around the world to find matches to four "continental" populations: Europeans, sub-Saharan Africans, Native Americans, and East Asians.[1]

When students opened the results, they were startled by the ancestral origins buried in their genes. Although twelve-year-old Sheila Guerrios identifies as Hispanic, she was told that her ancestors were 63 percent European, 25 percent sub-Saharan African, and 12 percent East Asian. "I didn't know I had European family," said Guerrios. "That's weird to me." Another Latina student, Rossay Gomez, had a similar reaction. Her ancestral roots broke down into 69 percent European and 31 percent sub-Saharan African. "I thought I was going to be Native American," she said, highlighting the indigenous component of Latino identity. Two African American students also received unexpected results. Testing showed one to have 19 percent European origins, while the other's ancestry was 23 percent European.

Many Americans have a burning desire to know who their great-great-great-grandparents were. Genealogy is one of the most popular hobbies in the United States. After pornography, ancestry Web sites are the most commonly visited on the Internet. Genealogy used to involve pressing your oldest relative for anecdotes passed down from prior generations or paging through dusty archives for a biographical record of a distant ancestor. Today, more and more Americans are using DNA technology to trace their family and geographic origins. A cottage industry of about forty online companies like DNAPrint Genomics located across the globe currently offers DTC genetic ancestry testing.[2] Half a million people have purchased at-home DNA test kits like those used to predict future health, except these reconstruct the past. Employing techniques developed in forensic genetics and human genomic research, the firms track family history through many generations of maternal and paternal lines of descent or identify a person's origins within one or more of either four or five continental groups or a more precise geographic location. Celebrities unveiling secrets in their lineage disclosed by DNA has become a common scene on both educational and reality TV alike, helping to boost public interest in this business. Like all the other genetic technologies I have discussed so far, ancestry testing, too, has race at its center.

We Have a Race for You!

Using DNA to identify biological relatives or to piece together family histories is nothing new. Paternity tests used in custody battles, infidelity cases, and "Who's your daddy?" reality shows are already a familiar technology. Some companies help people augment genealogies created using conventional methods by validating records, filling in gaps, or extending familial branches with the use of genetic information. A growing segment of the business, however, has nothing to do with finding family members. Instead, some ancestry testing companies help people trace their *racial* roots. They use DNA samples to assign percentages of a person's lineage to large continental populations that mirror conventional races. AncestrybyDNA, the leading racial test used by a number of companies, examines 176 genetic markers to split a customer's ancestral origins among four "major population groups"—indigenous American, East Asian, sub-Saharan African, and European—originally described as contemporary terms for "anthropological lineages that extend back in time tens of thousands of years."[3] The AncestrybyDNA Web site now states that it estimates what it calls "BioGeographical Ancestry (BGA)," defined as "the term given to the biological or genetic component of race."[4] So a customer might learn that she is 50 percent European, 30 percent sub-Saharan African,

12 percent indigenous American, and 8 percent East Asian. Although these groupings are not explicitly labeled "races," they map on to the familiar racial groupings: white, black, Native American, and Asian. Most people learning these results accordingly interpret them as racial designations. They seem like pieces of a racial puzzle, adding together fractions of pure races that represent different-colored ancestors who mated at some point in the past.

How do companies figure out this composite racial makeup? They use the flawed racial science growing out of population genomics described in chapter 3. Recall that some molecular biologists use allelic frequencies of certain polymorphic genes as probable markers of geographic origin and statistically divide this genetic variation into continental groupings that look exactly like races to the untrained eye. These ancestry informative markers, known as AIMs, are not exclusive to each continental group; companies use differences in the statistical frequencies of alleles to identify AIMs that can distinguish among continental populations.[5] They then analyze AIMs found in customer DNA and compare them to the statistically generated worldwide groupings. From this comparison, they infer the percentages of "admixture" in the customer's genotype that comes from each of the continental groups.

Telling customers that they are a composite of several broad anthropological groupings reinforces three central myths about race: that there are pure races, that each race contains people who are fundamentally the same and fundamentally different from people in other races, and that races can be biologically demarcated. The concept of dividing a person's genotype into racial components assumes that each component is racially pure. We can only imagine someone to be a quarter European if we have a concept of someone who is 100 percent European.

The ancestry testing companies base their breakdown on a made-up baseline of what constitutes 100 percent genetic purity for each race. The measure for 100 percent racial purity, in turn, is derived from a statistical analysis of allele frequencies in DNA data sets. But as I pointed out before, the sampling methods used were inadequate and unscientific. Researchers never sampled the world's populations in either a systematic or random fashion. Small, isolated groups that do not represent most of the continental populations were preferred over groups that were more likely to have migrated and mixed with others. Vast portions of continents were never sampled. By using these samples of convenience as proxies for human genetic variation the

AIMs technology arbitrarily differentiates among continental groups and treats people within these groups as if they were basically all the same. Ancestry testing companies assume that there is a "biological or genetic component of race" that can be isolated from its social meaning. Their claim that race is determined by DNA and therefore can be identified with statistical accuracy by a genetic test perpetuates the false definition of race as a biological category.

Some scientists believe that the disclosures of racial intermixing and genetic kinship generated by these tests will help overcome racism. "Each of us has around 6.7 billion relatives," writes Aravinda Chakravarti, a molecular biologist at Johns Hopkins. "The global picture of relatedness that is emerging from DNA studies stands to shatter many of our beliefs about ourselves."[6] Noting that "we are all Africans under the skin," Spencer Wells, head of the Genographic Project, a research partnership between National Geographic and IBM "to better understand our human genetic roots," hopes that these genomic revelations will "help people to overcome some of the prejudices they might have."[7] This recognition of racial intermixture might hasten a new racial paradigm in the United States: some sociologists predict that as more minorities, especially Asians and Latinos, marry whites and have biracial children, the next generation may be accepted as "honorary whites."

It is not clear, however, why genetic technologies would have any greater impact on racial inequality than the knowledge of racial intermixture we had without them. Surely, Southern slave owners were well aware that the children they fathered with enslaved women were racially mixed and intimately related to them. Yet their response was to pass laws guaranteeing that their offspring would have the status of slaves. As a little girl growing up in a liberal university community in the 1960s, I used to cherish the fantasy that the intimate hybridity of my own biracial, multiethnic family constituted a blow against the racial order. But that was before I formed my own moral allegiance to black people based on a sense of common struggle against racial oppression. Looking back, I can see that my childhood fantasy was not only unenlightened but privileged by my middle-class existence, largely disconnected from the majority of black residents in other parts of the city who may have lived out their entire lives without ever experiencing even one moment when race did not matter or any chance of gaining some benefit from "mixed ancestry." There is no evidence that the genetic technologies that have emerged since then have the power to overturn the racial hierarchy that still exists today.

Recovering African Origins

Most African Americans are not interested in finding out what race they are—they are pretty sure they are black and that there is racial mixing somewhere in their heritage. Instead, they want to know more about their ancestors in Africa. Alex Haley's best-selling 1976 novel, *Roots: The Saga of an American Family*, and the miniseries that followed were based on Haley's own quest to track down his ancestral family in Africa. Haley's narrative sparked hope among African Americans that it was possible to use slave-ship manifests, plantation records, and family lore to trace their lineage back to their motherland. But blacks in the United States who have tried to reconstruct their family trees with conventional genealogical tools almost always meet a brick wall erected by the slave trade. With the right genealogical tools, most African Americans can trace at least one side of their family to the 1870 federal census taken after the Civil War, the first to list blacks as citizens rather than property.[8] Finding slave ancestors living before then is considerably harder. Birth certificates, marriage documents, and other evidence of their life in America are scarce. To track their lineages all the way back to a point in Africa prior to their enslavement is virtually impossible. Records were rarely kept of the names, much less the origins, of Africans who were captured by slave sellers and forcibly shipped to the Americas.

A popular segment of commercial genetic genealogy offers to fill this gap. Instead of tracing ancestry to the broad category "sub-Saharan Africa," some ancestry-testing companies pinpoint which part of the continent the forebears of Americans of African descent came from. Using mitochondrial DNA (mtDNA) or Y chromosome sequences, the companies compare the genotypes of their African American customers to a database of DNA profiles sampled from groups living in Africa today to match those with common maternal or paternal ancestors. Mitochondrial DNA, the energy producer located in each cell, is directly transmitted unchanged from mothers to their sons and daughters. Only daughters pass it on, so it is used to trace maternal lineage along one line of descent. From mtDNA, it is possible to identify a person's mother, the mother's mother, the mother's mother's mother, and so on. Y-chromosome DNA is passed down from fathers exclusively to sons and its markers are used to trace the paternal line in similar fashion. (In 1998, Y chromosome analysis was used to settle the debate over whether or not Thomas Jefferson fathered children with his enslaved companion Sally Hemings. The results confirmed that an individual carrying the Jefferson Y-chromosome fathered Eston Hemings, born to Sally Hemings in 1808.)

The companies that specialize in recovering black people's African roots cannot possibly identify individual ancestors who lived in Africa prior to their capture by slave traders. Thus, ancestry testing cannot reveal the identity of a black customer's great-great-great-great-grandmother, but it may tell a black customer that her mtDNA shows common ancestry with the present-day Bamileke people of Cameroon. Columbia sociologist Alondra Nelson describes these genetic tools as "ethnic lineage instruments through which undifferentiated racial identity is translated into African ethnicity and kinship."[9] Distinct from either family-focused genealogies or racial apportionments, this specialized service provides the sources for discovering a new aspect of black identity and connection to the peoples of Africa.

One of the most popular companies, African Ancestry, can match the genotypes of African Americans to more than four hundred ethnic groups in Africa. To date, more than 100,000 customers have received a Certificate of Ancestry specifying their newfound lineage. The company was founded in 2003 by Rick Kittles, an African American medical geneticist at the University of Chicago who was responsible for a string of breakthrough genetic studies on skin color, prostate cancer, and diabetes. But Kittles is best known for spearheading the use of cutting-edge DNA technology to help African Americans trace their roots in Africa. Kittles boasts that he has stored over 25,000 African lineages collected from thirty countries—the most comprehensive database of its kind in the world.

"That really was something that happened out of the blue for me," Kittles told me when we met at his office in Hyde Park in 2009. As one of the few black geneticists with training in anthropology, Kittles was hired in 1997 to handle genetic forensics on the New York African Burial Ground project. In 1991, during excavation for a new federal office building in lower Manhattan, workers unearthed skeletal remains in what turned out to be a graveyard that had been paved over. Further investigation revealed that, from 1640 to 1795, the site had been the burial ground for thousands of free and enslaved African Americans outside the perimeter of the New Amsterdam settlement that was to become New York. Controversy ensued when black

New Yorkers, including Mayor David Dinkins and state senator David Paterson, protested the government's careless handling of the graves and failure to consult with the black community. Within two years, Dr. Michael Blakey, a physical anthropologist at Howard University, assumed oversight of the scientific investigation and the remains were transferred to his lab in Washington, D.C. There, Kittles conducted genetic analysis of the bones to determine whether their origins were African, European, or American Indian.

"There was an ABC reporter who did a story on the burial ground, and he came to my lab and said, 'If you can do that for the bones, why can't you do it for me?'" Kittles recalled. Kittles matched the reporter's DNA to a group in Nigeria, and the procedure was mentioned on a D.C. local news program. In June 2000, Kittles was attending the physical anthropology meetings in Columbus, Ohio, when a journalist from the *Boston Globe* approached to ask him how much he charged for the ancestry-testing service. Kittles's off-the-cuff response of $300 made it into her article.[10]

The *Boston Globe* piece was picked up by the Associated Press, and blacks around the country read about Kittles's technique for tracing their ancestry back to Africa. By the end of the day, the torrent of inquiries from the media, potential customers, and firms interested in a business deal forced Howard to shut down its switchboard. When talks with Howard's president about starting a business there fell apart, Kittles set up his own Web site, registered as AfricanAncestry.com, where people could sign up for information. The company took off in 2002, when a friend introduced Kittles to Gina Paige, an experienced entrepreneur with a marketing degree from the University of Michigan, who soon became his business partner. They officially launched the company in 2003 during Black History Month, with Paige as the president, responsible for marketing and finances, and Kittles as the scientific director.

The chance to regain their lost ancestry has struck a deep chord with many black Americans in particular. You hear the theme "knowing where you're from is critical to knowing who you are" echoed over and over by people who have done the testing. One of African Ancestry's customers, Rodney Wilson, explains how tracing his origins in Africa filled a void in his identity: "I can remember as a child in elementary school going to cultural days every year and not being able to participate. And other kids would have their flags up and native dishes from their countries and representing Ireland and China and various countries from South America. As an African American

I was never able to engage in any of these activities. I wasn't able to claim any kind of native land."[11] For African Americans, tracing their ancestry to a group in Africa constitutes a type of racial justice, righting a wrong inflicted by racism and gaining a valuable possession that most Americans have always been able to claim if they wanted to.

Kittles and Paige spend a good part of the year on the road, promoting their service to gatherings of African Americans across the country, from churches to community groups, conventions, cultural events, and genealogy clubs. If everyone pitches in $10 toward the testing kit, an extended family can learn the results at a reunion, and Kittles may even make an appearance. African Ancestry has attracted the media spotlight; Kittles has been interviewed by the press multiple times and has appeared on *60 Minutes*, CNN, and countless local news programs. His profile is also enhanced by the many celebrity endorsements that grace the Web site. Well-known customers include actor Blair Underwood, singer India Arie, statesman Andrew Young, and director Spike Lee.

The biggest splash came in 2005 when *Grey's Anatomy* star Isaiah Washington had his DNA tested. Kittles announced at a videotaped event that the actor's maternal lineage traced to the Mende people of Sierra Leone, as Washington shouted ecstatically and other celebrities clapped and cheered. "I always wanted to go to Sierra Leone," he exclaimed. Finding his Mende roots transformed Washington's life. The following year, Washington made a triumphant journey to a Sierra Leone. Shocked by the devastation caused by civil war, Washington established the Gondobay Manga Foundation, dedicated to improving living conditions in Sierra Leone.[12] He also donated $25,000 toward restoration of a crumbling castle on nearby Bunce Island, where slave traders brought Africans before shipping them across the Atlantic. After being inducted as a village chieftain, Washington became an official citizen of Sierra Leone.[13]

"I wanted to bring African Americans closer to Africa and vice versa," Kittles told me. "I wanted them to say, 'Wow, I'm connected with the Bamileke in Cameroon, I want to go to Cameroon and learn about them. I want to lobby for the Cameroon people. They are part of our extended family.'" To Kittles, Washington's projects in Sierra Leone are perfect illustrations of the affiliation with Africa that genetic testing can launch. Indeed, many African Americans see their genetic test results as a springboard to action related to Africa—learning more about the country their ancestors came from, making trips there, and donating to the country's development. Some customers have even followed Washington's example and petitioned their African country of origin for dual citizenship.

The celebrity factor skyrocketed when Henry Louis Gates Jr., director of the W.E.B. Du Bois Institute for African and African American Research at Harvard, entered the scene. His four-hour series *African American Lives*, which aired in 2006 on most public television stations nationwide, disclosed the genealogical histories of eight famous black Americans, including composer Quincy Jones, actor Chris Tucker, actress and comedian Whoopi Goldberg, and media superstar Oprah Winfrey. At first, the program centered on fascinating details of their family trees that Gates had tracked down using archival documents like census records, birth and death certificates, land deeds, and wills. Gates tells the celebrities amazing stories about their ancestors who survived slavery to become property owners, politicians, teachers, and artists. He hands Winfrey the deed to eighty acres her great-great-grandfather, a freed slave named Constantine Winfrey, earned from a white Mississippi landowner by picking what amounted to five thousand pounds of cotton. Watching the series, I was riveted by the monumental feats within the stories of ordinary yet heroic black folks. I was inspired by the message of their accomplishments: we shall overcome.

In the last episode, DNA takes center stage. As each celebrity waits anxiously, Gates dramatically unveils his or her African origins and racial breakdown as determined by genetic ancestry testing. Gates used Kittles's database of African lineages to trace their ancestral connections to continental ethnic groups. Gates extols the power of finding your "roots in a test tube." "Now for the first time in three centuries," he said, "we can begin to reverse the Middle Passage." Eleven million people watched as Gates revealed to Winfrey that her ancestors belonged to a tribe in Liberia, famously dashing her long-held belief that she was descended from the Zulu people. The 2008 sequel, *African American Lives* 2, picks up the genetic theme with a new set of black celebrities. "They won Oscars, Grammies, and Olympic Gold, but nothing means as much to them as discovering their roots," announced the promotion for the series.

Two years later, in February 2010, Gates co-produced and hosted another star-studded, four-hour television series, *Faces of America*, exploring the genealogies of a more diverse celebrity ensemble, including actor Meryl Streep, cellist Yo-Yo Ma, and comedian and talk-show host Stephen Colbert. Like his other programs, the first episodes revealed aspects of the guests' immigrant forebears unearthed by conventional genealogical research, while the final segment, titled "Know Thyself," revolved around DNA analysis. Gates and his ninety-six-year-old father even had their entire genomes sequenced—"the first African Americans, the first father and son, and the oldest individual" to do so.[14] The writer Louise Erdrich, an Ojibwa, refused to take the test because she views her DNA as belonging to her entire extended family, who told her "it's not yours to give," providing a refreshing break from the unquestioned assumption throughout the rest of the program that genetic testing is an unadulterated good.

Gates also informed his guests about the racial breakdown of their ancestors, estimated by admixture testing using AIMs. Some are a mélange, like actor Eva Longoria, whose recent forebears were Mexican cattle ranchers in what is now Texas. Gates tells her that she is 70 percent European, 27 percent Native American, and 3 percent African—what gets translated into Latina. Others are deemed to be racially "pure": Yo-Yo Ma is "100 percent Asian," while Meryl Streep is "100 percent European." When Gates appeared on *The Colbert Report* in February 2010 to promote the series, he gave Stephen Colbert a preview of the results: "I found out that you have absolutely this much African ancestry," Gates told the comedian, making a zero with his fingers. "I have done *African American Lives 1*, *African American Lives 2*, *Finding Oprah's Roots*, and *Faces of America*, and you, Stephen, are the whitest man I have ever tested." Gates made it clear that he regards so-called biogeographical ancestry as the equivalent of a test for race.

Riding the popularity of his PBS specials, Gates launched his own ancestry testing brand, African DNA, in 2007.[15] Gates claims that his company offers a more complex portrait of African American ancestry than other services by revealing its non-African components, by consulting with prominent historians and anthropologists, and by adding the option of purchasing a "Genealogy Package" that includes family tree research. For customers interested in tracking their African roots, the service provides nothing novel. Gates partners with the industry leader, Family Tree DNA; the companies have headquarters near each other in Houston, and Family Tree DNA supplies customer support for both firms. African DNA runs customer samples through the already-existing database of Family Tree DNA, which Kittles is quick to point out has fewer African samples than his proprietary collection. While owners quibble over the details, all of these genetic genealogy companies share the same underlying philosophy: that genetic technology can tell us the identities of our ancestors, revealing an important part of our own identities in the process.

Ancestry, Not Race

What intrigued me most about Kittles was the paradox I saw in his work. On the one hand, he is an outspoken critic of a genetic definition of race and has written several prominent articles debunking claims of natural racial boundaries found in our genes. On the other hand, he is a leader in the scientific investigation of distinctive African genetics, starting a business that deals in African DNA and conducting studies on genetic contributions to the high rate of prostate cancer among African American men. Kittles likes to point out that being black in America is a social category that disregards the European lineages that most African Americans possess. "All I have to do is walk out in the street and try to get a cab and I'm reminded I'm a black man, no matter how many European chromosomes I have," Kittles said at a talk I attended. "One of the first things I looked at when I did my [genetic ancestry] profile was what we call the Duffy null, meaning the West African allele—an allele common in West Africa. And I don't have any copy of it. I have a buddy, a fellow researcher at Penn State named Mark Shriver, who did the same thing. He's white and he had one of those Duffy nulls. So he kids me all the time to say he's more black than I am, and I say, 'Yeah, but at least you can get a cab.'" How can the Duffy null be a "West African allele" if Kittles, a black man, does not have it, and Mark Shriver does, despite being white? It seems to me that mismatch poses a problem for genetic testing of African ancestry as well as for a genetic definition of race.

When I met with Kittles, I brought along a 1997 article Kittles coauthored with his mentor Shomarka Keita, an Oxford-trained physician and biological anthropologist, when he was still in graduate school. Titled "The Persistence of Racial Thinking and the Myth of Racial Divergence," the article demolishes the concept of human races—that "visible human variation connotes fundamental deep differences within the species, which can be packaged into units of near-uniform individuals." Kittles and Keita observe that, although anthropologists had discredited this type of racial thinking, it was creeping back into genetic research. "Received racial categories, concepts, and constructs should *not* be used as starting points in analyses," they conclude.[16] Their message was quite prescient, written before the human genome was mapped and the surge in racial science that followed.

I asked Kittles how he could square these statements against studying race as a genetic category with his current research and ancestry-testing business focused on the genetics of African Americans.

"Because when I use African Americans in my research, I am studying them as a demographic group that has some level of shared environment and some level of shared ancestry, but is not defined as a race," Kittles responded. "A race is a subspecies. There are no subspecies of humans. So there are no human races. When I say African Americans have shared West African ancestry, that means that the bulk of their gene pool comes from populations that emerge from west and central Africa."

But aren't the genetic markers used to trace African American lineage to a region in Africa verifying a common racial affiliation? I asked. To the contrary, he answered: "Those markers are *deconstructing* race. What they are being used for is to show that within these so-called social races, there is a lot of biological diversity. Within the macro ethnic group called African Americans, there are lineages from a plethora of very diverse continental groups. There are also significant European chromosome lineages, as well as significant Native American lineages in certain areas. So that says that African Americans are not a monolithic racial group. They are a genetically diverse social group." Likewise, Kittles emphasized the genetic diversity of Africans. "I don't see a lot of shared ancestry across Africa," he pointed out.

Kittles told me he devoted his career to studying African Americans because, as a black scientist, he wanted to apply his skills to serve "my own people." He first encountered the dearth of genetic data on African ancestry as a PhD student in biology at George Washington University in the 1990s. Unable to excite any of his professors about Africa, he stumbled on thousands of Finnish DNA samples at an NIH lab, where geneticists thought (incorrectly) that Finnish DNA would reveal the genetic sources of common diseases like diabetes. Kittles ended up writing his dissertation on Finnish population structure, applying what was then a fledgling technique using Y-chromosome and mtDNA to dissect the ancestry of Finns, and he was the first to publish a study showing that at least half of the group's male ancestors came from Asia, while most of the women came from Europe.

After graduate school, Kittles embarked on a mission to fill the gap in genetic research on African ancestry. "I think the work that's been done in whites obviously is important, but there could be alternative data. It is a better design for biomedical researchers to be more inclusive," Kittles said. Kittles credits his successful career in African genetics to the National Human Genome Center (NHGC) at Howard University, where Kittles landed fresh out of graduate school in 1998. Modeled after the Human Genome Project, the goal of the NHGC was to expand the resources for investigating population-based genome variation in African Americans and its applications in mapping disease susceptibility genes. According to Kittles, the center's founder, Georgia Dunston, grew its research base by developing shrewd collaborations with Francis Collins, then head of federal human genome research. "Georgia was very savvy," Kittles says. "Francis wanted to study diabetes and prostate cancer in blacks. So she says, 'If you give me a certain number of millions of dollars a year, we will be the coordinating center and collect DNA samples, whether from West Africa or from urban areas in this country.'"

In 1998, Howard received a huge federal contract to be the coordinating center for a study of the genetics of type-2 diabetes in West Africans called the Africa-America Diabetes Mellitus (ADM) Study Network, with the goal of storing DNA from four hundred affected sibling pairs. Along with the biological specimens, which include blood and urine samples, investigators gathered extensive epidemiological, family history, and medical data at five sites, two in Ghana and three in Nigeria. The ADM project was followed by the African-American Hereditary Prostate Cancer study. Under Kittles's direction, a team of African American investigators from the Howard NHGC center collected samples and pedigree data from 150 families affected by prostate cancer from seven different sites around the country, providing a "model for how research should be done in the black community," Kittles told me. In May 2003, Dunston announced a partnership with Chicago-based First Genetic Trust to establish GRAD (Genomic Research in the African Diaspora), a biobank derived from people of African descent. Dunston hoped to amass DNA samples from 25,000 volunteers within five years. "I think we'll learn some lessons on how life works by looking inside the cell, rather than outside the individual," Dunston told the AARP when the organization honored her with a 2004 Impact Award.[17]

Although NHGC relies on claims about the distinctiveness of African genomes to obtain institutional and financial support, its key researchers sought to construct "a new paradigm" in research on health disparities and human genomic variation that rejects biological definitions of race.[18] The center houses epidemiological and clinical as well as genetic materials to study the complex interplay between environmental and genetic factors that cause disease. Ethical concerns about the potential use of genetic information for racist ends, studied in the GenEthics unit developed by Charmaine Royal, were at the center of NHGC's agenda from its inception. The new paradigm pursued by the Howard geneticists attempts to resolve the tension between making genetic research more inclusive and rejecting the biological definitions of race that African American scientists in particular are navigating.

Am I Jewish?

> Have you ever wondered if you have Jewish or 10-Israel Ancestry? Many people from non-Jewish families ask themselves, "Am I Jewish?" "Am I Hebrew?" or "Were my ancestors of the Lost 10 Tribes of Israel?" Believe it or not, you can actually have your DNA tested by FamilyTreeDNA (a simple cheek swab) and start a journey of answering some of those burning questions.[19]

A Web site that promotes the controversial theory that ten of the original Jewish tribes are missing from Scripture looks to genetic ancestry testing to validate claims of Jewish authenticity by tracing genetic roots to one of the tribes. Thousands of people have purchased DNA testing kits to answer the question it poses: "Am I Jewish?" Family Tree DNA, the world's largest genetic genealogy company, is one of several online firms employing the same methods used in African ancestry tracing to determine if customers have Jewish ancestral roots. The company offers to test a person's mtDNA or Y-chromosome DNA and compare the results to its database containing genetic profiles for Ashkenazi and Sephardic Jews, as well as the hereditary priestly castes, Levites and Cohanim or Cohens. In the summer of 2010, two different international research teams reported genome-wide analyses comparing the genotypes of a variety of Jewish population groups, including Ashkenazi, Iraqi, and Syrian Jews, and non-Jewish groups from across the globe. Both teams found distinctive Jewish genetic patterns that revealed a shared ancestry with contemporary Middle Eastern groups—Palestinians, Bedouins, and Druze.[20]

Underlying the question "Am I Jewish?" is the deeper issue of what defines Jewishness. Jewish genetic tests presuppose that modern-day Jews are a biological grouping that descended from ancestral tribes marked by distinctive genetic patterns or haplotypes. According to this view, those genetic markings were passed down generation after generation and can now determine who is an authentic Jew.

David Goldstein, the Duke molecular biologist who has played a prominent role in personalized medicine, also dabbles in genetic genealogy. As Kittles traces genetic lineages back to tribes in Africa, Goldstein explores the genetic history of Jews. Like Kittles, Goldstein became fascinated with his own people's genetics as a graduate student. While working with Stanford population geneticists Marcus Feldman and Luca Cavalli-Sforza, Goldstein discovered a "genetic signature" on the Y chromosome for the Cohen caste and calculated that it first appeared about three thousand years ago, during King Solomon's reign. He also resolved a puzzle of Jewish genetics: why Jewish men across the globe are genetically similar to each other while Jewish women show more genetic differences. Goldstein theorizes that Jewish men, who might have been traders along the Silk Road or the Arabian Peninsula, traveled long distances by themselves to found small Jewish communities. Once settled, they married local women who converted to Judaism, welcomed additional male travelers from their Jewish homeland, and then married exclusively within the religion.[21]

Goldstein weighs in on another heated controversy concerning Jewish ancestry: whether the Lemba, a Bantu-speaking group in Southern Africa, are descended from one of the Lost Tribes of Israel. Lemba oral tradition holds that the group's ancestors fled Judea by boat three thousand years ago. Comparing Y-chromosome samples from Lemba men, non-Lemba Bantu speakers, and Semitic groups, Goldstein made a remarkable discovery: not only did two-thirds of Lemba Y chromosomes come from a Semitic source, but nearly one in ten was the haplotype that identifies the Cohanim. Although it is impossible to confirm the Lemba ancestry narrative, Goldstein concluded that the origin of the Semitic Y chromosomes among the Lemba is probably Jewish.[22] When Goldstein first reported his findings at a 1999 talk, the story made the front page of the *New York Times* and set off a tidal wave of interest in the African tribe as well as in using DNA to resolve disputes over ethnic identity.[23]

Although Goldstein believes that ancestry is central to Jewish identity, he never says that Jews constitute a race or that genetics is a racial identity test.

In fact, Goldstein states plainly, "Jewishness is not a matter of DNA."[24] By contrast, Jon Entine, author of *Abraham's Children: Race, Identity, and the DNA of the Chosen People*, is adamant that race is essential to Jewish identity. "Jews need to acknowledge the relevance of race to Jewishness or risk trapping themselves in a Philip Roth novel, constantly at battle with neuroses about their identity," he writes. "Unlike Christianity and Islam, Judaism is grounded in more than faith. It is the only major tribal religion to have survived. Jews have seen themselves and have been seen by Gentiles as a race for their entire history, except for the post-Holocaust period."[25] To Entine, race gives Jewish identity a definitive meaning that comports with historical understandings.

But race also seems relevant because it makes it possible to test for Jewish authenticity using genetics. Genetics serves not only as a tool for discovering Jewish history, it provides proof of inclusion or exclusion in the Jewish race. In the passage quoted above, Entine was responding to a review of his book by Hillel Halkin, who wrote sympathetically in *Across the Sabbath River* about a Tibeto-Burmese ethnic group in northeast India whose members believe they descended from a Jewish tribe called Manasseh.[26] Entine objected to Halkin's embrace of groups who claim to be Jews but do not meet a racial test for Jewishness, which for Entine depends on genetic ancestry. The question of whether or not DNA testing can confirm Jewish ancestry is intimately tied to the question whether or not Jews are a race, defined as a biological grouping.

While Entine advocates genetics to authenticate Jewish identity, some Jews object to a biological test. In 2009, Britain's Supreme Court was called to decide the legality of a racial definition of Jewish identity.[27] The case involved a twelve-year-old boy, identified in court papers as "M," who applied to the exclusive JFS, founded in 1732 as the Jews' Free School by North London's Jewish community. JFS is one of seven thousand publicly funded religious schools, which admit students of all backgrounds but are permitted to give preference to applicants within their own faiths. Although M is an observant Jew, he was rejected under the Orthodox definition of Judaism: because M's mother converted to Judaism in a Reform synagogue, she did not meet the Orthodox requirements for Jewish identity. Following the classic test based on the mother's status, M was not considered a Jew either. M's parents sued to challenge, as their lawyer put it, the "exclusion of children who are devout in their Jewish faith, but considered by some to be not quite

Jewish enough."[28] They argued that M should not be treated as less Jewish than a nonbelieving atheist whose mother happened to be a Jew.

The court ruled that the school's test for Jewishness violated Britain's Race Relations Act because it relied on race rather than religion. In other words, the school had discriminated against M based on race. The decision angered many British Jews who felt that the court had no business foisting its meaning of Jewish identity on them. Others, such as Rabbi Danny Rich, chief executive of Liberal Judaism, applauded the more inclusive definition tied to faith rather than heredity, noting that the Orthodox rule "excludes 40 percent of the Jewish community in this country."[29] Bioethicist and religion scholar Laurie Zoloth observes that there is a persistent tension between the idea that blood and genes determine identity and the idea that one can choose identity and membership. It is significant that Jewish tradition permits the full legal transformation of self by converting to Judaism. "Traditional rules of conversion place obligations on both the community and the new member of the community, which transcend genetic ties and endorse the new, non-genetic identity as the actual one," Zoloth writes.[30]

Authenticating Native American Blood

Another segment of the genetic genealogy industry caters specifically to Native Americans to validate their claims of Indian ancestry with DNA testing. "Do you need to confirm that you are of Native American descent?" asked one such company called Genelux in an advertisement placed in the weekly *Indian Country Today*. "Whether your goal is to assist in validating your eligibility for government entitlements such as Native American Rights or just to satisfy your curiosity, our Ancestry DNA test is the only scientifically rigorous method available for this purpose in existence today."[31] Many other companies make precisely the same boast.

These companies apply two distinct approaches to determining Indian authenticity, one focused on kinship ties, the other on tribal blood.[32] The first traces an individual's line of descent to establish family kinship to another tribal member. Kinship is a social, legal, and cultural concept of relatedness that need not entail genetics at all; yet ancestry testing companies reduce kinship exclusively to a genetic determination.[33] To Kimberly TallBear, a science and technology professor at Berkeley and a member of the Sisseton-Wahpeton Oyate in South Dakota, defining kinship as a genetic connection drastically differs from the traditional Indian meaning. She distinguishes between the genetic understanding of kinship and one that supports "a land-based, group identification as the locus of Native American identity." Although both may be characterized as "blood" relationships, the traditional Native American understanding involves ties to the entire tribe and not just to individuals through a genealogical line. "The molecular knowledge produced by DNA tests does not account well for group kinship that is central to tribes," concludes TallBear.[34]

A second approach tests an individual's genotype to determine whether it carries distinctive Native American traits. Instead of connecting individual tribal members, this method provides evidence of whether or not someone is biologically Native American based on the assumption that there is a pure Native American "biogeographical ancestry." The genetic yardstick consists of "Native American markers," a set of haplotypes that genomic scientists claim are distinctive enough to separate Native Americans from other human beings.[35] Some companies use AncestrybyDNA, the technique that calculates the percentage of an individual's ancestry from four continental groups, including indigenous Americans. This "tribal blood" approach to Native American ancestry converts the social requirements for membership into a racial test. It treats Native Americans as a biological race of people, with membership hinging on having a high enough proportion of indigenous genes.

Genetic testing that calculates the percentage of Native American ancestry sounds suspiciously like the blood quantum rules that arose in the late nineteenth century when the U.S. government needed a test for membership in "federally recognized tribes."[36] Defining who is American Indian by the amount of Indian "blood" was first implemented in treaties with the federal government and reinforced by the General Allotment Act of 1887 (the Dawes Act), which divided tribal communal land into individually held parcels. Federal distribution of lots to every Indian head of household required that tribes construct official rolls of "tribal members." The principal test became degree of Indian blood—the number of Indian parents and grandparents the person seeking enrollment had. Since then, most of the 562 federally recognized tribes require that a certain degree of "blood quantum," ranging from full Indian blood to 1⁄32 Indian blood, be proven as a condition for membership.[37]

Bitter battles over blood quantum enrollment criteria continue to divide tribes today, with race and genetic testing at the forefront. Marilyn Vann, an engineer in Oklahoma City, can document her family's Cherokee roots with certificates, tribal enrollment cards, and land deeds. Her late father, George Musgrove Vann, was raised in Cherokee country, participated in tribal culture, and received 110 acres from the federal government as reparation for land stolen from the tribe when it was forcibly removed from Georgia to Oklahoma in 1838 along the Trail of Tears.[38] But when Vann applied for citizenship in the Cherokee Nation in 2001, she received a letter stating her father was ineligible because he was black and not Cherokee. According to Cherokee enrollment rules, members must be directly descended from someone listed on the 1907 Dawes Roll. Instead, her father was a descendant of black slaves of the Cherokee, later known as the Freedmen, who relocated with the tribe to Oklahoma.

An 1866 treaty between the United States and the Cherokee Nation granted the emancipated slaves full tribal citizenship, including voting in tribal elections and benefits granted by the Bureau of Indian Affairs, but federal clerks who came to Oklahoma to register Indians on the Dawes Roll put many blacks on a separate Freedmen list.[39] A constitutional amendment approved by the Cherokee Nation of Oklahoma in 2007 to expel the Freedmen from citizenship is currently under review by U.S. federal and tribal courts.

Contests over Indian authenticity are more fierce than ever because of the huge fortunes at stake. Some tribes have earned millions in annual revenues after Congress passed the Indian Gaming Regulatory Act in 1988, allowing them to build their own casinos. In 1990, the federal government awarded the Seminoles $56 million in compensation for land seized in Florida almost two hundred years ago. The tribe, the first to open a high-stakes bingo hall in 1979, acquired the Hard Rock Café brand for approximately $1 billion in 2004 and has been constructing hotels and casinos under the popular name. Tribes have been able to raise the standard of living of their members by distributing profits through college scholarships, health care, and low-interest loans. The blood quantum rules initially worked against tribal interests by reducing the numerical strength of tribal nations. Now some tribes are embracing them to conserve tribal resources. It is a matter of simple mathematics: per-capita payments divided among tribal members shrink as the number of tribal members increases. So tribes are becoming stricter about defining and enforcing enrollment criteria, even turning to genetic tests.[40]

Genetic ancestry testing is a two-edged sword: on the one hand, it gives tribes a convenient way to screen out pretenders who cannot demonstrate a sufficient amount of Indian ancestry; on the other hand, it offers an increasingly popular means for outsiders to discover an Indian fragment of their lineage and stake a genetic claim to tribal wealth. While many Freedmen have a history of identifying as tribal members, some interlopers take advantage of genetics out of a newfound and purely monetary interest in the tribe.

Consumer Beware!

In chapter 3, I discussed the problems inherent in trying to differentiate major population groups based on allele frequency. These limitations are only intensified when inferred population structure is then applied to individual ancestry. The type of statistical analysis conducted by genomic scientists translates poorly to describing the ancestry of a single individual because they are looking for differences in frequencies, not absolute genetic distinctions between groups. Genetic genealogy companies that make biogeographic designations are guilty of what bioethicist Pilar Ossorio calls the "tendency to transform statistical claims into categorical ones."[41] Because of the huge amount of genetic diversity within groups, two siblings may end up with different racial proportions, depending on the gene variants each one happens to have. Most consumers do not grasp that they are paying for a crude, probabilistic guess rather than a definitive test.

The inadequacy of reference databases, composed of present-day populations whose genetic information is readily available, poses another significant problem. For example, some companies use the publicly available database from the International HapMap's initial phase, which sampled only four populations: Yoruba from Nigeria, Han Chinese, Japanese, and Utah residents of Northern and Western European descent. The computer program forces results into a composite of whatever reference samples the company has. If the company is missing reference samples that match the customer's DNA, the computer will compensate with the closest match—which may be from a population that has nothing to do with the customer's actual ancestry. Companies give erroneous results to any customer whose ancestry includes groups that are not represented in their database.

A 2010 white paper released by the American Society of Human Genetics explains how easily an error can occur:

> Consider genetic ancestry testing performed on an individual we will call Joe, whose eight great-grandparents were from southern Europe. The HapMap populations are used as references for testing Joe's genetic ancestry. The HapMap's European samples consist of "northern" Europeans. In regions of Joe's genome that vary between northern and southern Europeans (such regions might include the lactase gene), the genetic ancestry test using the HapMap reference population is likely to incorrectly assign the ancestry of that portion of the genome to a non-European population because that genomic region will appear to be more similar to the HapMap's Yoruba or Han samples than its (northern) European samples.[42]

When Joe receives the results, he is thus led to believe that he has newfound African or Asian ancestry.

Likewise, if the database contains no samples from East Africa, the computer will designate someone whose ancestors are from that region as part West African and part Western European.[43] Because the HapMap database initially contained no Native American samples, a customer's Native American ancestry was classified as Asian, considered the most genetically similar group. Tests based on databases that do include Native American samples do not account for the genocide of tribes whose genes contributed to current genetic profiles but who cannot be sampled because they no longer exist.[44] One method using an advanced mathematical procedure called "principal-component analysis," when applied to a person whose parents are from geographically disparate origins, will assign an incorrect ancestry that falls somewhere in between. For example, an individual with one East Asian and one European parent looks to the computer like someone from Central Asia.[45]

Although some companies have collected their own unpublished databases, none can guarantee completely accurate inferences. And because firms keep their proprietary data—both their databases and computer algorithms—confidential, it is impossible to verify the reliability of their results. Because both the proprietary databases and statistical models differ, customers who submit their DNA to several companies get a different composite profile from each one. The ancestry estimations seem to confirm that races are biological when all they are really able to do is demonstrate the futility of trying to determine racial identity with genes.

The mtDNA and Y chromosome lineages used by ethnic ancestry companies raise a whole other set of issues. To begin, these lineages cover only one line of descent; they include one female or male ancestor per generation. But every individual has hundreds of ancestors going back just a few centuries and thousands over the course of a millennium.[46] Eight generations alone encompass 256 ancestral lines. Since the time of the slave trade, about fourteen generations ago, any individual has accrued more than 16,000 ancestors. So maternal and paternal lineage tests reflect only a tiny fraction of a customer's entire genetic ancestry. The test may determine that the customer has a distant relative who shares a common ancestor with someone living in Sierra Leone. But this same customer has thousands of other distant relatives who are not accounted for in the mtDNA or Y chromosome test and who may share common ancestors with people from other parts of Africa or beyond. Since each of those ancestors contributed equally to the customer's genetic history, there is no reason to focus identity on only one or two branches of the ancestral tree.

In learning that his mtDNA could be traced back two millennia to Jews living in the Middle East, Steven Pinker felt a "primitive tribal stirring," despite being a "secular, ecumenical Jew." But he quickly put this information in the context of his total ancestry: "my blue eyes remind me not to get carried away with delusions about a Semitic essence. Mitochondrial DNA, and the Y chromosome, do not literally tell you about 'your ancestry' but only half of your ancestry a generation ago, a quarter two generations ago, and so on, shrinking exponentially the further back you go." If you go back far enough, Pinker points out, "there aren't enough ancestors to go around, and everyone's ancestors overlap with everyone else's, and the very concept of personal ancestry becomes meaningless."[47]

How can a DNA test possibly trace the lineage of a modern-day African American back to its precise tribal origins in Africa centuries ago? A major limitation of this type of genetic genealogy is the assumption that ethnic groups have remained genetically discrete and immobile for four centuries. In fact, peoples across Africa have migrated and mixed extensively. Bert Ely, a molecular biologist at University of South Carolina and principal investigator for the African-American Roots Project, a nonprofit group that studies black genealogies, believes that half of the mtDNA in its company databases is shared by at least twenty groups dispersed throughout the continent.[48] He found that DNA from a modern-day person from Ghana, for example, matched groups currently living in more than a dozen African countries.

Because "mtDNA will never have the resolution to specify a country of origin," says Mark Thomas, a geneticist at University College London, "what they are selling is little better than genetic astrology."[49]

This may be why so many customers who buy DNA test kits from several ancestry-tracing companies end up with just as many differing results. This problem was dramatically demonstrated in a 2007 CBS *60 Minutes* segment featuring an African American woman from Harlem named Vy Higgensen who discovered her "new kin" with the help of DNA.[50] When she learns from African Ancestry that she descended from the Mende people of Sierra Leone, she shouts, "I am thrilled! It puts a name, a place, a location, a people!" But her joy soon turns to uncertainty when she receives a second opinion from a company called Relative Genetics informing her that her DNA traces to the Wobe of the Ivory Coast instead. Her confusion mounts when a third company, Trace Genetics, finds a putative match to Senegal's Mendenka tribe. While doing research for an article on commercial genetic genealogy, journalist Ron Nixon had his own mtDNA tested by several companies.[51] Every company identified his maternal ancestors as African, but while African Ancestry said they were from the Mende and Kru in Liberia, Family Tree DNA traced them to a host of ethnic populations, including the Songhai and Bambara in Mali, various groups in Mozambique and Angola, and the Fulani, who live in eight African countries. The African-American Roots Project and DNA Tribes only added to the ethnic smorgasbord.

Changes in the gene pool of African ethnic groups also mean that the genetic profiles of present-day Africans contained in company databases are not identical to the profiles of these groups during the time of the slave trade. It has never made sense to define these groups as discrete genetic entities at *any* point in history. Ethnic groups like the Yoruba and Mende are not natural groupings of genetically related people; they were created by geopolitical forces that arose out of European colonialism. "Slaves from these cultural groups were known by the name of the particular kingdoms or sub-cultural groups to which they belonged, for example Effon/Ekiti, Ijesa, Ife, Ijebu, Egba, and Oyo," writes Charles Rotimi.[52] So while Kittles boasts that African genetic genealogy appreciates the diversity among African ethnic groups, it fails to appreciate the cultural diversity within large populations like the Yoruba, Mende, Hausas, and Fulani.

Kittles the scientist acknowledges all of these limitations in an article he published in 2004 with Mark Shriver, his frequent collaborator at Penn

State.[53] I asked him how he could justify handing out certificates of ancestry when DNA technology cannot really pinpoint the location or ethnicity of anyone's ancestors at the time of the slave trade. He argued that, although he analyzes only a tiny fraction of ancestry, "that small fraction has so much power in telling you a little bit about your ancestry." He explained, "So it's like this: you are in a dark room, and the door has been cracked, and some light is coming in, and that's what this is providing. The value is knowing something, versus knowing nothing." To Kittles, providing this light on Africans' diverse heritage is far more honest than companies that "take African Americans' money to tell them their ancestors are from Africa." "I try not to oversell it," Kittles told me. "I try not to say this is going to be the rapture." But to view the rejoicing of some of his customers, their interpretation of the test results seems to come pretty close.

The ethics of the marketplace have skewed the ethics of scientists who give their imprimatur to companies' misleading claims.[54] Financial incentives pressure scientists to inflate company assertions about what genetic testing can reveal as well as its importance in defining identity. As we saw with marketing race-based pharmaceuticals, turning racial identity into a commercial product raises the monetary stakes for scientists who stand to benefit from "paid consultancies, patent rights, licensing agreements, stock options, direct stock grants, corporate board memberships, scientific advisory board memberships, media attention, lecture fees, and/or research support," notes a 2007 *Science* article.[55] In a candid moment, Kittles told me that, at one point, he thought about "getting out of this whole thing" because of the marketing aspect. "It's not about science anymore," he said. "It's about business, and business has no scruples."

Making Racial Claims

The implications of genetic genealogy are far graver than the elation or disappointment a customer might feel after learning about her tribal lineage or racial origins. It is even more significant than consumer fraud. I believe that the explosion in genetic ancestry testing is perpetuating a false understanding of individual and collective racial identities that can have widespread repercussions for our society. Genetic genealogy has tremendous power to influence the way we define race, determine who belongs to various racial groupings, and understand racial connections. Test results are being used not only as a means to explore personal identity, but also as a basis for claiming membership in racial groups in order to qualify for government benefits and entitlements. Although genetic ancestry testing companies are fond of calling their service "recreational genomics," their impact on relationships of power can be quite serious.

Matt and Andrew Moldawer always considered themselves to be white, but they decided to undergo genetic ancestry testing when they applied to college. The results allocated the twins' racial composition as 9 percent Native American and 11 percent northern African, in addition to their European heritage. Their father saw this as useful information for obtaining more financial aid. "Naturally when you're applying to college you're looking at how your genetic status might help you," he said.[56] *New York Times* journalist Amy Harmon reports that the Moldawers' racial conversion is not exceptional. "Prospective employees with white skin are using tests to apply as minority candidates, while some with black skin are citing their European ancestry in claiming inheritance rights," she writes. At one point DNA Print Genomics encouraged viewers of its Web site to buy the test "whether your goal is to validate your eligibility for race-based college admission or government entitlements." Harmon also cites a Christian, John Haedrich, who used test results showing his Jewish genetic ancestry to sue for Israeli citizenship without converting to Judaism.[57]

The genetic definition of race strips it of its political meaning. Applying for benefits under a genetic alias distorts the purpose of affirmative action policies designed to remedy institutionalized racism. Although the Moldawers had experienced the privileges of being white their entire lives, they wanted to use a genetic definition of race to obtain benefits meant to compensate minorities for disadvantages they endured because of their race. In response to a letter from someone whose brother claimed to be "part Indian" to get college grants, the *New York Times*'s "Ethicist" columnist Randy Cohen replies, "If a DNA test had shown your brother to be 20 percent Cherokee, so what? He did not live as a Cherokee. His Cherokee forebears did not affect his behavior. And so he ought not check the Cherokee box on any application."[58] Whites who discover for the first time after taking a DNA test that they have a tiny fraction of African, Asian, or Native American ancestry should not be entitled to claim any benefit because they have not experienced the racial disadvantages that affirmative action redresses. Their race has not changed on account of a genetic test.

Harmon also cites the case of Pearl Duncan, a black Manhattan writer, who established with archival records that she descended from a Scottish slave owner in Jamaica who was her mother's great-great-grandfather. She confirmed that she had 10 percent British ancestry with a DNA test. With this evidence in hand, Duncan contacted her Scottish cousins to request one of the family's eleven castles acquired with the fortune made from her African ancestors' forced labor.[59] Unlike the Moldawers, Duncan is not seeking an entitlement based on her race. She is insisting on property she is owed because of her family lineage and that was denied her because of her race. Rather than categorizing her by race, the genetic test showing British ancestry is supplemental evidence of a more important family lineage tracing back to the Scottish slaveholder in Jamaica.

African Americans have employed genetic ancestry technologies to produce evidentiary support for similar legal claims in court cases seeking reparations for slavery. Genetic testing features prominently, for example, in a federal lawsuit filed by eight African Americans against Lloyd's of London, FleetBoston Financial Corp., and R.J. Reynolds Tobacco Company for violating international law by committing acts of genocide against African people. A federal judge in Illinois had previously dismissed another slavery reparations lawsuit in part because the plaintiffs failed to allege any direct ties between themselves or their ancestors to the defendant corporations or their predecessors.[60] To establish their connection to the slave trade, these plaintiffs offered DNA evidence tracing their genealogies to regions in Africa that produced the people whose enslavement profited the companies. One of the plaintiffs, reparations activist Deadria Farmer-Paellmann, asserted that genetic testing showed that her ancestors were members of the Mende tribe in Sierra Leone who had been "kidnapped, tortured, and shipped in chains to the United States."[61] In these cases, genetic technology is used by African Americans to strengthen claims for racial justice within a legal framework set by the courts.

In addition to supporting legal claims with DNA evidence, black Americans are incorporating genetic ancestry into their racial identities and everyday institutions. In a video on African Ancestry's Web site, Mark Thompson, a radio host and community leader in Washington, D.C., explains that, for African Americans, genetic genealogy is "not just a business, it's not just a tremendous personal experience that is cathartic, but it is also very political because we were kidnapped, and we have the right to know the homes that we were kidnapped from." Thompson describes the current interest in making a genetic connection with Africa as following the tradition of W.E.B. Du Bois, who made "a real effort to link the African American struggle with the struggle of African people worldwide."[62]

Thompson also sees the trend as a way of contesting the negative messages about blackness and Africa that abound in dominant American culture. "Our images of Africa were the Tarzan movies. Our image of Egypt was Elizabeth Taylor. Moses was Charleton Heston," he says. "In this day and time, especially when all of these young people are besieged through the media saturation of negative images, why not have our young people know that they came from Africa?" The Association for the Study of African American Life and History (ASALH), founded by Carter G. Woodson in 1915, partnered with African Ancestry to instill in young African Americans "a real sense of pride that we can connect to the great kings and queens of Africa, to the great accomplishments and civilizations and scholarship in Africa," says Silvia Cyrus, ASALH's executive director. "Our young people need to know that they are connected to the greatest continent on this planet." For his part, Henry Louis Gates Jr. developed an ancestry-based curriculum for public school children that centers on studying their own DNA. "My plan," he announced, "is to revolutionize the way we teach history and science to inner-city black and minority kids."[63] Gates donates a portion of the profits earned by his genetic genealogy company to the Inkwell Foundation, dedicated to transforming inner-city education by teaching genetics and ancestry tracing.

Many African Americans seem less interested in finding an ancestral link to actual relatives who lived three hundred years ago than to making connections with people living in Africa right now. In her research, Alondra Nelson emphasizes that blacks deploy ancestry testing as part of a broader cultural project by which they seek to reconcile themselves to the destructive legacy of slavery. Rather than base their identity solely on genetic data, they treat test results as a resource that they incorporate into a more complicated process of "affiliative self-fashioning," translating them "from the biological to the biographical."[64] The way African Americans creatively incorporate genetic genealogy into their lives is a far cry from a reductionist notion that genetics determines one's race or one's fate. They embed test results in the family history they have already begun to construct, interpret them to fit their political and spiritual viewpoints, and integrate them into their collective customs. The work of constructing an identity rooted in African ethnicity starts with the "Certificate of African Ancestry"; it is not determined by it.

Yet despite these extra-genetic dimensions, treating genetic genealogy as the linchpin of identity and affiliation helps to reinforce the emerging understanding of citizenship rooted in biological sameness and difference. Educating blacks, as well as other Americans, about the history, politics, and cultures of Africa is an important antidote to the widespread ignorance that exists about the continent. But there is no reason why instilling a greater appreciation for Africa should hinge on genetic kinship. African Americans can choose a country in Africa to learn about, visit, and support based on any number of factors. Some love the music of Mali or the artwork of Ghana. Some are inspired by the South African people's victory over apartheid. Still others are intrigued by the ancient kingdoms of Kush, Hausaland, or Zimbabwe. These seem to me to be better reasons to embrace a group in Africa than the results of a DNA test that are not even guaranteed to be accurate. I want to know what inspired W.E.B. Du Bois, at age ninety-three, to travel to Ghana at the invitation of President Kwame Nkrumah to edit the *Encyclopedia Africana*, become a Ghanaian citizen, and live in his adopted African homeland until his death in 1963, despite never tracing his genetic lineage there. His political vision, which cannot be replaced by a DNA test, might serve as a basis for connection to contemporary African nations.

I know what it means to feel deep affiliation with an African nation. My mother left Jamaica in her twenties to teach in Liberia and was a Liberian citizen when I was born. She and my father traveled to Liberia when I was three months old, and I spent the first part of my childhood there. My two sisters were born there, and one chose to return to help the nation rebuild from its devastating civil war, lasting from 1989 to 2003. She now calls Liberia her home. A group of women in my church raises funds to support her work there. Our commitment to the people of Liberia does not depend on having a genetic connection to them. It is a separate issue for African Americans to find their actual ancestors. Unfortunately, the traces of lost relatives irreparably erased by the slave trade cannot be retrieved by any amount or refinement of DNA technology. Instead, African Americans are paying for a false sense of connection to a contemporary ethnic group in Africa, a connection that could be established through other, more authentic means.

Not only does ancestry testing create ties to ethnic groups on the African continent, it also is a way to cement black community ties here in the United States. One customer wrote to Kittles, "I found out that I am blood related to Isaiah Washington, Maya Angelou, Andrew Young, Ethan Thomas and Rev. Jesse Jackson, who all are descendants of the Mende people of Sierra Leone." Genetic genealogy is becoming incorporated into traditional black institutions and customs as cause for collective celebration. For example, African Ancestry partnered with Mt. Ennon Baptist Church, a large black church outside Washington, D.C., for a whole series of genetic events during the month of February 2008. The pastor and his wife launched the program by revealing their ancestral roots during church service. Then "Where Are You From?" workshops were held during Bible study and in the chapel following each service during the month. The campaign culminated with a Community Testing Day when the entire congregation was offered ancestry testing at a special price and could get their cheeks swabbed right in the church building. The following Sunday, African Ancestry invited them to "receive your ancestry results and connect with your friends and family in a whole new way during the Church Anniversary Celebration."[65]

Relying on a genetic tie to Africa to bond American blacks ignores black people's distinctive collective experience of creatively resisting racial oppression in the United States. By the turn of the twentieth century, black Americans had developed a sense of political unity apart from biological bonds or differences that laid the foundation for later civil rights struggles.[66] "Black nationhood is not rooted in territoriality," writes John Gwaltney, "so much as it is in a profound belief in the fitness of black culture and in the solidarity born of a transgenerational detestation of our subordination."[67] Blacks have escaped the constraints of racial ideology by defining themselves apart from inherited traits, understanding group membership and solidarity as a political affiliation. Whites defined enslaved Africans as a biological race. Blacks in America have historically resisted this racial ideology by defining themselves as a political group. African Americans gauge blackness primarily by a commitment to black people, not via a biological test.[68]

Our common struggle to end racism and improve what Michael Dawson calls our "linked fate" in America is a more historical and morally legitimate source of black solidarity. It is that political unity that still fills African Americans with feelings of pride and connection when they gather to sing "Lift Every Voice and Sing," known as the "Black National Anthem," penned by James Weldon Johnson in 1900: "Sing a song full of faith that the dark past has taught us / Sing a song full of hope that the present has brought us / Facing the rising sun of our new day begun / Let us march on til victory is won."[69] The shared march toward victory over racism has been the defining feature of black solidarity.

Perhaps ancestry tracing is a high-tech tool for creating black solidarity that modernizes the fight for racial equality with genetic science. But defining identity in genetic terms creates a biological essentialism that is antithetical to the shared political values that should form the basis for unity. In his definitive book *We Who Are Dark: The Philosophical Foundations of Black Solidarity*, Tommie Shelby, a philosopher at Harvard, argues for disentangling the call for an emancipatory black solidarity from the call for a collective black identity. He makes the critical distinction between a "black solidarity based on the common experience of antiblack racism and the joint commitment to bringing it to an end" versus "a form of black unity that emphasizes the need to positively affirm a 'racial,' ethnic, cultural, or national identity," which he believes impedes the collective struggle for racial justice.[70] Genetic genealogy is being used by some African Americans, as well as people from other groups, as the basis for a collective identity that lacks the shared political values and goals needed to fight racial oppression.

I also like the distinction between ancestry and heritage made by Christopher Rabb, founder of the social media outlet Afro-Netizen, at a Chicago discussion on black genealogy I attended. "Ancestry is who you are; heritage is who you choose to become," he said. Although our choice of identity is constrained by our ancestry, we have considerable freedom to decide how much importance to give our genetics, family history, and social relationships. I know, as do most African Americans, that my ancestry traces to Africa and Europe. Although I could easily track down my father's Welsh and German forebears, I have never felt the slightest inclination to do so. In fact, knowing my European ancestors frankly seems irrelevant to my identity. My father and I had a very close relationship, which has had a profound influence on my life; he was an anthropology professor who believed deeply in the common humanity of all people, and his spirit hovers over me as I write this book. But as a young girl, I chose a heritage rooted in the struggles of black people around the world to defeat racism and demand treatment as equal human beings. African Americans do not need to take a genetic test to claim that rich heritage. Nor should we displace the political unity it provides with genetic lineages to separate tribes in Africa.

Kimberly TallBear takes a stand on genetic testing by Indian tribes that parallels mine on African ancestry testing. "Native American ancestry DNA should never inform enrollment policy," she declares. "To do so would attack the very historical and political foundation upon which contemporary tribal governance and land rights are based." Although she finds that parentage tests might occasionally be useful in rare cases where biological parentage is uncertain, they should not be given more weight than other types of evidence, such as the testimony of enrolled relatives. It is a mistake to allow "techno-scientific knowledge" to supplant the cultural, social, and political factors that tribes have historically used to determine tribal membership. Genetic tests cannot resolve the "philosophical and political disagreements within tribes about who should count as a Dakota or Pequot or Cree," she writes.[71] I similarly believe that genes cannot replace—and reliance on them may even threaten—the philosophical and political foundations of black solidarity.

TallBear also warns that the shift in the meaning of Native American identity from tribe to race has dire implications for federal obligations to tribes. If claims to tribal lands and self-governance are determined by the absence or presence of DNA,

> anti-tribal interests will have strong ammunition to use against tribes whom they already view as beneficiaries not of treaty payments, but of special race-based rights. Second, groups without historical-colonial relationships, heretofore racially identified as other than Native American, may increasingly claim tribal authority and land based on DNA. . . . Thus, DNA markers—when there is something tangible to gain—may be used to legitimate claims that contradict and potentially contravene prior tribal claims based on historical treaties, law, and policy, even if the groups that use Native American DNA analysis do not intend to undermine existing tribal claims and law.

I do not regard my genetic makeup as important to my identity and was initially puzzled that so many of my brothers and sisters treat genetic ancestry as critical to theirs. In consciously crafting my identity as a black woman, I deliberately downplayed genetics to affiliate with African Americans whose ancestors were enslaved in the U.S. South, though none of my ancestors were. Working on this book helped me to understand the emotional connections that ancestry testing creates for many African Americans and to appreciate the creative ways they have incorporated it into their lives. But I believe the search for identity and political consciousness in genetics is not only misguided but dangerous during the rise of a new racial science that is establishing a genetic definition of race. These technologies promote a racial identification that depends more on common biology than on the common struggle for social justice. It is now more critical than ever to recognize the difference.

PART IV

The New Biopolitics of Race

11

Genetic Surveillance

In September 2009, the United Kingdom Border Agency began a pilot program that uses DNA to check the nationality of asylum seekers. The Human Provenance Pilot Program verifies the nationality of people who said they had fled war-torn Somalia by asking them to submit to genetic screening. Immigration officials want to make sure the asylum seekers are not actually from another country, such as Kenya. Human rights advocates and scientists protested the program, pointing out that it conflated ancestry and nationality and ignored the possibility that people with origins in one country might have moved to another. British geneticist Sir Alec Jeffreys, who pioneered human DNA profiling, is one of the program's chief critics, charging the Border Agency with "making huge and unwarranted assumptions about population structure in Africa."[1]

If the British use of ancestry testing to determine nationality is alarming, consider how some law enforcement agencies in the United States are using these same tests to predict the race of criminal suspects. The Florida-based company DNAPrint Genomics developed a technology called DNA Witness for use by law enforcement officers to determine "genetic heritage" from DNA samples left at a crime scene.[2] DNA is typically used by the police to match a crime scene specimen to the genetic profile of a known suspect. But instead of confirming the identity of a perpetrator the way a fingerprint would, the new DNA phenotyping *predicts the race* of an unknown suspect. Using the same techniques as for ancestry testing, DNA Witness reports the percentage of genetic makeup among four population groups—sub-Saharan African, Native American, East Asian, and European. If the test shows more than 50 percent European ancestry, Euro-Witness further refines the admixture to Northwest European, Southeastern European, Middle Eastern, and South Asian. Its objective is to narrow the potential suspect pool to a more focused group of likely candidates. In other words, by identifying the race of the person who left the crime scene sample, police can concentrate on suspects who belong to the same race. The company reported it has sold its product to medical examiners' offices, special task forces, sheriffs' departments, and district attorneys' offices in cities around the country, including New York, Los Angeles, and Chicago.[3]

DNA Witness claims to provide a "molecular" eyewitness to a crime that is more scientific than human recollection "because it is a quantitative analysis of inherited genetic markers." As a testament to its powers of identification, DNAPrint points to the case of Derrick Todd Lee, an African American convicted of rape and murder in Baton Rouge, Louisiana. In the early 2000s, a serial killer responsible for the murders of seven women was on the loose in the Baton Rouge area. A multi-agency homicide task force had initially relied on eyewitness testimony and FBI profiling to construct a "Caucasian" sketch of the suspect. But when DNAPrint was hired to test DNA evidence left at some of the crime scenes, its analysis determined instead that the suspect had 85 percent sub-Saharan and 15 percent Native American ancestry. The task force then shifted its focus to include Lee, a known sex offender, and eventually charged him with the crimes. Based on a DNA match, Lee was convicted of two murders and sentenced to death.

The Derrick Todd Lee case may seem like airtight validation of genetic ancestry testing as a tool for identifying criminal suspects. But whether used for recreational or law enforcement purposes, this procedure is subject to all the scientific flaws I have discussed in previous chapters—confusing probabilistic population estimates with definitive individual identification, inadequate reference databases, errors made by computer statistical models, and the fundamental flaw of equating genetic ancestry with race. The racial categories police use to identify people are social ones that have no definitive biological indicators and correspond poorly with patterns of identifiable genetic variation.[4] The Baton Rouge police were lucky that Lee's appearance matched the racial expectations in our society for a person who has African and Native American ancestry. Many people are surprised when they learn their ancestry from DTC genetic genealogy companies because the test results don't match their racial identity. There is no scientific test to discern what a person who is part African, part European, part Native American, and/or part East Asian will look like or what racial category he or she will fit into.

Once a suspect is located, though, DNA testing can confirm a match to the crime scene evidence—so what's the harm? For one thing, predicting the race of a suspect based on DNA gives the police added license to round up people identified as belonging to that race in order to isolate the perpetrator with further DNA testing. Police around the country engage in "DNA dragnets" that collect samples from large numbers of individuals who fit a loose physical description, such as "tall black male," but do not otherwise warrant suspicion. For example, in 1994, police in Ann Arbor, Michigan, asked more than six hundred African American men to submit genetic samples during the search for a serial rapist. The police chief warned that anyone who failed to "volunteer" DNA automatically became a suspect.[5] DNA phenotyping gives scientific cover to racial profiling in the form of DNA dragnets. As bioethicist Pilar Ossorio points out, law enforcement officers will regard phenotyping technologies as more useful when they predict a suspect from a minority population simply because a minority profile more drastically narrows the suspect pool.[6] Although racial dragnets already occur, basing them in DNA forensic technology gives them extra legitimacy, despite the scientific flaws that plague racial phenotyping.

Let's imagine that, after narrowing the suspect pool with racial phenotyping, the police find someone whose DNA matches the crime scene sample. The police have now added race as additional proof against the suspect: they have already assumed that the race of the suspect is the same as the race of the perpetrator. This assumption will be reinforced if the case goes to trial, when prosecutors typically state "random-match probability"—the odds that DNA from a randomly selected person (i.e., someone other than the defendant) has the same genetic profile as the crime scene DNA—in terms of race.[7] For example, in a leading California case involving a black man charged with first degree murder and attempted rape, the prosecutor stated the random match probability as 1 of 96 billion Caucasians, 1 of 180 billion Hispanics, and 1 of 340 billion African Americans.[8] Of course, all of these odds are greater than the world's population, so race adds no probative value to the DNA evidence. Used in criminal investigations and trials, racial identification technologies take the myth that race is a genetic category to a more sinister level. They encourage the government to categorize people into biological races in order to identify criminal suspects.

The Surveillance State

As racial phenotyping illustrates, the use of DNA by law enforcement has expanded beyond the confirmation of a particular suspect's involvement in a crime. There is yet another form of genetic surveillance being conducted by government agents that threatens to envelop whole communities. In California, anyone arrested for a felony, from murder to shoplifting to writing a bad check, can be compelled by state law to provide a DNA sample—without ever being convicted of a crime, without a warrant or court order, without even providing probable cause to suspect the person actually committed a crime. There is no right to request an attorney first. The person subjected to the genetic seizure may be completely innocent. All that is required is that a police officer decides to make an arrest. The genetic profile is run through state and federal databases and compared against DNA collected in unsolved crimes. If no match turns up, the person arrested may be released, with no criminal charges ever brought. But this innocent person has become a permanent suspect: his or her genetic information remains in the government databases for future screening for criminal involvement.[9] Most of the DNA forcibly retrieved, analyzed, and stored by state agents in this unprecedented exercise of power comes from blacks and Latinos.[10]

California's decision to take DNA upon arrest is part of a rapidly expanding collection of genetic information by federal and state governments for law enforcement purposes. With 8 million offender samples, the U.S. federal government has stockpiled the largest database of DNA seized from its citizens of any country in the world.[11] Because of rampant racial bias in arrests and convictions, the government DNA databases being amassed nationwide effectively constitute another race-based biotechnology emerging from genetic science. Unlike the voluntary technologies discussed in prior chapters, which claim to help people cure their diseases, improve their children, and find their identities, forensic DNA repositories are gathered by the state without consent and are maintained for the purpose of implicating people in crimes. They signal the potential use of genetic technologies to reinforce the racial order not only by incorporating a biological definition of race but also by imposing genetic regulation on the basis of race.

Genetic testing was first introduced as a type of supplemental evidence to help convict criminal suspects by comparing their DNA to crime scene samples. Data banking extended the purpose of DNA from confirming the guilt or innocence of particular suspects to detecting unknown suspects from crime scene evidence. The theory was that law enforcement officers could catch repeat offenders by running genetic information gleaned from semen, blood, saliva, or hair left by the perpetrator through a database containing genetic profiles of prior lawbreakers. The DNA profiles function as "genetic fingerprints" that can help match the crime scene sample with one in the DNA database. A match—called a "cold hit"—might save police months of investigation or help them catch a criminal who would have otherwise eluded detection. Initially, DNA was collected only from violent felons and sex offenders on the theory that they were the most likely to commit crimes again and to leave genetic evidence at the scene. The heinous nature of their crimes justified the state's interference in their privacy.

But the net has widened drastically in the last two decades. Not only have the categories of people subject to DNA seizure increased, but also the government's use of the banked DNA has widened. Throughout the 1990s, Congress dramatically enhanced the federal government's authority to collect, analyze, and permanently store DNA samples. The DNA Identification Act of 1994 provided funding for law enforcement agencies to amass DNA into a giant federal repository, the FBI's Combined DNA Index Systems (CODIS). Congress passed the Antiterrorism and Effective Death Penalty Act of 1996 and the Crime Identification Technology Act of 1998 to allocate federal funds for developing and upgrading DNA collection procedures. The federal DNA databank contains not only samples gathered by federal agents but also genetic profiles submitted by state law enforcement agencies to the FBI. All states, in turn, have access to the CODIS computerized data. By linking federal and state databases, law enforcement officers around the country can conduct interstate investigations, matching DNA evidence to suspects at the local, state, and national levels.

In the last decade, Congress passed a series of laws that gradually cast the federal DNA net even farther. The USA Patriot Act, passed in 2001 in the wake of the 9/11 attacks, extended the scope of federal DNA collection to terrorism-related crimes. Then came the Justice for All Act of 2004, which widened the reach to all federal felonies and any additional crimes of violence or sexual abuse. On January 5, 2006, President George W. Bush signed into law a stunning extension of government power, without anyone apparently noticing. Buried in the pages of the popular Violence Against Women Act reauthorization bill, the DNA Fingerprint Act of 2005 authorizes U.S. agents to take and store DNA from anyone they arrest or detain and permits CODIS to retain profiles from arrestees submitted by the states that collected their DNA. This includes citizens who have not been convicted or charged with any crime and immigrants detained on suspicion of Immigration and Naturalization Services violations.[12]

A similar escalation has been taking place at the state level. All fifty states now extract DNA from at least some classes of offenders and send it to the CODIS system. Forty-seven take a sample from anyone convicted of a felony, and some states include misdemeanor offenders. In eighteen states, people who are only arrested are forced to submit DNA, even if they are never convicted of a crime. Thirty-five states have extended their genetic collection law to children. Twenty-nine states retain the actual DNA samples and not just the genetic profiles derived from them, allowing investigators to mine them for additional genetic information in the future. Half of states allow the stored DNA samples to be used for purposes other than law enforcement, such as biomedical research. Some local law enforcement agencies also maintain their own "offline" databases, which have even less judicial oversight and procedural safeguards than the official state systems. The district attorney's office in Orange County, California, for example, maintains a repository of fifteen thousand DNA samples extracted from criminal suspects using unorthodox methods. When Charlie Wolcott was arrested in May 2009 for allegedly trespassing on railroad property, a deputy district attorney offered to drop the charges if Wolcott agreed to submit a DNA sample and pay a $75 fine.[13]

The state of California has amassed the third-largest DNA database in the world, behind only the U.S. and British governments.[14] In 1998, the California legislature passed the DNA and Forensic Identification Database and Databank Act to permit police to retrieve DNA from anyone, including children, convicted of a felony, sex offense, or arson. By 2004, the California database had grown to 220,000 offender profiles. But law enforcement clamored for a wider net to stockpile even more DNA samples for the database. That year, Proposition 69, the DNA Fingerprint, Unsolved Crime, and Innocence Protection Act, was placed on the ballot. The ballot initiative took a giant leap beyond the current law: it broadened the scope of individuals in the state who are subject to warrantless DNA seizures to anyone, even children, arrested on suspicion of committing any felony.[15] The measure also applies retroactively to authorize collection of DNA from all five hundred thousand Californians in prison or on probation or parole with a felony record. The state's DNA database is expected to mushroom to more than two million samples over the next five years.

There was also strong opposition to the law. A range of advocacy groups, including the California ACLU, the League of Women Voters, the Privacy Rights Clearinghouse, the Children's Defense Fund, and the American Conservative Union, objected to the initiative's inclusion of arrestees for undermining the principle of presumptive innocence, branding children with lifetime suspicion, and wasting taxpayer money on the monumental costs of DNA processing.[16] According to the California Department of Justice, of the approximately 332,000 people arrested for felonies in California in 2007, more than 101,000 were not convicted of any crime.[17] That means that roughly 30 percent of arrestees are subject to DNA seizure without a determination of guilt.

In 2009, the ACLU of Northern California filed a class-action lawsuit charging that Proposition 69 is unconstitutional because it subjects innocent Californians to "a lifetime of genetic surveillance" that constitutes an unreasonable search under the Fourth Amendment.[18] The named plaintiff, Lily Haskell, was arrested at a peace rally in San Francisco and forced to provide a DNA sample even though she was quickly released without being charged with a crime. "When your DNA is taken after an arrest at a political demonstration, it can have a silencing effect on political action," Haskell says. "Now my genetic information is stored indefinitely in a government database, simply because I was exercising my right to speak out."[19] But the ACLU lost its case. A California federal judge ruled that the ACLU failed to show that individual privacy rights outweighed the government's compelling interest in DNA profiling that works to "swiftly and accurately" solve past and present crimes.[20]

In California, the scope of genetic surveillance extends to another category of people never even suspected of a crime. In what is known as "familial searching," investigators question relatives of people whose DNA is stored in the government databank, pressuring them to submit genetic samples to avoid being implicated in a crime. People become candidates for inclusion just because they are related to someone who has already been profiled. Familial searching is used when a crime scene sample fails to produce a perfect cold hit, but there is a partial match between it and a DNA profile stored in the database.[21]

Forensic genetic analysis examines the arrangement of alleles located at a selected number of specific points, or loci, on the DNA molecule where the sequence varies greatly among individuals. The sequence at each position is converted to a series of numbers and forms a DNA profile. The FBI standard is to use thirteen genomic locations to create a profile. If two samples match at all thirteen loci, the odds are extremely high that both came from the same person. If all thirteen loci do not match, police sometimes conduct a "low stringency" search that looks at matches of eight to twelve loci. A partial match of fewer than thirteen loci by someone in the database suggests that a relative of that person might fully match all thirteen loci and therefore be the person who left DNA evidence at the crime scene. The police then track down close relatives, such as siblings, parents, or children of the partial matches and ask them to provide a cheek swab to compare with the crime scene evidence. Although submitting a sample is voluntary, refusing to submit one looks suspicious. Colorado and New York have also approved these DNA dragnets based on family relationships.[22]

Familial searching has proven effective in catching perpetrators when a near match to crime scene evidence is found in a DNA database. In 2004, Craig Harman of Surrey County in England became the first person in the world to be convicted of a crime following identification through this technique. Police investigating the death of a motorist who was killed when a drunken man hurled a brick through his car windshield drew a blank when they ran DNA found on the brick through the national UK database. But they tracked down Harman after his brother's DNA profile in the government database partially matched the blood left on the brick. Familial searching made U.S. news six years later for helping police capture the notorious Grim Sleeper, a serial killer who had murdered at least ten black women in South Los Angeles since 1985. As in the British case, crime scene DNA produced no perfect match among the genetic profiles stored in the California database. Then the state forensics lab found a partial match with a recently convicted felon named Christopher Franklin, leading them to Christopher's father, Lonnie Franklin Jr., whose matching DNA, drawn from saliva on a discarded slice of pizza, clinched the case. Lonnie Franklin Jr. was charged in July 2010 with ten counts of murder.[23]

These law enforcement successes have to be weighed against the expansion of state surveillance beyond people who have committed or are arrested for crimes to millions of people who are not even suspected of wrongdoing but fall under suspicion simply because they happen to have a relative whose profile is in a database. This suspicion by association seems like a throwback to the old English "corruption of blood," which stripped inheritance rights from the descendants of anyone convicted of a felony and was explicitly abolished by the U.S. Constitution. "The idea of holding people responsible for who they are rather than what they've done could challenge deep American principles of privacy and equality," warns George Washington University law professor Jeffrey Rosen.[24]

Government DNA data banking began as a targeted procedure to assist law enforcement in identifying perpetrators of a narrow set of crimes. It has swelled into a form of state surveillance that ensnares innocent people or petty offenders who have done little or nothing to warrant intrusion into their private lives. Databanks no longer detect suspects—they create suspects from an ever-growing list of categories. Even so, the public shows little alarm about the massive retention of genetic information because the balance between protecting individual privacy and keeping the streets safe seems to fall in favor of more law enforcement. DNA profiling is a far more precise and objective method of identifying suspects compared to less sophisticated law enforcement techniques, such as eyewitness identification or smudged fingerprints found at a crime scene.[25] Far from feeling threatened by this gigantic storehouse of genetic data, many Americans see it as a surefire way of catching criminals and ensuring that only guilty people are convicted of crimes. Storing an innocent person's DNA seems a small price for such a great public good.

Americans are very familiar with news stories about wrongfully convicted people who have been released from prison based on newly tested DNA evidence. The Innocence Project in New York City, the Center on Wrongful Convictions at Northwestern University's School of Law, and other advocacy groups have used DNA to free more than two hundred innocent people who languished in prison for decades, sometimes on death row, for crimes they did not commit. In May 2010, for example, Raymond Towler walked out of the Ohio prison where he had served thirty years of a life sentence for rape after a judge vacated his conviction based on DNA evidence.[26] As is common in wrongful conviction cases, Towler was misidentified by the victim and witnesses. DNA exonerations deservedly make national news and have brought needed public attention to grave flaws in the way criminal justice is dispensed.

But DNA databanks do not always generate such stories of delayed justice. The countless cases where they have either yielded no benefit or produced erroneous identifications received little attention from the media. Nor does the public hear from the thousands of innocent people whose DNA was seized and stored against their will. Most Americans have simply not noticed the creeping expansion of government authority or think it will never affect them. They have not considered the implications of the new form of state surveillance for civil liberties or for social justice.

The Fallacy of DNA Infallibility

Television programs like *CSI* and news reports of people released from prison based on DNA evidence have helped to elevate DNA as the pinnacle of scientific proof. Many Americans believe that a match between a genetic profile in a database and crime scene evidence is airtight proof of guilt. The belief that DNA is infallible weighs heavily in favor of expanding state DNA databases.

But DNA is *not* infallible. The genetic material in government databanks has to be retrieved, transferred, transported, identified, labeled, analyzed, and stored by human hands, and there is opportunity for error at every stage.[27] In 2003, the Houston police department's crime lab was shut down and hundreds of convictions were called into question when an independent audit uncovered widespread problems in the way DNA evidence was handled, including "poor calibration and maintenance of equipment, improper record keeping and a lack of safeguards against contamination of samples."[28] Josiah Sutton, convicted at sixteen, spent nearly five years in prison for a rape he could not have committed because a Houston lab technician mistakenly reported that Sutton's DNA profile matched a semen sample found in the car where the rape occurred. Such problems are not limited to Houston's lab; mishandling of DNA samples has been found at state crime labs throughout the country.[29]

Even simple mistakes can lead to tragic consequences for people whose DNA is in the system. Lazaro Soto Lusson, a twenty-six-year-old jailed at a Las Vegas detention center for an immigration violation, faced life in prison for two sexual assaults he never committed because of a DNA mix-up.[30] It all started when his cellmate accused him of rape and police investigating the allegation took DNA samples from both men. When entering the genetic data into a computer, a clerk at the Las Vegas police crime lab accidentally switched the names on the samples. The mislabeled samples were then run through the Las Vegas database to see whether they matched DNA collected from any unsolved crimes. A match between the DNA erroneously linked to Soto Lusson and two child sexual assaults led prosecutors to charge him with multiple felonies. It was only a week before trial, after Soto Lusson had already spent more than a year in prison, that a DNA expert detected the labeling error and all charges against Soto Lusson were dropped.

False positives, when crime-lab analysts conclude that the crime scene DNA matches a database profile when it really does not, are more commonplace than most people think.[31] As in the Soto Lusson case, false positives often result from human error. People processing DNA samples sometimes contaminate them with someone else's DNA, mislabel them, or enter the wrong information about them in a computer.

False positives may also result from mistakes in interpretation of the genetic data. The DNA samples do not speak for themselves. When DNA is typed, an analyst must interpret a computer-generated graph to determine the alleles located at thirteen specific loci and record the information to create a genetic profile. If an error is made in this analysis, the genetic profile designated for the crime scene sample may be wrong and could be falsely matched with someone in the database. Timothy Durham of Tulsa, Oklahoma, was convicted of raping an eleven-year-old girl and sentenced to 3,200 years in prison on the basis of a misinterpreted DNA test, despite having produced eleven witnesses who placed him at a skeet-shooting competition in Dallas, Texas, at the time of the crime. Postconviction reanalysis of the initial test, which had been conducted by a lab called GeneScreen, revealed that the lab had failed to separate the male and female alleles in the sample, generating an erroneous genotype that implicated Durham. According to Durham's lawyers at the Innocence Project, "The tests by GeneScreen were riddled with quality control problems that, when ignored, turn solid science into junk." Durham served four years in prison, where he was beaten severely by other inmates for being convicted of child molestation, before being exonerated.[32]

Finally, false positives can occur from sheer coincidence. It's possible for the crime scene DNA to match the genetic profile of someone in the database even though the crime scene specimen was left by someone else. This is because the "genetic fingerprints" that crime labs create, containing only thirteen markers (and fewer, in the case of partial DNA profiles), are not absolutely unique.[33] While running samples through the state database in 2001, Arizona crime-lab analyst Kathryn Troyer happened upon two unrelated felons, one black and one white, whose genetic profiles matched at nine of the thirteen loci.[34] Over the years, Troyer noticed dozens of similar matches that contradicted the astronomical odds of random matches touted by law enforcement officials and prosecutors. A systematic study of the 65,000 profiles in the Arizona database turned up 122 pairs that matched at nine of thirteen loci. Twenty pairs matched at ten loci. A similar search of 220,000 profiles in the Illinois database ordered by the state supreme court in 2006 discovered 903 pairs matching at nine or more loci.

William Thompson, a criminologist at University of California at Irvine and the leading expert on DNA forensic error, points out another serious danger with cold-hit searches. Defendants are reluctant to tell jurors that the incriminating DNA match arose from a database search because that entails admitting that their profile was in the database, thereby revealing a prior conviction or arrest. But without knowing how the match was discovered, "jurors will assume (incorrectly) that the DNA evidence confirms other evidence that made the defendant the subject of police suspicions and hence will underestimate the likelihood that the defendant could have been incriminated by coincidence." This potential for misinterpretation of a coincidental match "puts innocent people who happen to be included in a database at risk of false conviction," Thompson concludes.[35]

Not only can DNA be inadvertently mislabeled and misinterpreted, it can be deliberately planted or tampered with to throw suspicion on the wrong person. There have already been cases of suspects who interfered with DNA evidence or paid others to take DNA tests to escape prosecution. Law enforcement officers have implicated people in crimes by planting guns, drugs, and other evidence at a crime scene or on a suspect's property, and they can just as easily plant DNA evidence to frame innocent people.[36] Some of the wrongfully convicted people who have been exonerated in recent years were initially convicted with evidence doctored by police officers and prosecutors. Terry Harrington, an African American teenager, was being recruited by Yale for a football scholarship when he was falsely convicted by an all-white jury of killing a retired police officer and sentenced to life without parole. A postconviction investigation revealed that prosecutors deliberately based their case against him on a witness who was a known liar and perjurer while withholding evidence that pointed to a white suspect who was the brother-in-law of the local fire chief. Harrington served twenty-five years of his life sentence before the Iowa Supreme Court overturned his conviction. His lawsuit against the prosecutors, who claimed immunity from suit, went to the U.S. Supreme Court to determine whether there is a constitutional right "not to be framed," but was settled before a decision was issued.[37]

Forensic analysts, too, have "fabricated test results, reported results when no tests were conducted or concealed parts of test results that were favorable to defendants," reports the Innocence Project.[38] In June 2010, the former chief crime scene investigator in Douglas County, Nebraska, David Kofoed, was sent to prison for planting blood evidence that led two cousins to be wrongfully convicted of the shotgun slayings of a rural couple.[39] To bolster the prosecution's case, Kofoed put a speck of blood from one of the murder victims in a car authorities believed was used by the cousins the night of the killings. It was the only evidence that tied the cousins to the double murder. Charges against them were eventually dropped when new evidence pointed to two Wisconsin drifters, who were eventually convicted of the crime.

Why would we expect DNA to eliminate the corruption in law enforcement that led to the framing of these innocent people? To the contrary, it gives dirty lab analysts, police officers, and prosecutors a simpler and more effective tool to use in building a false case against someone they want to see behind bars.

Not Worth the Cost

The public will never know how many innocent people are sitting in prison because they were convicted with faulty DNA evidence. Despite recent DNA exonerations, most of the power to use genetic technology lies in the hands of law enforcement. Even where justice ultimately prevailed, "courts and law enforcement imposed obstacles to conducting DNA testing and then denied relief even after DNA proved innocence," writes Brandon Garrett in a study of the first two hundred people to be exonerated by DNA testing.[40] In 2009, the U.S. Supreme Court ruled that prisoners have no constitutional right to postconviction testing of DNA samples that might prove their innocence.[41] Defendants may have been pressured to plead guilty based on a prosecutor's threat to enter into evidence crime scene DNA that was not theirs but might nevertheless convince a jury of their guilt because of a coincidental or erroneous match to their DNA. The Grammy-winning rap artist Lil Wayne, for example, pleaded guilty to felony attempted weapon possession, for which he received a one-year sentence, when police used an unreliable DNA test based on fewer than thirteen markers to tie him to a gun found on his tour bus in Manhattan in 2007.[42] Although his attorneys could have challenged the DNA profile at trial, Lil Wayne was apparently unwilling to take the chance. Already, prosecutors wield a tremendous amount of power to get innocent people to accept a plea deal rather than risk being falsely convicted or being convicted of a more serious crime. Because juries usually believe DNA is infallible, its availability greatly strengthens a prosecutor's hand because no amount of exculpating evidence may be able to overcome its persuasive power. Erroneous eyewitness testimony is damaging, but erroneous DNA evidence is absolutely devastating.

Against all of these dangers, genetic surveillance has not been proven to make us any safer. As bioethicist George Annas notes, "there is no independent, peer-reviewed study of the overall effectiveness of DNA data banks in solving crimes."[43] In their 2011 book, *Genetic Justice*, Sheldon Krimsky and Tania Simoncelli find, based on a mathematical index, that DNA data banking can be effective when used in connection with violent crimes, but when extended to those involved in petty crimes or innocent individuals, "the marginal benefits of loading names into a national database decline rapidly." They add, "Moreover, there is no evidence that posting people's DNA profiles on a national database will deter or prevent crime."[44] Public opinion is based instead on isolated anecdotes and unsupported assertions by government officials, such as the statement by Senator Jon Kyl (R-Arizona), one of the authors of the 2005 DNA Fingerprint Act, "We know from past experience that collecting DNA at arrest or deportation will prevent rapes and murders that would otherwise be committed."[45]

At the same time, there is evidence that the unbridled expansion of DNA databanks is hindering law enforcement. Crime laboratories throughout the nation are overwhelmed by backlogs of genetic samples that are not being processed in a timely fashion. Even before Proposition 69 went into effect, an emergency report issued by a bipartisan criminal justice panel warned of a backlog of some 160,000 untested DNA samples in California's state lab, causing delays of six months or more. A U.S. Department of Justice audit released in March 2009 reported a nationwide backlog of 600,000 to 700,000 convicted-offender samples, despite more than $1 billion in federal spending on forensics labs.[46] While DNA is being seized from innocent people to fill government databanks, DNA gathered from crime scenes goes unanalyzed. In one California case, a rapist attacked two more victims before the state lab had time to analyze a rape kit containing his DNA. Backlogs increase the risk of errors in DNA processing as lab technicians "are pressured to cut corners" to keep up with the ever-growing influx of samples.[47]

Because we are more likely to learn about cases in which DNA exonerates innocent people rather than incriminates them, or when DNA helps catch someone who is guilty rather than fails to identify a perpetrator, it is hard to evaluate the net gain or loss in terms of criminal justice. But we can avoid the errors entailed in DNA databanks without losing DNA's value for uncovering wrongful convictions or confirming a suspect's guilt. Testing the DNA of someone suspected, charged, or convicted of a crime to see if it matches crime scene evidence does not require expansive government storage of DNA profiles. The usefulness of DNA evidence to confirm the guilt or innocence of a particular suspect is no excuse for using DNA to broaden the categories of people treated as *permanent* suspects by indefinitely maintaining their genetic profiles.

Although DNA testing has shed light on the injustice of false convictions, it cannot solve the underlying problems that lead innocent people to be convicted in the first place. Most wrongful convictions result from deep biases in the criminal justice system that make poor, minority defendants vulnerable to police abuse, misidentification, and inadequate representation. One of the main causes of wrongful convictions is false confessions that were coerced by the police. According to the Innocence Project, "In about 25 percent of DNA exoneration cases, innocent defendants made incriminating statements, delivered outright confessions or pled guilty."[48] Coerced false confessions were factors in fifteen of thirty-three exonerations won by the Center on Wrongful Convictions.[49] In one of the center's defeats, a jury convicted Juan Rivera of the rape and murder of an eleven-year-old girl based on a confession he made after four days of intense interrogation (the final session lasted twenty-six hours without a break)—despite DNA testing that positively *excluded* Rivera as the source of crime scene evidence, including semen recovered from the child.[50] It makes no sense to correct a problem created by law enforcement's abuse of power by handing over even more authority to law enforcement in the form of DNA collection. The way to reduce wrongful convictions is to remove the biases based on race and class that corrupt our criminal justice system. Extending the reach of state surveillance does just the opposite.

These weaknesses in the state's use of DNA data banking as a tool for reducing crime make it harder to justify the resulting breach of individual privacy. We recognize that the government violates our civil liberties if it taps our telephones or secretly searches our homes without court permission. Collecting and storing our DNA is also a serious intrusion into our private lives because DNA is a part of the body; taking it without consent violates our bodily integrity. In addition to this material aspect, DNA contains sensitive personal information that can be used to identify us and our family members and can be matched with other private records, including medical files.[51]

We tolerate the state forcibly extracting highly personal data from people convicted of serious crimes because these offenders have a diminished right to privacy as a result of their antisocial conduct. But as the categories of people who are compelled to submit their DNA broaden, it becomes less clear why the state should have so much power over them. Once compelled DNA collection goes beyond murderers, rapists, and armed robbers, law enforcement's need for suspects' DNA lessens and suspects' right to retain control over their private information strengthens. Although people convicted of heinous crimes may forfeit their claim to privacy, there is no such justification for seizing genetic samples from someone who has, say, forged a check. State agents should have to get informed consent in order to take or test DNA from someone who has not been convicted of a serious crime.

Although U.S. courts have been slow to recognize this threat to civil liberties, in 2008 the European Court of Human Rights unanimously held that UK government storage of DNA for purposes of criminal investigation infringed privacy rights protected by Article 8 of the European Convention.[52] The court was especially troubled by the indefinite retention of genetic information taken from children and adults who were never convicted of a crime, stigmatizing them as if they were convicted criminals. This equation of the innocent and the guilty disregards the presumption of innocence accorded to citizens in a democracy. Massive government collection of DNA transforms the relationship between citizens and their government in ways that contradict basic democratic principles. Government becomes the watchdog of citizens instead of the other way around. Huge segments of the population are perpetually under suspicion although they are guilty of no wrongdoing. Citizens can no longer rely on the state to safeguard their privacy by forgetting their past behavior; evidence about them is stored forever.[53] The state has the authority to take citizens' private property—in this case, their genetic information—without due process. Those are the features of a totalitarian state, not a liberal democracy.

Jim Crow Databases

These violations are exacerbated by the racial inequities that plague every part of the U.S. criminal justice system. The most stunning aspect of this injustice is the mass incarceration of African American men. Radical changes in crime control, drug, and sentencing policies over the last thirty years produced the explosion of the U.S. prison population from 300,000 to 2 million inmates.[54] The United States has the highest rate of incarceration in the world at a magnitude unprecedented in the history of Western democracies. The gap between black and white incarceration rates has increased along with rising inmate numbers. Black men are eight times as likely as white men to be behind bars. One in nine black men aged twenty to thirty-four is in prison or jail. Most people sentenced to prison today are black. Sociologist Loïc Wacquant argues that "[m]ass incarceration is a mischaracterization of what is better termed *hyperincarceration*" because this excessive confinement is finely targeted (at "one particular category, *lower-class African American men trapped in the crumbling ghetto*") rather than spread among the masses.[55]

The targeted imprisonment of black men is translated into the disproportionate storage of their genetic profiles in state and federal databases. We can look to the United Kingdom to gauge the likely racial impact of our own federal database now that it has surpassed theirs in size: 40 percent of all black men and 77 percent of young black men ages fifteen to thirty-five, compared with only 6 percent of white men, were estimated to have genetic profiles in the UK national DNA database in 2006.[56] Stanford bioethicist Hank Greely estimated in 2006 that at least 40 percent of the genetic profiles in the U.S. federal database were from African Americans, although they make up only 13 percent of the national population.[57] Krimsky and Simoncelli arrive at a similar estimate of 41 to 49 percent of CODIS profiles.[58]

The extension of DNA collection by the federal government and a number of states to people who are only arrested (as opposed to charged or convicted) brings many more whites into the system, but it is also on its way to creating a nearly universal database for urban black men. These men are arrested so routinely that upwards of 90 percent would be included in databases if the collection policy is strictly enforced.[59] A controversial Arizona law signed by Governor Jan Brewer in April 2010, giving police broad authority to detain anyone suspected of being in the country illegally, is held up as a model for immigration enforcement policy in other states.[60] When combined with congressional authorization of DNA sampling from all federal detainees, these immigration laws will cause the number of Latino profiles in state and the CODIS systems to skyrocket.

Police routinely consider race in their decision to stop and detain an individual.[61] A *New York Times*/CBS News Poll conducted in July 2008 asked: "Have you ever felt you were stopped by the police just because of your race or ethnic background?" Sixty-six percent of black men said yes, compared to only 9 percent of white men.[62] The U.S. Supreme Court has authorized police to use race in determining whether there is reasonable cause to suspect someone is involved in crime. Legal scholar Michelle Alexander calls the Court's license to discriminate the "dirty little secret of policing."[63] In recent decades, a conservative Supreme Court has eroded the Warren Court's protections against police abuse in ways that promote the arrest of blacks and Latinos—relaxing, for example, the standard for reasonable suspicion—and has blocked legal channels for challenging racial bias on the part of law enforcement.

Numerous studies throughout the country demonstrate that the police engage in rampant racial profiling.[64] There is overwhelming evidence that police officers stop motorists on the basis of race for minor traffic violations, such as failure to signal a lane change, often as a pretext to search the vehicle for drugs. One of the first confirmations was a 1992 *Orlando Sentinel* study of police videotapes that discovered that, while blacks and Latinos represented only 5 percent of drivers on the Florida interstate highway, they comprised nearly 70 percent of drivers pulled over by police and more than 80 percent of those drivers whose cars were searched.[65] A study of police stops on the New Jersey Turnpike similarly found that, although only 15 percent of all motorists were minorities, 42 percent of all stops and 73 percent of all arrests were of black drivers.[66] In Maryland, only 21 percent of drivers along a stretch of I-95 outside of Baltimore were African Americans, Asians, or Latinos, but these groups made up nearly 80 percent of those who were stopped and searched.[67] An Illinois state police drug interdiction program known as Operation Valkyrie targeted a disproportionate number of Latinos, who comprised less than 8 percent of the Illinois population but 30 percent of the drivers stopped by drug interdiction officers for petty traffic offenses.[68]

Police officers also make drug arrests in a racially biased manner. Although whites use drugs in greater numbers than blacks, blacks are far more likely to be arrested for drug offenses—and therefore far more likely to end up in genetic databases. The latest National Survey on Drug Use and Health, released in February 2010, confirms that young blacks aged eighteen to twenty-five years old are less likely to use illegal drugs than the national average.[69] Yet black men are twelve times more likely than white men to be sent to prison on drug charges.[70] This staggering racial disparity results in part from the deliberate decision of police departments to target their drug enforcement efforts on inner-city neighborhoods where people of color live. Indeed, the increase in both the prison population and its racial disparity in recent decades are largely attributable to aggressive street-level enforcement of the drug laws and harsh sentencing of drug offenders.

A crusade of marijuana arrests in New York City in the last decade provides a shocking illustration.[71] Since 1997, the New York Police Department has arrested 430,000 people for possessing tiny amounts of marijuana, usually carried in their pockets. In 2008 alone, the NYPD arrested and jailed 40,300 people for the infraction. Even more alarming is the extreme racial bias shown in whom the police target for arrest. Although U.S. government studies consistently show that young whites smoke marijuana at the highest rates, white New Yorkers are the least likely of any group to be arrested. In 2008, whites made up over 35 percent of the city's population but less than 10 percent of the people arrested for marijuana possession. Instead, the NYPD has concentrated arrests on young blacks and Latinos. Police arrested blacks for marijuana possession at seven times the rate for whites and Latinos at four times the rate for whites.[72]

The racist marijuana policing strategy is based on the routine police practice of stopping, frisking, and intimidating young blacks and Latinos. According to Harry Levine, the City University of New York sociologist who exposed the arrest campaign, "In 2008, the NYPD made more than half a million recorded stop and frisks and an unknown number of unrecorded stops, disproportionately in black, Latino and low-income neighborhoods."[73] That year, the Center for Constitutional Rights, a New York human rights organization, released a report showing that 80 percent of people stopped and frisked by the NYPD were black or Latino. Among those stopped, "45 percent of blacks and Hispanics were frisked, compared with 29 percent of whites, even though white suspects were 70 percent more likely than black suspects to have a weapon."[74] According to the *New York Times*, between January 2006 and March 2010, officers made nearly 52,000 stops in an eight-block area of Brownsville, an African American neighborhood in Brooklyn.[75] The overwhelming majority of these victims of police harassment have done nothing wrong. But when police discover contraband, it is usually a small amount of marijuana. Although New York City is the "marijuana arrest capital of the world," other cities, like Atlanta, Baltimore, Denver, Houston, Los Angeles, Philadelphia, and Phoenix, are also arresting and jailing huge numbers of blacks and Latinos for marijuana possession.[76]

The widespread arrests of young blacks and Latinos for marijuana possession and other petty offenses, such as truancy, skateboarding, and playing loud music, have devastating consequences. A first-time offender who pleads guilty to felony marijuana possession has a permanent criminal record that can block him or her from getting a student loan, a job, a professional license, food stamps, welfare benefits, or public housing.[77] Even if they avoid prison on a first offense, those who are arrested a second time risk a harsh sentence for being a repeat offender. We can now add a lifetime of genetic surveillance to the long list of "collateral consequences" created by discriminatory arrests. The laws permitting police to take DNA from arrestees also give them added incentive to stop and arrest people in order to obtain more profiles for the growing pool of suspects.

With more and more states extending DNA collection to juveniles, these racially skewed arrests also mean that minority children run a greater risk of having a genetic profile in government databases. It is appallingly easy for black and brown children to be arrested in this country. Although only 16 percent of the nation's youth population, black children make up 28 percent of juvenile arrests, 30 percent of youth adjudicated for delinquency charges, 37 percent of youth placed in secure detention, and 58 percent of youth sent to state prison.[78] More than half of black male teenagers in some inner-city high schools are arrested each year.[79] Astonishing proportions of black and Latino youth appear on the police lists of probable gang members in some cities. In Denver and Los Angeles, for example, nearly half of the cities' young black men were marked as suspected gangbangers.[80]

Perfectly law-abiding black children are looked upon with suspicion by police. A group of eleven black and Latino high school students were traveling across New Jersey on a college trip to Howard University in 2009 when they were surrounded by ten police cars and hauled off the bus in handcuffs as police pointed rifles at them.[81] Someone had called 911 to falsely report the presence of a weapon when the group stopped for lunch at a highway exit. Police arrested more than thirty black and Latino teens who were peacefully walking down a Brooklyn street in broad daylight in 2007 to attend a wake for a friend who had been murdered.[82] A New York ACLU class action lawsuit filed in 2010 alleges that New York City children are often arrested by police officers stationed in schools for minor misbehavior like shoving or back-talking. In one case, a twelve-year-old girl was hauled out of class in handcuffs for doodling on her desk.[83] On top of the brutality inflicted on these children, states that allow their DNA to be gathered have added the stigma of making them permanent suspects in government databases.

Racial disparities in DNA databanks make communities of color the most vulnerable to state surveillance and suspicion.[84] The disproportionate odds faced by blacks and Latinos of having their DNA extracted and stored will, in turn, intensify the racial disparities that already exist in the criminal justice system. People whose DNA is in criminal databases have a greater chance of being matched to crime scene evidence. While a guilty person may have no right to complain, that is no excuse for unfairly placing certain racial groups at greater risk of detection. Blacks and Latinos have greater odds of being genetically profiled largely because of discriminatory police practices.[85] Moreover, people whose profiles are entered in DNA databases become subject to a host of errors that can lead to being falsely accused of a crime.

As the federal government and a growing number of states extend the scope of DNA collection to innocent people, they are imposing this unmerited risk primarily on blacks and Latinos.

The problem is not only that all of these harms are visited disproportionately on people of color; the dangers of state databanks are multiplied when applied to blacks and Latinos because these groups are already at a disadvantage when they encounter the criminal justice system. They have fewer resources than whites to challenge abuses and mistakes by law enforcement officers and forensic analysts.[86] They are stereotyped as criminals before any DNA evidence is produced, making them more vulnerable to the myth of DNA infallibility. "The experience of being mistaken for a criminal is almost a rite of passage for African-American men," writes journalist Brent Staples.[87] One of the main tests too many Americans apply to distinguish law-abiding from lawless people is their race.

Many, if not most, Americans believe that black people are prone to violence and make race-based assessments of the danger posed by strangers they encounter. One of the most telling reflections of the presumption of black criminality is biased reporting of crime by white victims and eyewitnesses. Psychological studies show a substantially greater rate of error in cross-racial identifications when the witness is white and the suspect is black. White witnesses disproportionately misidentify blacks because they expect to see black criminals. According to Cornell legal scholar Sheri Lynn Johnson, "This expectation is so strong that whites may observe an interracial scene in which a white person is the aggressor, yet remember the black person as the aggressor."[88]

In numerous carefully staged experiments, social psychologists have documented how people's quick judgments about others' criminal acts are influenced by what is called implicit bias—positive or negative preferences for a social category, such as race or gender, based on unconscious stereotypes and attitudes that people do not even realize they hold. Whites who are trying to figure out a blurred object on a computer screen can identify it as a weapon faster after they are exposed to a black face. Exposure to a white face has the opposite effect. Research subjects playing a video game that simulates encounters with armed and unarmed targets react faster and are more likely to shoot when the target is black.[89] The implicit association between blacks and crime is so powerful that it supersedes reality: it predisposes whites literally to see black people as criminals. Most wrongful convictions occurred after witnesses misidentified the defendant.[90] Databanks filled with DNA extracted from guilty and innocent black men alike will enforce and magnify the very stereotypes of black criminality that led to so many wrongful convictions in the first place.

Consider a 2005 study by Princeton sociologists Devah Pager and Bruce Western finding that whites just released from prison fared better in the New York City job market than blacks with identical résumés but no criminal record. Employers' preference for white offenders over law-abiding blacks shows not only the inequitable opportunities available to blacks but also that blackness itself is seen as a mark of criminality. "Being black in America today," writes Pager, "is just about the same as having a felony conviction in terms of one's chances of finding a job."[91] This stark racial bias in employment puts into practice the racist notion that blacks are meant to labor in prisons and not in decent jobs. Myths of black criminality are so embedded in the white psyche that it seems perfectly natural to many Americans that blacks are disproportionately stopped for traffic infractions, arrested for drug offenses, swept off the streets for "gang loitering," and sent to prison.

Collecting DNA from huge numbers of African Americans who are merely arrested, with no proof of wrongdoing, embeds the sordid myth of black criminality into state policy. As databanks swell with DNA from black people who are arrested or convicted on petty offenses, and as their relatives also come under suspicion in states with familial searching, the government effectively treats every black person in many communities as a criminal suspect. It seemingly also legitimizes the myth that blacks have a genetic propensity to commit crime. The public, who already implicitly associates blacks with violence, may link together research claiming that genes cause gangbanging and aggression to the disproportionate incarceration of African Americans and the banking of their genetic profiles to reach the false conclusion that blacks are more likely to possess these crime-producing traits—or even that most blacks possess them. Americans will become even more indifferent to racial injustice in law enforcement if they are convinced that black people belong behind bars anyway because of their genetic predisposition to crime.

Despite the racial harms, civil rights advocacy groups have done little to challenge the threat posed by government genetic surveillance. Their silence stems from a tension surrounding DNA testing. Although it is serving an unjust criminal justice system, DNA technology is also responsible for the biggest success in criminal justice reform. As of 2010, more than half (151 out of 254) postconviction DNA exonerations involved African Americans.[92] There could not be a better public-relations campaign for DNA than the compelling stories of falsely imprisoned people released after DNA testing. As criminologist Simon Cole notes, "At first glance, post-conviction DNA exonerations appear to be a powerful example of the use of technoscience to offset social inequality."[93] Civil rights organizations that may have ordinarily opposed the expansion of DNA databanks because of their intrusion into communities of color instead embrace DNA technology because of its ability to exonerate victims of the system.

One of the rare times President Barack Obama has addressed a criminal justice issue was in order to endorse expanded law enforcement collection of DNA. In March 2010, the president was interviewed on the tough-on-crime television show *America's Most Wanted* by host John Walsh about the virtues of building a national database with DNA profiles of anyone arrested, regardless of guilt. "We now have eighteen states who are taking DNA upon arrest. England has done it for years. It's no different than fingerprinting or a booking photo," Walsh stated. "It's the right thing to do," Obama replied. Walsh conveniently neglected to mention that the European Court of Human Rights declared the British practice of retaining DNA from people arrested but not convicted a violation of their privacy rights. Agreeing with Walsh that "every one of the states [should] have DNA compliance," Obama declared, "That's how we make sure that we continue to tighten the grip around folks who have perpetrated these crimes."[94] Shortly after the president's appearance, the House of Representatives approved a bill sponsored by Representative Harry Teague, a Democrat from New Mexico, allocating $75 million to states that agree to collect DNA upon arrest and send the genetic profiles to the federal CODIS system. Not a single Democrat voted against the measure.[95]

Champions for racial justice who support expanded DNA databanks or are silent about them make the mistake of embracing DNA technology without analyzing its full role in the criminal justice system. Although DNA testing can correct injustices when used narrowly to confirm a suspect's guilt or innocence, the massive genetic surveillance we are witnessing threatens to reinforce the racial roots of the very injustices that need to be corrected. Wrongful convictions are a symptom of our Jim Crow system of criminal justice, which is systematically biased against blacks and Latinos. Creating a

Jim Crow database filled with their genetic profiles only intensifies this travesty of justice.

The Test

If you believe that the government should forcibly collect DNA from innocent people, I have one question for you: are you willing to voluntarily submit *your* DNA to the police and the FBI? The test for the legitimacy of collecting DNA from anyone arrested should be our willingness to undergo such collection ourselves. If the government has no right to store genetic profiles of everyone, then it has no right to store them from people who have not been found guilty of any crime.

There are good reasons why most people do not want their genetic profiles kept in law enforcement databases. It breaches their privacy and subjects them to the risk of false incrimination. Why, then, do citizens in a liberal democracy tolerate the expansion of state genetic surveillance? I believe the answer to this puzzle lies in the race of the people most threatened by this creeping state power. Legislatures can get away with passing these laws because white voters do not think such legislation threatens their own privacy rights. They want the crime-solving benefits of databases without paying the personal price of having their own genetic information retained by government agents. I suspect that if, after the Columbine shootings, state legislatures proposed laws that gave police authority to retrieve DNA from children who matched the teen killers' demographics—white children from middle-class homes—the public outcry would quickly doom its passage. But this kind of racial profiling has been deemed acceptable as long as it is largely confined to racial minorities.

The escalation of genetic surveillance is part of a broader practice of experimenting with solutions to social problems at the expense of minority group freedom. Arguments that white Americans should relinquish any part of their liberty for the sake of creating a more egalitarian society are renounced as reverse discrimination. Yet proposals to restrict black and brown people's liberties to improve public welfare span the arenas of crime control, immigration policy, and welfare reform. This is why 70 percent of white Americans support the Arizona immigration law passed in 2010, while only 30 percent of Latinos do so.[96]

Criminal law has resolved the tension between liberty and order by protecting the liberty of white citizens while enforcing order against minorities. Georgetown legal scholar David Cole argues that the criminal justice system affirmatively exploits this inequality: "Absent race and class disparities, the privileged among us could not enjoy as much constitutional protection of our liberties as we do; and without those disparities, we could not afford a policy of mass incarceration we have pursued over the past two decades." Government DNA databases that increasingly target people of color similarly exploit America's racial divide, continuing to "sidestep . . . the difficult question of how much constitutional protection we could afford if we were willing to ensure that it was enjoyed equally by all people."[97]

Some officials, scholars, and pundits have called for a universal database that holds genetic information on every resident of the United States. A universal database, promoters argue, would have the dual benefit of catching criminals more effectively while eliminating the racial bias that currently distorts collection procedures.[98] Opponents, such as bioethicist George Annas, rightly protest that such widespread surveillance of citizens would turn a "free country into a 'nation of suspects.'"[99] But black people in America have long belonged to a "nation of suspects," reinforced with the explosion of the prison population since the 1970s and more recently with the escalation of government DNA data banking. This is not a reason to stretch the civil liberties violations to more people. It is reason to be concerned about the police state that already exists in inner-city neighborhoods. The burgeoning government databases of genetic profiles reflect a broader political trend at work in America today. The latest science and technology that has redefined race as a genetic reality could facilitate brutal government policies that threaten disaster for this nation—what I call the new biopolitics of race.

12

Biological Race in a "Postracial" America

Race is central to every aspect of the new science and technology that is emerging from genomic research—computer-generated portraits of the molecular structure of human populations, biomedical studies searching for genetic cures, personalized medicine tailored to each individual's genotype, reproductive technologies for improving children's genetic makeup, genetic genealogy for tracing ancestral roots, and forensic DNA testing that helps law enforcement catch criminals. This science and technology is redefining race as a natural division written in our genes. Yet at the very moment that science, government, and business are promoting race-based genomics, the idea that we are living in a "postracial" America is gaining traction. Race is becoming more significant at the molecular level precisely as it appears less significant in society. On the one hand, scientists claim genetic confirmation of racial categories and treat race as a proxy for genetic difference, leading genetic testing and pharmaceutical companies to push products for race-specific genotypes. On the other hand, many policy-makers endorse a color-blind approach that rejects race-conscious remedies for inequality, while many pundits declare the Obama presidency to be evidence that race no longer affects the U.S. social order.

As the ideology that race is important to genetics but not society is spreading, we are also witnessing the escalation of a particularly brutal form of state control over large numbers of people on the basis of race. The expansion of massive government DNA databases that make millions of poor blacks and Latinos permanent suspects is a vivid example of unprecedented state surveillance of people of color. Government agents treat the men, women, and children of color locked in prison cells and detention centers in this country as less than human, subjecting them to routine abuse, inadequate medical care, and even torture.[1] Scientists, politicians, and corporations are constructing a genetic understanding of race that not only obscures the continuing social significance of race but promotes post–civil rights mechanisms for preserving racial inequality. This pernicious convergence of race-based genomic science and biotechnology at a time of escalating brutality that cuts short the lives of minorities defines the new biopolitics of race in America.

One of the remarkable aspects of racial science and technology today is the support it has from opposing ends of the political spectrum. Both conservatives who espouse a color-blind ideology and liberals who believe in a postracial America have embraced both the science validating racial difference at the genetic level and biotechnological solutions for inequality at the social level. Even some activists who oppose racism have adopted the view that race-based genetics and technology can be used as a tool for alleviating health inequities, building solidarity, and fixing the criminal justice system. Examining the grounds for this bipartisan endorsement of racial science shows the flaws in its logic and the urgent need for a political understanding of race that does not rely on biology.

Conservative Color Blindness and Genetic Race

Color blindness and race consciousness compete as major schemes for determining the proper treatment of race in social policy.[2] In the political arena, advocacy for color-blind policies is based on the assertion that racism has ceased to be the cause of social inequities, while race-conscious policies are promoted as a necessary means for remedying persistent institutional racism. Color blindness emerged as a conservative strategy after the civil rights movement succeeded in toppling the Southern Jim Crow system and forms of de jure segregation in the North. A backlash movement intent on crushing black empowerment and preserving white dominance latched on to the concept of color blindness as an ideological tool of retrenchment. As sociologist Eduardo Bonilla-Silva notes in his classic *Racism Without Racists*, "Much as Jim Crow served as the glue for defending a brutal and overt system of racial oppression in the pre–Civil Rights era, color-blind racism serves today as the ideological armor for a covert and institutionalized system in the post–Civil Rights era."[3]

Whites rationalized the persistent gaps between white and minority health, wealth, and status as products of unbiased market operations, not continuing social injustice. Racial minorities have fallen behind whites on every measure, the theory goes, as a result of their own failings, which make them unable to compete with whites in purportedly fair social, political, and economic arenas. Color-blind ideology posits that because racism no longer impedes minority progress, there is no need for social policies to account for race.

Pretending that the civil rights movement attained perfect equality ignores the lingering effects and systemic incorporation of three centuries of official white supremacy as well as newly minted forms of racial discrimination. Civil rights advocates explicitly argued that eliminating the structural roots of racial inequality required paying attention to race. As U.S. Supreme Court Justice Thurgood Marshall explained, "It is because of a legacy of unequal treatment that we now must permit the institutions of this society to give consideration to race in making decisions about who will hold the positions of influence, affluence, and prestige in America."[4] Color blindness strips the government of any justification for taking affirmative steps to correct past injustices.

Color-blind ideology has been increasingly implemented in government policy by a conservative majority on the U.S. Supreme Court as well as in some state legislatures. California (by voter referendum), Michigan, Nebraska, and Washington banned consideration of race in public employment, education, and contracting; Florida prohibits using race as a factor in state university admissions. A series of Supreme Court decisions in the last two decades struck down race-conscious measures to desegregate schools and workplaces. In *City of Richmond v. J.A. Croson Co.*, decided in 1989, the Court ruled that the former capital of the Confederacy practiced reverse discrimination against whites by adopting a set-aside program to steer some of its construction dollars to minority-owned firms—"even when, without the program, less than one percent of construction contracts went to minorities in a city over 50 percent African American," as legal scholar Ian Haney Lopez pointed out.[5] Justice Clarence Thomas articulated the preposterous perspective that equates official Jim Crow segregation with state efforts to end its legacy, noting "a 'moral and constitutional equivalence' between laws designed to subjugate a race and those that distribute benefits on the basis of race in order to foster some current notion of equality. . . . In each instance, it is racial discrimination pure and simple."[6] In June 2007, the Court continued to dismantle affirmative-action programs in its 5–4 decision striking down voluntary plans to desegregate elementary schools in Seattle, Washington, and Jefferson County, Kentucky.[7] The Court reiterated the position that the Constitution requires the government to be color-blind by paying no attention to race in policy-making. "The way to stop discrimination on the basis of race is to stop discriminating on the basis of race," Chief Justice John Roberts declared.

One of the techniques the Court uses to mask the existence of institutionalized racism is to treat its impact on people of color as a mere statistical correlation. The Court reasons that disadvantages experienced disproportionately by minorities represent a higher statistical chance of harm rather than reflect institutions that systematically discriminate against them. Because people of color just happen to be more frequent victims of neutral policies, this view holds, there is no reason to invalidate the policies for being racially discriminatory. One of the most outrageous applications of this statistical masquerade was the Court's decision in *McCleskey v. Kemp* to reject the claim that Georgia's death penalty was biased against blacks. Despite evidence that Georgia executed blacks who murdered whites at twenty-two times the rate for blacks who killed blacks, the Court concluded that these figures proved "at most . . . a discrepancy that appears to correlate with race."[8] Both majority and dissenting justices recognized that such disparities were so endemic to U.S. criminal justice that making racial discrimination unconstitutional would threaten the entire system. As Justice Brennan wrote in dissent, the reason for *McCleskey*'s holding was "a fear of too much justice."[9] Reducing this systemic bias against blacks to a correlation with race served as an excuse for maintaining a discriminatory institution.

More recently, the Court used this statistical sleight of hand to disguise discrimination against black firefighters in New Haven, Connecticut. In its 2009 decision *Ricci v. DeStefano*, the Court ruled that the city discriminated against its white firefighters when it threw out a written promotional exam that failed to select any blacks. The majority opinion by Justice Anthony Kennedy continuously downplays the discriminatory impact of the test by calling it a "statistical disparity" between whites who were promotable because they passed the test and blacks who were denied the opportunity for promotion because they did poorly. In contrast, the dissenting opinion by Justice Ruth Bader Ginsberg emphasizes the "backdrop of entrenched inequality" in firefighting that New Haven officials were trying to reverse. By casting the test results as merely a statistical disparity, the majority ignored the long-standing institutionalized barriers against black firefighters in New Haven. This statistical erasure of racism goes hand in hand with genomic studies of population structure, pharmacogenomics, and genetic ancestry testing, all of which treat race as a statistical proxy for genetic difference. Like this ploy of conservative color blindness, the new genomic methodology hides the political nature of race by recasting race as nothing more than a probabilistic grouping.

If systemic racism no longer occurs, then color blindness requires some other explanation for the startling racial disparities that continue to mark every socioeconomic, health, and political indicator. As I described in the first part of this book, biological difference was the predominant explanation for racial inequality for several centuries. When eugenics was discredited after World War II, most conservatives developed an alternative explanation based on black cultural depravity and individual irresponsibility. The new racial science based on genetics has rejuvenated the biological rationale in a modern version that does not appear to be racist—and is all the more insidious for it. Some conservative proponents of social color blindness have eagerly embraced the new genetic definition of race as well as genetic explanations of health disparities and racial medicine.

Sally Satel, for example, distinguishes between the proper use of race as a biological category in medicine and the improper use of race as a social category in policy. In a 2004 essay for the Manhattan Institute advocating the FDA's approval of the first race-specific drug, Satel argued that the story of BiDil proved that "race and medicine can mix without prejudice." She writes, "Social race is the phenomenon constructionists have in mind. . . . Biological race, however, is what BiDil's developers are concerned with—that is, race as ancestry."[10] Similarly, journalist Jon Entine writes, "We talk a lot about diversity in the United States, as long as we wink and smile that this diversity is not real, just superficial, a cultural patina. But in some aspects of our humanity, it is very real, and such differences can have huge consequences in everything from sports performance to success in the classroom."[11] According to this view, genetic race is scientific truth while social race is just politically correct ideology. It reverses the understanding of race as a social construction, which recognizes that there is no biological basis for dividing humanity into races but affirms that race is a political reality. Instead, racial differences are seen as real at the molecular level, but merely constructed in society.

With the new distinction between biological and social race, conservatives now have a way to speak about racial difference while maintaining a color-blind approach to social policy. They find it acceptable to refer to race explicitly as long as it has a biological meaning because that use of race is purportedly scientific and unbiased. It is no longer necessary to use code-word proxies for race, such as "welfare" and "illegal immigration." Instead, race can be used outright as a proxy for genetic difference. When racial-justice advocates refer to the political meaning of race, however, it is interpreted as an expression of racism morally equivalent to forms of overt white supremacy. By this view, addressing racist policies that make blacks more vulnerable to imprisonment is a violation of color blindness, but suggesting that blacks are genetically predisposed to crime is simply considering a scientific hypothesis.

Genomic science, conservatives argue, frees us from political correctness so we can act on racial differences in genetics that determine our health. In this ingenious twist of political logic, those who criticize racial biomedicine because of its social impact are seen as interfering with health out of loyalty to racial ideology. These critics are not politicizing science, however. Instead, it is conservatives who are using science to *de*politicize race.

Liberal Postracialism and Genomic Science

When Mark Warner gave the keynote address at the Democratic National Convention in August 2008, he emphasized that an Obama victory would represent America's return as a beacon of science and technology. "In four months we will have an administration that actually believes in science," he declared.[12] As president, Barack Obama quickly fulfilled his inaugural promise to "restore science to its rightful place" by issuing in March 2009 a memorandum of "scientific integrity" to complement his executive order restoring federal funding for embryonic stem cell research. For many liberals, the election of Barack Obama as president represented twin victories: it ushered in an era both of postracial social equality and of respect for science. Obama's pledge to make sure "scientific data is never distorted or concealed to serve a political agenda" was an important repudiation of the ideological influences that had characterized the Bush administration's handling of scientific research.[13] But a naive faith in the ability of science to transcend racism has blinded many liberals to the potential for a genetic definition of race to deepen America's racial divide.

Liberal Americans have bought into the new racial science in part because it is science. Many believe in the inherent progress of science and have faith that scientists conducting research on race and genetics must be advancing knowledge in an objective, rational, and ultimately beneficial way. The majority of scientists who are conducting this research are probably themselves liberals who oppose racism. Those who express blatantly racist views about black genetic inferiority are the exception and are swiftly and publicly excoriated by the scientific establishment, as was the case for both William Shockley and James Watson. The researchers who are developing the new race-based genomics are more oblivious to racism than active proponents of it. Indeed, some are people of color who see their work as furthering the interests of minorities in a field long dominated by whites. When I asked Esteban Burchard what the goal of his genetic research on asthma was, he compared himself to "a feminist back in the eighties" who criticized the NIH for funding studies of heart disease exclusively in men and fought for greater research spending on women's health.

The liberal faith in scientific objectivity has generated an approach to the genetic definition of race that sounds remarkably similar to the conservative one. Like conservatives, liberals separate racial science from racial politics to retain a supposedly scientific concept of race as a genetic category. Liberal scientists erect a wall between their objective study of racial difference in the lab and racial politics at play in the outside world. Burchard is well aware that his research on a race-specific gene for asthma could lend support to claims that blacks and Latinos are biologically inferior. In fact, David Duke, the former wizard of the Ku Klux Klan and a vocal advocate of white supremacy, cited Burchard's research on his Web site. But Burchard sees such aberrational misuses of his findings as no reason to discontinue what he believes is fundamentally sound research. "Scientifically, I do not think we should fear what other folks are going to do, like David Duke and other politicians," Burchard told me when we spoke at his lab in 2008. "They may try to distort our findings to fit their neat little box but that should not deter us from doing good science."

In an article claiming to demonstrate the genetic validity of race, Burchard and co-author Neil Risch criticize the opposing claim, "that there is no biological basis for 'race,'" as merely ideological. "In our view, much of this discussion does not derive from an objective scientific perspective," they wrote dismissively. "This is understandable, given both the historic and current inequities based on perceived racial or ethnic identities, both in the US and around the world, and the resulting sensitivities in such debates."[14] Like Entine and Satel, Burchard and Risch distinguish between "perceived" racial identities that cause controversy at the social level and the "real" racial ancestries they study at the molecular level.

While describing their racial research as objective and free from political bias, Burchard and Risch insinuate that scientists who reject the biological understanding of race are fudging the truth for the sake of political harmony. By dismissing opposing claims as politically motivated, the new racial scientists need not deal seriously with the evidence that refutes a genetic definition of race. Nor do they ever question the extent to which their own scientific methods are shaped by political interests and the unquestioned presumption that biological races exist. Like those who drafted the UNESCO statements, Burchard and Risch argue that past harms that occurred from studying race stemmed from labeling races as superior or inferior rather than from the biological definition of race itself. "Great abuse has occurred in the past with notions of genetic 'superiority' of one particular group over another," they write. "The notion of superiority is not scientific, only political."[15] This view strips the biological concept of race of its political moorings so it can be seen as purely scientific, separate from its political deployment by racists who blatantly espouse notions of racial domination.

The liberal position, then, is to see the concept of biological race as a neutral scientific fact that can be put to good or bad use and to advocate for safeguards against its misuse by racists. Many liberals call for rules mitigating the damage caused when the findings of racial science fall in the wrong hands but see any criticism of the research itself as an impermissible interference in science. "Rather than allowing past mistakes in scientific research and medicine and a legacy of discrimination and exploitation based on race and ethnicity to shackle science," writes law professor Michael Malinowski, "the law should be applied to ensure that research, including race, ethnicity, and ancestry-based population genetics, advances responsibly."[16] The first step in this reasoning is to cast the horrors of racial science as aberrant errors of the past. The next step is to dismiss them as impediments to progress rather than heed them as evidence of a fundamental flaw in racial science.

Liberals' failure to criticize racial science also stems from the political wars during the Bush administration over virtually every scientific issue, from stem cell research to evolutionary theory to climate change. One of the chief complaints about George W. Bush was that he undermined health, education, and rights by "placing ideology ahead of science," as Hillary Clinton and Cecile Richards wrote in a *New York Times* op-ed.[17] Because liberals viewed Bush as fanatically anti-science, opposing him seemed to require an equally fanatical pro-science stance. A striking example is liberal support for the California Stem Cell and Cures Act, creating a constitutionally guaranteed right to conduct stem cell research in the state. In November 2004, California voters approved a landmark ballot initiative, Proposition 71, authorizing the state to allocate $3 billion in public funds for stem cell research in both the public and private sectors. Proposition 71 created "the largest state-sponsored medical research program in U.S. history," according to Richard Hayes, executive director of the Center for Genetics and Society.[18] This unprecedented allocation of public money to the biotech industry relied on an alliance between liberals who support stem cell research and conservatives who advocate state support for private business.[19] Both Silicon Valley venture capitalists and Hollywood celebrities poured millions of dollars into the "Yes on 71" campaign.

California Democrats saw the initiative as a volley against Bush, who had pledged to ban federal funding of research on certain embryonic stem cell lines. "An anti–Prop 71 position has almost become synonymous with Christian conservatism," observed Tali Woodward in the *San Francisco Bay Guardian*.[20] When the *New York Times* ran a story on the battle over the initiative, it noted that funding for opposition to the referendum came from the Catholic Church and religious conservatives.[21] Less attention was paid to a small but passionate alliance of pro-choice progressives who were appalled at the fortune promised to private enterprise, the lack of government oversight to ensure public accountability, and the potential harms to young women who might be pressured to donate eggs for research. Debra Greenfield, a bioethicist who campaigned against Proposition 71, noted that the measure "create[s] a 'right' to conduct stem cell research in the California constitution without a corresponding right to receive the benefits of research."[22]

The progressives who opposed Proposition 71 were frustrated that others were unable to disentangle their support for stem cell research and opposition to Bush from this corporate goldmine. It seems that opposition by liberals to the Bush administration's ideological limits on scientific research led to a similar reluctance to scrutinize the merits of race-based genomic and biomedical research.

Just as being pro-science became a liberal litmus test, critics of scientists doing racial research were accused of being both scientifically and politically backward. A 2008 National Human Genome Research Institute meeting included a tense discussion of a paper by University of Chicago geneticist Bruce Lahn reporting evidence that mutations of two genes that regulate brain growth are more common among Eurasians than sub-Saharan Africans. Celeste Condit, a professor of speech communication at the University of Georgia who studies the social impact of genetic research, charged that the paper seemed to have a "political message embedded" in it: it implied that genes contribute to racial differences in brain size and therefore might explain racial differences in IQ. Lahn lobbed back that Condit was "putting words in [my] mouth," and later accused some scientists as being "almost like creationists" in their refusal to acknowledge evolutionary pressures on brain size.[23]

Similarly, in a 2008 essay in the *Humanist*, Kenneth Krause, an editor of *Secular Nation*, makes a "progressive argument" in favor of acknowledging gene-based racial distinctions. Claiming that recent evolutionary selection produced racial differences in intelligence and disease, Krause writes, "We should never confuse the social construct with the scientific reality. Denial is the least mature and, certainly, the least progressive response to fear."[24] Proponents of racial science portray criticism as an unscientific, Luddite-type stance that plays into a political agenda. They don't see that their unwillingness to consider empirical evidence that contradicts the faith in biological races stems from its own form of political correctness.

Genes, Race, and the Emerging Biopolitics

Why does it matter if scientists hunt for race-specific genes, drug companies label their products by race, or people define their racial identities in genetic terms? Is it so bad if we hold a variety of views about the meaning of race—some seeing it as a biological category, others as a social construct—as long as everyone rejects the view that one race is superior to others? I contend that the ideology of race as a natural division between human beings that is written in our genes will have devastating political consequences. It can serve as the linchpin of a new, already emerging biopolitics in which the state's power to control the life and death of populations relies on classifying them by race.

Finding racial differences at the molecular level seems to make sense of the post–civil rights paradox. Not only has there been little change in racial gaps in health and economic status since the civil rights movement, but some conditions have worsened for the most marginalized people. The rate of incarceration today is five times what it was in 1972, and there are more blacks under correctional control today than there were slaves in 1860. A 2010 Amnesty International report, *Deadly Delivery*, stated that "African-American women are nearly four times more likely to die of pregnancy-related complications than white women." This death gap has not improved for more than twenty years.[25] The wealth gap between blacks and whites increased fourfold between 1984 and 2007, widening even more during the current economic crisis. The rise in black unemployment, already nearly double the white rate, was seven times greater than for whites between July and August 2010.[26] Because blacks were targeted for predatory loans, foreclosures in their neighborhoods outpace foreclosures in white neighborhoods.[27]

Biological distinctions, seemingly validated by genomic science and technology, appear to explain why stark racial disparities persist despite the abolition of official discrimination on the basis of race and despite most white Americans' belief that racism has ceased to exist. So far, most claims of genetic differences between races have focused on health disparities. But researchers are increasingly asserting that genes also determine behavior. Consider the following genetic claims that made the news in 2010.

The public radio show *Marketplace* reported on the book *Born Entrepreneurs, Born Leaders: How Your Genes Affect Your Worklife* by Scott Shane, a professor of entrepreneurial studies at Case Western Reserve University, which contends that "your DNA accounts for one third of the difference between you and your co-workers in many aspects of work life, from job satisfaction to income level."[28] Shane proposed using genetic testing to determine job assignments and design training programs.

Newspapers reported that voting behavior is determined by genes after political scientist James Fowler analyzed the voting patterns of identical and nonidentical twins in Los Angeles for eight elections between 2000 and 2005 and concluded that 53 percent of the differences in voter participation could be explained by genetics. Later Fowler identified a specific variant of the D2 dopamine receptor gene that is partly responsible for the tendency to join political groups.[29]

Florida State University criminologist Kevin Beaver published a widely reported study claiming to show that young men with the low-activity form of the monoamine oxidase A (MAOA) gene—dubbed by the press as the "warrior gene"—were more likely to join gangs than were those who had the high-activity version of the MAOA gene. He concluded that "male carriers of low MAOA activity alleles are at risk for becoming gang members and, once gang members, are at risk for using weapons in a fight."[30] In June 2009, the Associated Press reported his study with the headline "Gang-Banging May Be Genetic." The story ran on the Internet accompanied by a large color photo of bare-chested, tattooed Latino men flanked by a security force holding automatic rifles. "Bad neighborhoods and lack of opportunity are usually blamed for boys joining violent street gangs," the article stated. "But a new study finds that the urge to join gangs might lie, at least in part, in their genes." Beaver endorsed this view, which explicitly downplayed a social explanation for gang involvement. "While gangs typically have been regarded as a sociological phenomenon, our investigation shows that variants of a specific MAOA gene, known as a 'low-activity 3-repeat allele,' play a significant role," he said.[31]

Beaver's research on gang involvement illustrates the trend toward finding genetic answers for sociological questions. It is increasingly common for studies examining social relationships to incorporate genetic testing of research subjects and then let genetics overwhelm the social approach. In September 2009, HealthDay ran a story, "Genetics Linked to Early Sexual Activity in Kids," on the genetic reason why children who grow up in a home without a biological father have sex at a younger age than children whose fathers are present.[32] It was based on a study published in *Child Development* by Jane Mendle, an assistant professor of psychology at the University of Oregon, that looked at a thousand cousins aged fourteen and older who took part in the National Longitudinal Survey of Youth. Mendle claims that children of absent fathers have sex earlier not because of their home or social environment, but because of the genetic traits they inherit from their parents, including "impulsivity, substance use and abuse, argumentativeness, and sensation-seeking." She says, "Our study found that the association between fathers' absence and children's sexuality is best explained by genetic influences, rather than by environmental theories alone."[33]

The two powerful and deeply flawed ideologies of biological race and genetic determinism are ascending in tandem in America today. Americans are so willing to accept genetic explanations for social relationships and behaviors because they believe that race is inherited biology. When scientists report genetic explanations for negative behaviors like gangbanging, absentee fatherhood, and teenage sex, which the public stereotypically associates with blacks and Latinos, it seems more plausible that genes cause social problems. These race-gene claims simultaneously confirm the myths that races exist in genes and that genes can tell us everything about ourselves. Together, they support a biological explanation for the widening racial chasm in health, incarceration, and social welfare.

There is also evidence that genetic theories of race can reinforce dehumanizing stereotypes already embedded in the minds of most Americans. The depiction of blacks as apelike is a vestige of the scientific theory that placed Africans on an evolutionary spectrum far from whites, who were deemed the most developed, and closer to less advanced primates. The association of blacks with apes gave scientific credence to stereotypes of blacks as innately lazy, aggressive, unintelligent, and licentious, and therefore undeserving of the privileges whites enjoyed. A team of social psychologists discovered that the college students they studied implicitly linked blacks with apes and that this linkage influenced the extent to which students condoned police violence against black suspects. In the first study, Stanford undergraduates were able to identify scrambled ape images faster after they had been subliminally exposed to a black male face and required more frames to identify the ape when primed with a white male face. In other words, the students' ability to see the ape was facilitated by a black face and inhibited by a white face. In another study, the team showed students at Penn State a video of police officers beating a suspect whom the students were told was either black or white. Students primed with ape words before watching the video were most likely to think the violence was justified, but only when they thought the suspect was black. Representations of innate racial difference, the team concluded, can lead to dehumanization—"a method by which individuals and social groups are targeted for cruelty, social degradation, and state-sanctioned violence."[34]

Privatization and Punishment

Dehumanization is the perfect word to describe the impact of current federal and state policies that are depriving communities of color of needed services, social programs, and economic resources while keeping them under brutally enforced control. We see not only entrenched inequality but also intensifying state coercion that treats whole populations as less than human and cuts short their lives on the basis of race. This new biopolitics of race played out on television screens around the world in the wake of Hurricane Katrina as thousands of poor blacks in New Orleans were left to drown and the survivors were rounded up en masse when the levees failed as a result of government neglect.[35] Surveying the Bush administration's disastrous response, Henry Giroux observes that the desertion of Katrina's victims "revealed the emergence of a new kind of politics, one in which entire populations are now considered disposable, an unnecessary burden on state coffers, and consigned to fend for themselves."[36] The dismayed reaction many white Americans expressed to the extent of death and destruction left by the storm also revealed widespread blindness toward glaring racial inequities such as those that already plagued the Gulf Coast before the hurricane hit. As the mayor of Winstonville, Mississippi, remarked following the disaster, "No one would have checked on a lot of the Black people in these parishes while the sun shined, so am I surprised that no one has come to help us now? No."[37]

In recent decades, the United States has led both industrialized and developing nations in drastically cutting social programs while promoting the free-market conditions conducive to capital accumulation. Critical to this process of state restructuring is the transfer of services from the welfare state to the private realm of market, family, and individual in tandem with government support of corporate interests. Scholars call this trend toward privatization, which has cut across Republican and Democratic administrations, "neoliberalism."[38] President George W. Bush's domestic agenda explicitly sought to establish an "ownership society" that would replace New Deal programs. The Cato Institute, a libertarian think tank that championed the ownership society, described the concept as follows: "An ownership society values responsibility, liberty, and property. Individuals are empowered by freeing them from dependence on government handouts and making them owners instead, in control of their own lives and destinies. In the ownership society, patients control their own health care, parents control their own children's education, and workers control their retirement savings."[39]

President Bush's proposals for creating an ownership society included restructuring Social Security to rely on individual investments, eliminating tax rules seen to penalize wealth accumulation and transfer, and changing class action laws to shield corporations from large tort damages awards.[40] Under Chief Justice John Roberts, nominated by Bush in 2005, the Supreme Court "ruled for business interests 61 percent of the time," compared with 42 percent of all decisions since 1953, the *New York Times* reported in December 2010, citing a study prepared for the newspaper by scholars at Northwestern University and the University of Chicago.[41] The ownership society and the privatization philosophy it reflects demand that individuals rely on their own wealth to meet their needs and discourages government aid for poor people who face systemic hardships, while encouraging government aid for big business. This philosophy also motivated policies under the Democratic Clinton and Obama administrations. Although free market fundamentalism gained popularity with the election of Ronald Reagan, corporate deregulation accelerated when Bill Clinton took office.[42] The Reagan administration stigmatized welfare by deploying the myth of the black "welfare queen" who had babies to fatten her benefit check, but it was the Clinton administration that worked with Congress to abolish the entitlement to welfare altogether.

Privatization also takes the form of bodily neglect of people who cannot afford medical care and other resources needed for good health. One of the quotes that began this book referred to hospital cuts in kidney dialysis care for indigent people in Miami, who were forced to "rely on emergency rooms for their life-sustaining treatment."[43] And the prison system has become the primary source of health care for increasing numbers of poor people of color. The Los Angeles County Jail psychiatric wing is the largest mental health care facility in the country, servicing thousands of mentally ill inmates charged with or convicted of minor offenses. A 2010 ACLU report revealed that inmates at the jail, most of whom are simply awaiting trial, risk severe beatings by deputies if they complain about the overcrowding and filth that has plagued the "modern-day medieval dungeon" for decades.[44] In

2009, a special panel of federal judges ordered California to reduce its prison population by 46,000 to ease an overcrowding crisis that had been mounting over twenty years because it was unnecessarily killing inmates. One California prisoner dies every eight days from the state's failure to provide adequate medical care, defined by the U.S. Supreme Court as "the minimal civilized measure of life's necessities." The state's appeal of the order, supported by eighteen other states, was argued before the Supreme Court in November 2010, delaying relief to dying prisoners even longer.[45]

As I discussed in chapter 9, the promotion of biological citizens who use genetic testing, reprogenetic technologies, and personalized medicine to manage their own health at the molecular level facilitates this shift of responsibility for public welfare from the state to the market and individual. Genetic technologies service a form of privatization that makes individuals responsible for their own welfare through the self-regulation of genetic risk. The prominence of race in the new biocitizenship promotes mutually supporting trends toward neoliberalism and racial apathy by making racial inequities seem like biological rather than social problems.[46] As the state and market jointly offer individualized technological solutions for social wrongs, race remains relevant to social policy even as citizens are admonished to be more color-blind. It is not that stark racial gaps in health, welfare, and opportunity no longer exist in this postracial America, but their roots in racial bias and systemic injustice are concealed by intensified attention to race at the genetic level. The pivotal role of race in creating biocitizens encourages their support for the interests of biocorporations and discourages demands for social change.

At the same time that the government is dismantling the social safety net, it has intensified its coercive intervention in poor communities of color. The intensification of corporate power and racial marginalization does not entail a unidimensional shrinking of government. It equally depends on the brutal, even barbaric, infliction of pain on the nation's most disenfranchised residents.[47] Over the last two decades, the welfare, prison, foster care, and deportation systems have clamped down on poor communities of color, increasing many people's experience of insecurity and surveillance.[48] In her 2010 book, *The New Jim Crow*, Michelle Alexander demonstrates that black incarceration functions like a modern-day Jim Crow caste system because it "permanently locks a huge percentage of the African American community out of the mainstream society and economy," replicating the subjugated status of blacks that prevailed before the civil rights revolution.[49] People are suffering not only because the government has abandoned them but also because punitive policies are making their lives more difficult.

These two trends—private remedies for systemic inequality and punitive state regulation of the most disadvantaged communities—are mutually reinforcing. Barbara Ehrenreich noted the correspondence between privatization and punishment when she wrote, "The pattern is to curtail financing for services that might help the poor while ramping up law enforcement: starve school and public transportation budgeting, then make truancy illegal. Shut down public housing, then make it a crime to be homeless. Be sure to harass street vendors when there are few other opportunities for employment."[50] Mass imprisonment of blacks and Latinos is a way for the state to exert direct control over poorly educated, unskilled, and jobless people who have no place in the market economy because of racism, while preserving a racial caste system that was supposed to be abolished by civil rights reforms. As the deadly conditions in the California prison system exemplify, the twin policies of bodily neglect and hyperpunishment converge in prison cells. On New Year's Eve 2010, Mississippi governor Haley Barbour suspended the prison terms of sisters Gladys and Jamie Scott, who were given life sentences for aiding a robbery in 1993 in which no one was hurt and only $11 stolen, on the condition that Gladys donate a kidney to Jamie, who barely survived on dialysis as a result of severe diabetes and high blood pressure. Governor Barbour explained that he made the decision not to render justice for the excessive penalty the sisters had endured but because "Jamie Scott's medical condition creates a substantial cost to the state of Mississippi."[51]

Naomi Klein's chilling exposé of violent state strategies employed to introduce radical free-market reforms abroad, *The Shock Doctrine*, reveals the horrific potential of state corporatism:

> Its main characteristics are huge transfers of public wealth to private hands, often accompanied by exploding debt, an ever-widening chasm between the dazzling rich and the disposable poor and an aggressive nationalism that justifies bottomless spending on security. For those inside the bubble of extreme wealth created by such an arrangement, there can be no more profitable way to organize a society. But because of the obvious drawbacks for the vast majority of the population left outside the bubble, other features of the corporatist state tend to include

> aggressive surveillance (once again, with government and large corporations trading favors and contracts), mass incarceration, shrinking civil liberties and often, though not always, torture.[52]

Readers may recognize the model Klein describes in foreign nations such as Chile under the dictator General Augusto Pinochet. But every word fits current conditions *within* the United States for poor minority communities. Take a look at what is going on in the nation's public schools, prisons, detention centers, and hospitals serving poor people of color. You will see not only stark inequalities but also a dehumanizing brutality committed by state agents, a brutality that hinges on bodily neglect, abuse, confinement, and even torture of black and brown people.

Torture is the only word that can describe the chilling footage of guards at a Panama City, Florida, youth detention center tormenting fourteen-year-old Martin Lee Anderson to death in January 2006. A security videotape, aired on television and the Internet, captured seven guards "covering Martin's mouth, forcing him to inhale ammonia and striking him repeatedly after he stopped running" on his first day at the boot-camp-style facility.[53] The tape shows a nurse standing by and deliberately watching the entire ordeal, as if giving medical sanction to the guards' deadly brutality. Violence, sometimes amounting to torture, is an accepted part of U.S. prisons and detention centers, which house mostly minority men, women, and children.[54] It is part of a long history of the U.S. government's systematic torture of people of color as a means of racial subjugation in this country and abroad. Torture has historically been an extreme but integral part of racial biopolitics. It graphically imposes a subordinated status on the bodies of its victims while classifying those bodies as fundamentally different.[55] I'm not claiming that anyone involved in today's racial science would support the torture of people based on their race; they would condemn it. Rather, I'm claiming that the legacy of race-based torture continues to this very day and must be taken into account in assessing the implications of a new racial science in the genomic age.

The physical and sexual abuse of inmates in U.S. prisons is rampant and tolerated by prison administrators and courts.[56] In addition to the violent enforcement of order meted out on a daily basis, prison guards sometimes inflict extreme measures to intimidate inmates. As one example, guards have used restraint chairs (a retraining device that locks a prisoner's legs, arms, and torso with belts and cuffs), aptly known as the torture chair or slave chair, not only to control violent inmates but to sadistically punish those who challenge prison rules. Male and female prisoners have been strapped to the chair completely naked, gagged, hooded, beaten, pepper-sprayed, and even left to die from asphyxia and blood clots.[57] In many states, when incarcerated women go into labor, they are routinely shackled to the hospital bed, their legs, wrists, and abdomens chained during the entire delivery. Shawanna Nelson, a twenty-nine-year-old African American woman imprisoned for credit card fraud and writing bad checks, was shackled to the sides of the bed in the Arkansas hospital where she gave birth. The chains prevented her from stretching or changing positions during the most painful contractions, permanently injuring her hip and ripping her stomach muscles.[58]

The horrors inflicted on children like Martin Lee Anderson who are confined to juvenile detention centers and adult prisons are especially heartbreaking. Incarcerated children around the country not only live in cramped, unsanitary quarters without adequate services and health care, but also are subjected to regular physical and sexual abuse. A federal inquiry begun after the 2006 killing of a fifteen-year-old by prison staff found that guards at four of New York's most hazardous youth prisons "routinely used physical force to discipline the youths, resulting in broken bones, shattered teeth, concussions and dozens of other injuries in a period of less than two years," according to a July 2010 *New York Times* report. Federal investigators discovered that staff also kept order among the child inmates by overmedicating them with powerful psychotropic drugs in "substantial departures from generally accepted professional standards."[59] Although New York officials belatedly agreed to limit the use of aggressive restraints and to expand mental health services for children in custody, they failed to address adequately the more basic injustice of relying on imprisonment to deal with children's behavioral and mental health problems.[60]

The abuse that took place in New York youth prisons is not exceptional. A 2010 study by the Justice Department's Bureau of Justice Statistics found that 12 percent of nine thousand children confined to juvenile facilities reported being raped at least once, mainly by employees.[61] In some detention centers, the rate of sexual abuse was as high as 30 percent or more. Even worse, children as young as twelve can be tried as adults in most states (and as young as seven in twenty states). More than ten thousand children are currently incarcerated with grown men and women where, according to a

Columbia University study, they are "five times as likely to be raped, twice as likely to be beaten and eight times as likely to commit suicide than adults in the adult prison system."[62]

The sketchy yet horrific accounts of immigrants dying in U.S. custody that *New York Times* journalist Nina Bernstein began to report in 2008 seemed to me to fit the pattern of racial biopolitics I was investigating in my research. In 2007, Boubacar Bah, a fifty-two-year-old tailor from Guinea, West Africa, who was jailed at the Elizabeth Detention Center in New Jersey for overstaying his tourist visa, died after emergency surgery for a skull fracture and multiple brain hemorrhages.[63] The internal records of the Corrections Corporation of America, the private company that runs many immigration jails, reveal that Bah was handcuffed and placed in leg restraints on the floor of the medical unit as he moaned and vomited after sustaining a head injury. Because his crying out was considered a "behavior problem," Bah was taken in shackles to a disciplinary cell, where he remained in solitary confinement for more than thirteen hours, despite a nurse's notation that he was "unresponsive on the floor with foamy brown vomitus noted around mouth."[64]

Hiu Lui Ng, a thirty-four-year-old computer engineer from Hong Kong who, like Bah, overstayed his visa, spent the final months of his life in detention centers in several New England states.[65] In April 2008, Ng began to experience excruciating back pain that became so severe over the following three months that he could no longer walk or stand. Ng's complaints about his suffering were dismissed by Rhode Island detention officials, who refused to give him a wheelchair or conduct an independent medical evaluation. Instead, guards "dragged him from his bed on July 30, carried him in shackles to a car, bruising his arms and legs, and drove him two hours to a federal lockup in Hartford."[66] Ng was finally taken to a hospital on August 1, where he died in custody five days later, his spine fractured by cancer that had spread through his body while immigration agents denied him treatment.

It would be a mistake to think of these instances of state-sanctioned inhumanity as aberrations. Indeed, in the last decade, torture was incorporated in official federal policy and accepted by alarming numbers of ordinary Americans. At the outset of the U.S. incursions in Afghanistan and Iraq, lawyers in the Bush administration embarked on a mission to set a sufficiently high standard for torture to render the military's harsh custodial conditions and interrogation methods permissible.[67] President Bush himself set the machinery in motion on February 7, 2002, when he issued a memorandum declaring that the Geneva Conventions did not apply to detainees in Afghanistan and Guantanamo.[68] The legal edifice erected to shield torture has direct antecedents in the colonial jurisprudence that justified the uncivilized treatment of African and Asian natives by U.S. and European imperialists under the racialized theory of savage war. As one British judge explained in 1910, the rule of law could be suspended in the African colonies because "a few dominant civilised men have to control a great multitude of the semi-barbarous."[69] Racial classifications that marked Africans and Asians as uncivilized savages justified the uncivilized use of torture in waging imperialist wars against them. Lawyers in the Justice Department similarly took the position that the United States was engaged in a "new kind of war" that necessitated replacing protections accorded prisoners under the Geneva Conventions with a de novo legal regime that permitted harsher treatment.[70]

The legal defense by the nation's highest officials of inhumane treatment in foreign detention centers and the steady dose of torture delivered by the entertainment media acclimated much of the American public to the infliction of pain and degradation on Muslim "enemy combatants." In the wake of 9/11, torture was championed by movies and television shows that depicted its successful deployment by patriotic heroes to avert disaster and promote the national interest. On virtually every episode of the top-rated Fox television drama *24*, boasting 15 million viewers on average, Special Agent Jack Bauer (played by Kiefer Sutherland) tortured terror suspects in order to extract information critical to national security.[71] Meanwhile, some Americans find entertainment in videos showing the gruesome deaths of enemy soldiers and civilians that are currently circulating on the Internet.[72] National polls showed that more than one third of Americans reported that torture was legitimate in some cases, and only one third believed the abuses at Abu Ghraib constituted torture.[73]

The biological definition of race seems acceptable today because past forms of blatant racial violence, such as lynching, are now institutionalized in new ways that make them invisible to many Americans. Scientists, politicians, and entrepreneurs can disassociate their promotion of inherent racial classifications from prior explicitly racist and eugenic incarnations because racial inequality no longer relies on overt white supremacy. We are told not to worry about the inhumane potential of dividing human beings into supposedly biological races because, in the words of Nikolas Rose, "In advanced liberal democracies at least, the biopolitics of identity is very different from that which characterized eugenics . . . [because it involves] choice, enterprise, self-actualization, and prudence in relation to one's genetic makeup."[74] But as I have shown, the political context of the new racial science actually involves coercion, surveillance, restraint, and brutality inflicted on minority populations. The reconfiguration of race in genetic terms provides a modern mechanism for legitimating this oppressive politics of race at a time when the United States claims to have repudiated the violent enforcement of a racial caste system. It is as if straining their eyes to see race at the molecular level blinds people to the glaring contradiction of barbaric state practices and gaping social inequities that perpetuate the racial order in our advanced liberal democracy.

Conclusion: The Crossroads

The toxic convergence of race, biology, and politics is rooted in the very origin of this nation. Race was instituted as a system of governance and "moral apology" for keeping Africans in chains and violently dispossessing Indian tribes. It resolved the contradiction between America's professed commitment to liberty and equality and its reliance on the barbaric system of chattel slavery and genocide, followed by the colonization of Mexicans, exclusion of Asian immigrants, and Jim Crow segregation. During the eugenics period, this country experienced an especially virulent form of racial biopolitics when the government forcibly sterilized thousands of people on the grounds that they were biologically unfit. The nation's fervor for eugenics was quelled by mass outrage at Nazi racial science and by the victories won by the civil rights struggle. Yet America is once again at the brink of a dangerous biopolitics of race, fueled by a new racial science based on cutting-edge genomics, a staunch refusal to acknowledge enduring racial inequality, and a free-market fundamentalism that, having virtually eliminated the social safety net, relies instead on technological solutions to social problems.

In the post–civil rights era, America once more needs a "moral apology" for institutionalized brutality, adopted by conservatives and liberals alike, that preserves the racial order at a time when society claims to have moved beyond race. The ideology, promoted by science, government, and big business, that race is written in our genes provides the modern-day mechanism for concealing the most heinous crimes that racism inflicts on poor communities of color and an excuse for keeping a backward racial caste system firmly in place. At the same time, all Americans are increasingly expected to become biocitizens who assume full responsibility for their own welfare through the self-regulation of genetic risk, consumption of gene-based goods and services, and donations of their genetic information to scientific research. As they adopt individualized technological solutions to social problems, their apathy grows not only toward racial repression but also toward a more insidious corporate control of their own identities and relationships. By embracing a racial ideology rooted in genetics, Americans are accepting a genetic ideology rooted in race that makes everyone responsible for managing their own lives at the genetic level instead of eliminating the social inequalities that damage our entire society.

This nation is at a crossroads. One path is the one I have just described: adopting the view that human beings are naturally divided into races at the molecular level and looking to genomic science and technology to bridge the enduring chasm between racial groups. The other road means affirming our shared humanity by working to end the social inequities preserved by the political system of race. Which path we choose to follow is not only a question of scientific evidence. It is a question of moral commitment. There is no neutral scientific position on this question. We have long had scientific confirmation that race is a political and not a biological category. The re-creation of biological race in genomic science today, like its invention by scientists in past centuries, results from an ideological commitment to a false view of humanity.

Some people draw a blank when asked to imagine the scientific study of human beings without classifying them by race. How would scientists organize research on human health and behavior if not by racial groupings? Would molecular biology, population genomics, and biomedical research implode from being deprived of an essential research tool?

There is an alternative way of approaching scientific research on human beings that makes full use of the wealth of information in our genomes. Instead of asking, "How do genes work differently in different racial groups?" scientists could ask, "How do genes work in human beings?" Scientists can investigate the genetics that makes us human, the genetic mechanisms that human beings share. The unquestioned use of race as an organizing principle for research has hindered scientific progress. I contend that, instead of hamstringing scientists, a focus on both human genetic diversity and its commonality, freed from false and antiquated notions of biological race, would liberate them. Studying the genetics of how cancer tumors grow in human bodies has been more fruitful than searching for race-specific genes that explain differential rates of cancer. In response to a June 2010 front-page article in the *New York Times* decrying the failure of genomic research to produce personalized medicines, Harvard molecular geneticist Michael Farzan reminded readers, "This is a narrow definition of the genome project's purpose." He wrote, "The project provided an enormously powerful tool to study general molecular mechanisms of health and disease, not just those resulting from genetic variation."[1]

Stanford bioethicist Mildred Cho told me that many genome scientists she has talked to are thirsting for a shift away from research focused on race and other types of genetic difference toward research that explores more basically how genes function in human beings—an area scientists now realize they know surprisingly little about. "I think that there has been some complaining in the scientific community and yearning for research that's less focused on these differences and more focused on the commonalities, about basic issues of how genes work. We don't really know that," she said, adding that some scientists believe that studies looking for race-linked genes associated with disease are "really kind of getting ahead of ourselves. We should be focusing on species commonality."[2]

This does not mean there is no place for race in scientific research. As I discussed in chapter 6, many scientists are investigating the biological pathways through which racism is translated into poor health and devising strategies to end racial inequities in health by changing social policies. These studies require using race, defined properly as a social category, as a research variable and a way of classifying research participants. If its function as a political system is recognized, race can have a valid use in scientific research to locate, understand, and eliminate the effects of racism.

What does the path of common humanity mean for all of us? We should reject the notion of biological citizenship based on genetic ties and consumption of biotechnologies that allow us to manage our welfare individually at the molecular level. Instead, we should forge a notion of citizenship based on our common moral commitment to end racism and other forms of social inequality that recognizes social change cannot be accomplished with a technological fix. It is the belief in fundamental human equality that inspires many people to fight collectively against racism and its dehumanizing practices. Locating the causes of inequality in social rather than genetic structure is liberating because it is much easier to change society than genes. It is more enlightened to understand the potential for political alliances apart from biological distinctions than to believe we are inevitably divided and shackled by immutable differences programmed in our genes.

There are models we can look to for envisioning an alternative approach to race in the twenty-first century. One is the tradition within black politics of skepticism about the claim of inerrant scientific progress in America, of basing solidarity on common struggle and not common biology, and of using race as a political tactic to document and contest racism. Through courage, political solidarity, and a deep faith in our common human dignity, black slaves, sharecroppers, and civil rights protestors defied their racial status as servants. Black resistance to oppression in America challenges the notion that biology is destiny. Other disenfranchised groups have similar legacies of political struggle that we can adapt to the genomic era.

Recognizing the relationship between neoliberalism, state authoritarianism, and a new biopolitics of race creates the potential for alliances among groups that see the dangerous potential in the escalating march from state support for public welfare toward market-based solutions and repression of people who are suffering most from this trend. Antiracist, disability rights, and economic, gender, reproductive, and environmental justice movements all have a stake in fighting the emerging racial biopolitics. These potential coalitions provide hope for a broad-based social movement that rejects biological definitions of race and citizenship in favor of the radical restructuring of our society into one that respects the humanity of all people.

Will Americans continue to believe the myth that human beings are naturally divided into races and look to genomic science and technology to deal with persistent social inequities? Or will they affirm our shared humanity by working to end the social injustices preserved by the political system of race? This is the most pivotal question facing this nation in the twenty-first century because the answer will determine the basic nature of the relationship between citizens and the government and with each other. One path is already leading to aggressive state surveillance, extreme human deprivation, and unspeakable brutality against whole populations on the basis of race. By obscuring this coercive control over poor communities of color, the new racial biopolitics permits the growth of a state authoritarianism and a corporatized definition of citizenship that endangers the democratic freedoms of all Americans. We must choose the other path of common humanity and social change if we are to have any hope for a more free and just nation.

Acknowledgments

During the six years I worked on *Fatal Invention*, I was fortunate to be surrounded by many friends, colleagues, and collaborators, without whose help I could not have completed this project. I am especially grateful to colleagues at my two academic homes, Northwestern University School of Law and the Institute for Policy Research, for their encouragement and advice. I had the opportunity to collaborate with the Black Women's Health Imperative, the Center for Genetics and Society, and Generations Ahead on a number projects related to my book and owe special thanks to Byllye Avery, Marcy Darnovsky, Richard Hayes, Eleanor Hinton Hoytt, Sujatha Jesudason, and Osagie Obasogie for their camaraderie and insights. I am also indebted to Jonathan Kahn for inviting me in 2003 to join a working group on "Colliding Categories: Haplotypes, Race, and Ethnicity" at the University of Minnesota Consortium on Law and Values in Health, Environment, and the Life Sciences, which sparked my interest in the genetic redefinition of race, and for many stimulating conversations since then. An invitation from my longtime friend Susan Wolf, director of the Consortium, to present a paper at a 2005 conference on "Proposals for the Responsible Use of Race and Ethnic Categories in Biomedical Research" led to my first publication in this book project. My participation in three conferences at MIT's Center for the Study of Diversity in Science, Technology, and Medicine, directed by David Jones, leading to two more publications, was also instrumental to my progress. My interviews with thirty scientists, scholars, bioethicists, and activists, many of whom are quoted in the book, were a key source of information.

I am very grateful for their agreement to speak with me about their work and ideas. Michael Byrd, Linda Clayton, and Christopher Kuzawa graciously took the extra time to read and comment on parts of the manuscript, and Vanessa Gamble, Alondra Nelson, Jennifer Reardon, and Keith Wailoo gave me additional helpful information.

In addition to the presentations mentioned above, I presented papers that contributed to this book at the American Anthropological Association, American Bar Foundation, American Political Science Association, American Society for Bioethics and Humanities, American Society for Reproductive Medicine, Arizona State University Consortium for Science, Policy, and Outcomes, Association of American Law Schools, Boston College Sociology Department, Columbia University Institute for Research in African American Studies, DePaul University College of Law, Emory University School of Law, Fordham University School of Law, Genetic Alliance, Harvard University Committee on Degrees in Studies of Women, Gender, and Sexuality, Howard University School of Law, Johns Hopkins University Berman Institute, John Marshall Law School, Michigan State University Philosophy Department, Middlebury College, Midwest People of Color Legal Scholarship Conference, Ohio State University College of Public Health, Philosophy Born of Struggle, Providence College, Scripps College, Stanford Bioethics Center, Trinity University Center for Bioethics and Human Dignity, Tuskegee University National Bioethics Center, UCLA Center for Society and Genetics, University of Baltimore School of Law, University of British Columbia School of Law and Department of Sociology, University of California at Berkeley School of Law, University of California at San Diego Ethnic Studies Department, University of Illinois at Chicago African American Studies, University of Iowa School of Law, University of Miami School of Law, University of North Carolina School of Law, Washington and Lee University School of Law, WeACT, and the 2010 World Forum on Human Rights. I greatly appreciate the invitations to share my work in progress, as well as the comments of participants too numerous to name.

I could not have completed this book without the superb research assistance of several Northwestern undergraduate and law students: Tera Agyepong, Haile Arrindell, Alan Bakhos, Nicholas Gamse, Katherine Halpern, and Jessica Harris. Stanford law student Eunice Cho lent remarkable research support while I was a visiting fellow at the Center for Comparative Studies in Race and Ethnicity. At the final stage, S.J. Chapman skillfully and tirelessly tracked down needed information under a tight deadline. The students who took my seminar, "Race, Biotechnology, and Law in the Gene Age," in spring 2009 and 2010 also helped me think through my arguments in *Fatal Invention*. I am most grateful to all of these wonderful students for their commitment and friendship. I wish to thank Shirley Scott for her devoted administrative assistance and the staff of the Pritzker Legal Research Center, especially Marcia Lehr, for providing every source I ever requested. This book also benefited enormously from careful editing by Sarah Fan of The New Press. Her insightful queries pushed me to make my arguments clearer and stronger. Audra Wolfe adeptly helped me make painful but needed cuts to the original manuscript, which was one hundred pages too long. I also thank David Halpern of The Robbins Office for his steadfast encouragement and advice.

I received generous support from several sources, which enabled me to take time from teaching to devote to research and writing, to hire research assistants, and to pay transcription and travel costs. David Van Zandt, dean of Northwestern University School of Law; a fellowship at the Institute for Policy Research, directed by Fay Lomax Cook; and the Kirkland & Ellis Fund provided financial assistance throughout my book project. Support was also provided by the National Science Foundation under grant 00551869; an RWJF Investigator Award in Health Policy Research from the Robert Wood Johnson Foundation, Princeton, New Jersey; the Dorothy Ann and Clarence L. VerSteeg Distinguished Research Fellowship; and a visiting fellowship at Stanford's Center for Comparative Studies in Race and Ethnicity. I am extremely grateful for this backing, which was essential to launch and complete my project.

Finally, I owe the utmost thanks to my family for their uncompromising love, which inspired and nurtured me throughout my work on *Fatal Invention*. My mother's death in May 2009 as I was drafting the manuscript reinforced for me the profound influence my parents had on my understanding of our common humanity and my commitment to racial justice. Words cannot express my gratitude for the childhood they gave me, which was spent celebrating the diversity and equal dignity of people around the world. My Sherman United Methodist Church family was also a much appreciated source of comfort and support.

Some of the material in this book is adapted from the following articles and chapters: "Race and the New Biocitizen," in *What's the Use of Race*, ed.

Ian Whitmarsh and David Jones (Cambridge, MA: MIT Press, 2010), 259; "Race, Gender, and Genetic Technologies: A New Reproductive Dystopia?" *Signs* 34 (2009): 783; "Torture and the Biopolitics of Race," in *Rethinking America: The Imperial Homeland in the 21st Century*, ed. Jeff Maskovsky and Ida Susser (Boulder, CO: Paradigm Press, 2009), 167; "Is Race-Based Medicine Good for Us? African-American Approaches to Race, Biotechnology, and Equality," *Journal of Law, Medicine & Ethics* 36 (2008): 537; "Torture and the Biopolitics of Race," *University of Miami Law Review* 62 (2008): 229; and "Legal Constraints on the Use of Race in Biomedical Research: Toward a Social Justice Framework," *Journal of Law, Medicine & Ethics* 34 (2006): 526.

Notes

Chapter 1: The Invention of Race

1. Joseph L. Graves Jr., *The Emperor's New Clothes: Biological Theories of Race and the Millenium* (New Brunswick, NJ: Rutgers University Press, 2002), 6.

2. Author interview of Charmaine Royal, Sept. 4, 2008, Durham, NC.

3. Michael Banton, *Racial Theories* (Cambridge, UK: Cambridge University Press, 1998), 17.

4. George D. Kelsey, *Racism and the Christian Understanding of Man* (New York: Scribner's, 1965), 21; Jacqueline Stevens, *Reproducing the State* (Princeton, NJ: Princeton University Press, 1999), 176–86.

5. Nell Irvin Painter, *The History of White People* (New York: Norton, 2010), 9.

6. Kelsey, *Racism and the Christian Understanding of Man*, 19–20; Ruth Benedict, *Race: Science and Politics* (New York: Viking, 1945); Jonathan Marks, "Race: Past, Present, and Future," in *Revisiting Race in a Genomic Age*, ed. Barbara A. Koenig, Sandra Soo-Jin Lee, and Sarah S. Richardson (New Brunswick, NJ: Rutgers University Press, 2008), 21.

7. Winthrop D. Jordan, "First Impressions," in *Theories of Race and Racism*, ed. Les Back and John Solomos (New York: Routledge, 2000), 35.

8. Ibid., 36, 50.

9. Thomas Sowell, *Race and Culture: A World View* (New York: Basic Books, 1994), 186–87; Painter, *History of White People*, 34–42.

10. Painter, *History of White People*, 42.

11. Gunnar Myrdal, "Racial Beliefs in America," in *Theories of Race and Racism*, 112.

12. David R. Roediger, *How Race Survived US History* (London: Verso, 2008), 4–5; William J. Cooper, *Liberty and Slavery: Southern Politics to 1860* (Columbia:

University of South Carolina Press, 2001), 9; Ariela J. Gross, *What Blood Won't Tell: A History of Race on Trial in America* (Cambridge, MA: Harvard University Press, 2008), 16–17.

13. Roediger, *How Race Survived US History*, 6.

14. Wilbur Joseph Cash, *The Mind of the South* (New York: Knopf, 1941), 116.

15. Gordon S. Wood, *Empire of Liberty: A History of the Early Republic, 1789–1815* (New York: Oxford University Press, 2009), 541.

16. Ira Berlin, *Slaves Without Masters: The Free Negro in the Antebellum South* (New York: Pantheon, 1974), 97.

17. Wood, *Empire of Liberty*, 541.

18. Ibid., 542.

19. W.E.B. Du Bois, *Darkwater: Voices from Within the Veil* (New York: Harcourt, Brace and Howe, 1920).

20. Cheryl Harris, "Whiteness as Property," *Harvard Law Review* 106 (1993): 1707.

21. Eric Foner, *Nothing but Freedom* (Baton Rouge: Louisiana State University Press, 1983); Rayford W. Logan, *The Betrayal of the Negro* (Cambridge, MA: Da Capo Press, 1965); C. Vann Woodward, *The Strange Career of Jim Crow* (1955; New York: Oxford University Press, 1974).

22. Frederick Douglass, "The Nation's Problem," in *Negro Social and Political Thought, 1850–1920*, ed. Howard Brotz (New York: Basic Books, 1966), 311, 312.

23. Angela Onwuachi-Willig, "A Beautiful Lie: Exploring *Rhinelander v. Rhinelander* as a Formative Lesson on Race, Identity, Marriage, and Family," *California Law Review* 95 (2007): 2393.

24. Gross, *What Blood Won't Tell*, 97; Earl Lewis and Heidi Ardizzone, *Love on Trial: An American Scandal in Black and White* (New York: Norton, 2001).

25. Onwuachi-Willig, "A Beautiful Lie," 2430.

26. Ibid., 2429.

27. Ibid., 2454.

28. See generally Gross, *What Blood Won't Tell*; Laura E. Gómez, *Manifest Destinies: The Making of the Mexican American Race* (New York: NYU Press, 2008).

29. Gross, *What Blood Won't Tell*, 31.

30. Ibid., 5.

31. Ibid., 46.

32. Ibid., 55.

33. Ian Haney Lopez, *White by Law: The Legal Construction of Race* (New York: NYU Press, 2006), 3.

34. See Appendix A, "The Racial Prerequisite Cases," in ibid., 163.

35. *In re: Ah Yup*, 1 F. Cas. 223 (C.C.D. Cal. 1878).

36. Bill Ong Hing, *Defining America Through Immigration Policy* (Philadelphia, PA: Temple University Press, 2004), 36–43.

37. Lisa Lowe, *Immigrant Acts: On Asian American Cultural Politics* (Durham, NC: Duke University Press, 1996), 19–20.

38. *United States v. Thind*, 261 U.S. 204 (1923).

39. Lopez, *White by Law*, 61.

40. *Ozawa v. United States*, 260 U.S. 178 (1922).

41. Lopez, *White by Law*, 104–5.

42. *United States v. Thind*, 261 U.S. 204 (1923).

43. Laura E. Gómez, *Manifest Destinies: The Making of the Mexican American Race* (New York: NYU Press, 2007), 139–41.

44. Karen Brodkin, *How Jews Became White Folks and What That Says About Race in America* (New Brunswick, NJ: Rutgers University Press, 1998); Andrew Jacobson, *Whiteness of a Different Color: European Immigrants and the Alchemy of Race* (Cambridge, MA: Harvard University Press, 1998).

45. F. James Davis, *Who Is Black? One Nation's Definition* (University Park: Penn State Press, 1991), 35.

46. Jonathan Marks, *What It Means to Be 98% Chimpanzee* (Berkeley: University of California Press, 2002), 68.

47. Laura E. Gómez, "Opposite One-Drop Rules: Mexican Americans, African Americans, and the Need to Reconceive Turn-of-the-Twentieth-Century Race Relations," in *How the United States Racializes Latinos: White Hegemony and Its Consequences*, ed. José A. Cobas, Jorge Duany, and Joe R. Feagin (Boulder, CO: Paradigm, 2009): 87, 98.

48. Gross, *What Blood Won't Tell*, 15.

49. Michael Omi and Howard Winant, *Racial Formation in the United States from the 1960s to the 1990s* (New York: Routledge, 1994), 55.

50. Omi and Winant, *Racial Formation in the United States*, 82; Betsy Guzmán, "The Hispanic Population: Census 2000 Brief," U.S. Census Bureau, May 2001.

51. Laura E. Gómez, "What's Race Got to Do with It? Press Coverage of the Latino Electorate in the 2008 Presidential Primary Season," *St. John's Journal of Legal Commentary* 24 (2009): 425, 433–35.

52. Sam Roberts and Peter Baker, "Asked to Declare His Race, Obama Checks 'Black,'" *New York Times*, Apr. 3, 2010, A9.

53. Deborah Posel, "What's in a Name? Racial Categorisations Under Apartheid and Their Afterlife," *Transformation: Critical Perspectives on Southern Africa* 47 (2001): 50.

54. D. Wendy Greene, "Determining the (In)determinable: Race in Brazil and the United States," *Michigan Journal of Race & Law* 14 (2009): 152.

55. Alfred L. Kroeber, *Anthropology: Biology and Race* (1923; New York: Harcourt Brace & World, 1948, 1963), 2.

56. A. Leon Higginbotham Jr., *In the Matter of Color: Race and the American Legal Process: The Colonial Period*, (New York: Oxford University Press, 1980), 44.

57. Gross, *What Blood Won't Tell*, 36.

58. Omi and Winant, *Racial Formation in the United States*, 58.

59. Gunnar Myrdal, *An American Dilemma: The Negro Problem and American Democracy* (New York: Harper and Brothers, 1944): 89.

60. E. Frankin Frazier, "The Pathology of Race Prejudice," in *The Negro Caravan: Writings by American Negroes*, ed. Sterling A. Brown (Salem, NH: Ayer, 1987), 904, 906.

61. Paul Gilroy, *Postcolonial Melancholia* (New York: Columbia University Press, 2005), 39.

Chapter 2: Separating Racial Science from Racism

1. Evelynn M. Hammonds, "Straw Men and Their Followers: The Return of Biological Race," Social Science Research Council, "Is 'Race' Real?" available at http://raceandgenomics.ssrc.org/Hammonds.

2. Jon Entine, *Abraham's Children: Race, Identity, and the DNA of the Chosen People* (New York: Grand Central, 2007): 242.

3. Jonathan Marks, "Race: Past, Present, and Future," in *Revisiting Race in a Genomic Age*, ed. Barbara A. Koenig, Sandra Soo-Jin Lee, and Sarah S. Richardson (New Brunswick, NJ: Rutgers University Press, 2008), 21.

4. See Emmanuel Chukwudi Eze, ed., *Race and the Enlightenment: A Reader* (Cambridge, MA: Blackwell, 1997).

5. Francois Bernier, "A New Division of the Earth," in *The Idea of Race*, ed. Robert Bernasconi and Tommy Lee Lott (Indianapolis: Hackett, 2000), 1.

6. Robert Bernasconi, "Who Invented the Concept of Race? Kant's Role in the Enlightenment Construction of Race," in *Race*, ed. Robert Bernasconi (Malden, MA: Blackwell, 2001), 15.

7. Quoted by Michael Banton, "The Idiom of Race: A Critique of Presentism," in *Theories of Race and Racism: A Reader*, ed. Les Back and John Solomos (New York: Routledge, 2000), 54, 58.

8. Ibid., 55.

9. Nell Irvin Painter, *The History of White People* (New York: W.W. Norton, 2010), 72–90; Bruce Baum, *The Rise and Fall of the Caucasian Race: A Political History of Racial Identity* (New York: NYU Press, 2006), 73–94.

10. Marks, "Race," 22.

11. Bernasconi and Lott, *Idea of Race*, 8–22.

12. Immanuel Kant, "On the Different Races of Man," in Eze, *Race and the Enlightenment*, 39, 40.

13. Quoted in Bernasconi, "Who Invented the Concept of Race?" 90.

14. Bernasconi and Lott, *Idea of Race*, 87–89.

15. Thomas F. Gossett, *Race: The History of an Idea in America* (New York: Oxford University Press, 1997), 58–60; Stephen Jay Gould, "American Polygeny and Craniometry Before Darwin: Blacks and Indians as Separate, Inferior Species," in *The "Racial" Economy of Science: Toward a Democratic Future*, ed. Sandra Harding (Bloomington: Indiana University Press, 1993), 84.

16. Gould, "American Polygeny and Craniometry Before Darwin," 95.

17. Ibid., 96–97.

18. Gossett, *Race*, 58.

19. Gould, "American Polygeny and Craniometry Before Darwin," 100.

20. Banton, "The Idiom of Race," 61.

21. Quoted in Joseph L. Graves Jr., *The Emperor's New Clothes: Biological Theories of Race and the Millennium* (New Brunswick, NJ: Rutgers University Press, 2002), 63.

22. Evelynn M. Hammonds and Rebecca M. Herzig eds., in *The Nature of Difference: Sciences of Race in the United States from Jefferson to Genomics* (Cambridge, MA: MIT Press, 2008), introduction to chapter 4, 106.

23. Ibid., 105, 106; Graves, *Emperor's New Clothes*, 74.

24. C. Vann Woodward, *The Strange Career of Jim Crow* (New York: Oxford University Press, 1966).

25. Carole A. Barrett, *American Indian History* (Pasadena, CA: Salem Press, 2003).

26. George M. Frederickson, *The Black Image in the White Mind* (New York: Harper & Row, 1971), 228–55.

27. C.W. Birnie, "The Influence of Environment and Race on Diseases," *Journal of the National Medical Association* 2 (1910): 243, reprinted in *Nature of Difference*, 129.

28. "The Impossibility of Acclimatizing Races," *Medical and Surgical Reporter* 5 (1861): 623, in *Nature of Difference*, 113.

29. Daniel J. Kevles, *In the Name of Eugenics: Genetics and the Uses of Human Heredity* (New York: Knopf, 1985), 8.

30. Francis Galton, *Inquiries into the Human Faculty* (New York: Macmillan, 1883), 24–25.

31. Francis Galton, *Eugenics: Its Definition, Scope and Aims* (London: Macmillan, 1905), 50.

32. Francis Galton, "Hereditary Talent and Character," *Macmillan's Magazine* 12 (1865): 318, 320.

33. Ibid., 321.

34. Francis Galton, *Hereditary Genius: An Inquiry into Its Laws and Consequences* (London: Macmillan, 1869; Rye Brook, NY: Elibron Classics, 2000), 339.

35. I discuss the regulation of black women's reproductive decision making at length in Dorothy Roberts, *Killing the Black Body: Race, Reproduction and the Meaning of Liberty* (New York: Vintage, 1999).

36. Stephen Trombley, *The Right to Reproduce: The History of Coercive Sterilization* (London: Weidenfeld & Nicolson, 1988), 49.

37. Quoted in Philip Reilly, *The Surgical Solution: A History of Involuntary Sterilization in the United States* (Baltimore, MD: Johns Hopkins University Press, 1991), 28.

38. Kevles, *In the Name of Eugenics*, 63.

39. Graves, *Emperor's New Clothes*, 115.

40. Kevles, *In the Name of Eugenics*, 45–47.

41. Ibid., 56; Reilly, *Surgical Solution*, 19–20.

42. Reilly, *Surgical Solution*, 18.

43. Harry Hamilton Laughlin, *The Legal and Administrative Aspects of Sterilization: Report of Committee to Study and Report on the Best Practical Means of Cutting Off the Defective Germ-Plasm in the American Population* (Cold Spring Harbor, NY: Eugenics Record Office, 1914).

44. Graves, *Emperor's New Clothes*, 119.

45. Stephen Jay Gould, *The Mismeasure of Man* (New York: W.W. Norton, 1981), 175–77.

46. William H. Tucker, *The Science and Politics of Racial Research* (Urbana: University of Illinois Press, 1994), 72–73.

47. Michael W. McConnell, "Originalism and the Desegregation Decisions," *Virginia Law Review* 81 (1995): 947, 1131, n. 856.

48. Stephen Jay Gould, "Carrie Buck's Daughter," *Constitutional Commentary* 2 (1985): 331.

49. *Buck v. Bell*, 274 U.S. 200 (1927).

50. Quoted in Reilly, *Surgical Solution*, 78.

51. Ibid., 64–65.

52. Mai M. Ngai, *Impossible Subject: Illegal Aliens and the Making of Modern America* (Princeton, NJ: Princeton University Press, 2005).

53. Madison Grant, *The Passing of the Great Race* (New York: Scribner, 1923), 60.

54. Tucker, *Science and Politics of Racial Research*, 93, quoting "The Great American Myth," editorial, *Saturday Evening Post*, May 7, 1921.

55. Jonathan Peter Spiro, *Defending the Master Race: Conservation, Eugenics, and the Legacy of Madison Grant* (Burlington: University of Vermont Press, 2008), 158.

56. Tucker, *Science and Politics of Racial Research*, 91–92, quoting Ellsworth Huntington, "Heredity and Responsibility," *Yale Review* 6 (1917): 305.

57. Lothrop Stoddard, *The Revolt Against Civilization* (New York: Scribner's, 1922), 30.

58. Quoted in James Graham Cook, *The Segregationists* (New York: Appleton-Century-Crofts, 1962), 65; see David M. Oshinksy, *Worse than Slavery: Parchment Farm and the Ordeal of Jim Crow Justice* (New York: Free Press, 1997), 229.

59. Cornelia Dean, "Nobel Winner Issues Apology for Comments About Blacks," *New York Times*, Oct. 19, 2007.

60. Reilly, *Surgical Solution*, 111–27; Stefan Kuhl, *The Nazi Connection: Eugenics, American Racism, and German National Socialism* (New York: Oxford University Press, 2002); Edwin Black, *War Against the Weak: Eugenics and America's Campaign to Create a Master Race* (New York: Dialog Press, 2008).

61. Graves, *Emperor's New Clothes*, 149.

62. Ibid., 149–50.

63. Jenny Reardon, *Race to the Finish: Identity and Governance in an Age of Genomics* (Princeton, NJ: Princeton University Press, 2005), 29–31.

64. Ibid., 27.

65. Mary L. Dudziak, *Cold War Civil Rights: Race and the Image of American Democracy* (Princeton, NJ: Princeton University Press, 2002).

66. Stuart B. Campbell, review of Carlton Putnam, *Race and Reason: A Yankee View* (Washington, DC: Public Affairs Press, 1961), *American Bar Association Journal* 48 (June 1962), 567.

67. Carleton S. Coon, *The Origin of Races* (New York: Knopf, 1962), ix–x.

68. Frank B. Livingstone, "On the Nonexistence of Human Races," in *The Concept of Race*, ed. Ashley Montagu (New York: Free Press, 1964), 46, 55.

69. Reardon, *Race to the Finish*, 61.

70. Ibid., 65.

71. Kevin Begos, "Against Their Will: North Carolina's Sterilization Program," *Winston-Salem Journal*, 2002, available at www.againsttheirwill.journalnow.com.

72. Danielle Deaver, "WFU Medical School Apologizes Again for Role," *Winston-Salem Journal*, Nov. 4, 2003.

73. Noting that "[t]he main reason more black children are living in poverty is that the people having the most children are the ones least capable of supporting them," a 1990 *Philadelphia Inquirer* editorial proposed reducing the number of children born to poor black women by implanting them with the long-acting contraceptive Norplant. Donald Kimelman, "Poverty and Norplant: Can Contraception Reduce the Underclass?" *Philadelphia Inquirer*, Dec. 12, 1990, A18.

74. Robert Cowan, "Lawmaker Urges Condoms for Border Control," Reuters, June 22, 2007.

75. The percentage of genetic sameness in the human species more recently has been estimated as 99.5 percent. See Charles N. Rotimi and Lynn B. Jorde, "Ancestry and Disease in the Age of Genomic Medicine," *New England Journal of Medicine* 363 (2010): 1551; Samuel Levy, Granger Sutton, Pauline C. Ng, Lars Feuk, Aaron L. Halpern, et al., "The Diploid Genome Sequence of an Individual Human" *PLoS Biology* 5 (2007): e254.

76. Deborah A. Bolnick, "Individual Ancestry Inference and the Reification of Race as a Biological Phenomenon," in *Revisiting Race in a Genomic Age*, 70, 73. See also Rotimi and Jorde, "Ancestry and Disease in the Age of Genomic Medicine," 1552 ("variation found outside Africa tends to be a subgroup of African genetic variation").

77. Rick A. Kittles and Kenneth M. Weiss, "Race, Ancestry, and Genes: Implications for Defining Disease Risk" *Annual Review of Genomics and Human Genetics* 4 (2003): 44.

78. Joseph L. Graves Jr., *The Race Myth* (New York: Dutton, 2004): 17.

79. Marcus W. Feldman, Richard C. Lewontin, and Mary-Claire King, "A Genetic Melting Pot," *Nature* 424 (2003): 374.

80. Richard S. Cooper, "Race and IQ: Molecular Genetics as Deux ex Machina," *American Psychologist* 60 (2005): 71.

81. Rotimi and Jorde, "Ancestry and Disease in the Age of Genomic Medicine," 1552.

82. David B. Goldstein and Gianpiero L. Cavalleri, "Understanding Human Diversity," *Nature* 437 (2005): 1241.

83. Richard C. Lewontin, "The Apportionment of Human Diversity," *Evolutionary Biology* 6 (1972): 381–98.

84. American Anthropological Association, *Response to OMB Directive 15* (Sept. 1997), available at http://www.aaanet.org/gvt/ombdraft.htm.

85. A.W.F. Edwards, "Human Genetic Diversity: Lewontin's Fallacy," *BioEssays* 25 (2003): 798.

86. Kittles and Weiss, "Race, Ancestry, and Genes," 37; Bolnick, "Individual Ancestry Inference," 72–73.

87. Marcus W. Feldman and Richard C. Lewontin, "Race, Ancestry and Medicine," in *Revisiting Race*, 93.

Chapter 3: Redefining Race in Genetic Terms

1. Constance Holden, "The Touchy Subject of 'Race,'" *Science* 322 (2008): 839.

2. Pamela Sankar and Mildred Cho, "Toward a New Vocabulary of Human Genetic Variation," *Science* 298 (2002): 1337.

3. Rick A. Kittles and Kenneth M. Weiss, "Race, Ancestry, and Genes: Implications for Defining Disease Risk," *Annual Review of Genomics & Human Genetics* 4 (2002): 33, 34.

4. Pamela Sankar, "Moving Beyond the Two-Race Mantra" in *Revisiting Race in a Genomic Age*, ed. Barbara A. Koenig, Sandra Soo-Jin Lee, and Sarah S. Richardson (New Brunswick, NJ: Rutgers University Press, 2008), 271.

5. David B. Goldstein and Gianpiero L. Cavalleri, "Understanding Human Diversity," *Nature* 437 (2005): 1241.

6. John Dupre, "What Genes Are and Why There Are No Genes for Race," in *Revisiting Race*, 49.

7. Deborah A. Bolnick, "Individual Ancestry Inference and the Reification of Race as a Biological Phenomenon," in *Revisiting Race*, 74; S.T. Kalinowski, "The Computer Program STRUCTURE Does Not Reliably Identify the Main Genetic Clusters Within Species: Simulations and Implications for Human Population Structure," *Heredity*, Aug. 4, 2010, 1, 5.

8. Noah A. Rosenberg, Jonathan K. Pritchard, James L. Weber, Howard M. Cann, Kenneth K. Kidd, et al., "Genetic Structure of Human Populations," *Science* 298 (2002): 2381.

9. Deborah A. Bolnick, "Individual Ancestry Inference and the Reification of Race as a Biological Phenomenon," in *Revisiting Race*, 76.

10. Ibid., 76–77; Kalinowski, "The Computer Program STRUCTURE Does Not Reliably Identify the Main Genetic Clusters Within Species," 4.

11. Sadaf Firasat, Shagufta Khaliq, Aisha Mohyuddin, Myrto Papaioannou, Chris Tyler-Smith, Peter A. Underhill, and Qasim Ayub, "Y-Chromosomal Evidence for a Limited Greek Contribution to the Pathan Population of Pakistan," *European Journal of Human Genetics* 15 (2007): 121.

12. Nicholas Wade, "Gene Study Identifies 5 Main Human Populations," *New York Times* Dec. 20, 2002, A1.

13. Noah A. Rosenberg, Saurabh Mahajan, Sohini Ramachandran, Chengfeng Zhao, Jonathan K. Pritchard, and Marcus Feldman, "Clines, Clusters, and the Effects of Study Design on the Inference of Human Population Structure," *PLoS Genetics* 1 (2005): 660, 668.

14. Rosenberg et al., "Genetic Structure of Human Populations," 2384.

15. Kenneth M. Weiss and Brian W. Lambert, "Does History Matter? Do the Facts of Human Variation Package Our Views or Do Our Views Package the Facts?" *Evolutionary Anthropology* 19 (2010): 92, 97.

16. Peter A. Chow-White, "The Informationalization of Race: Communication Technologies and the Human Genome in the Digital Age," *International Journal of Communication* 2 (2008): 1168, 1176.

17. Barbara A. Koenig, Sandra Soo-Jin Lee, and Sarah S. Richardson, "Introduction: Race and Genetics in a Genomic Age," in *Revisiting Race*, 3.

18. Jonathan Pritchard, "Where Does Diversity Come From? Genetic Approaches to Studying Human Evolution," lecture, Northwestern University Center on the Science of Diversity, Oct. 23, 2009.

19. Theodosius Dobzhansky, *Genetics and the Origin of Species* (New York: Columbia University Press, 1937), 138.

20. Theodosius Dobzhansky, "A Debatable Account of the Origin of Races," *Scientific American* 208 (1963): 169, 169–70.

21. Richard C. Lewontin, "Confusion About Human Races," Social Science Research Council, June 26, 2006, at http://raceandgenomics.ssrc.org/Lewontin/.

22. Author's interview of Charmaine Royal, Sept. 4, 2008, Durham, NC.

23. Author's telephone interview of Charles Rotimi, Nov. 19, 2008.

24. Author's telephone interview of Camara Jones, Nov. 18, 2008. See also Camara Phyllis Jones, "Invited Commentary: 'Race,' Racism, and the Practice of Epidemiology," *American Journal of Epidemiology* 154 (2001), 299, 300.

25. Deborah A. Bolnick, "Individual Ancestry Inference and the Reification of Race as a Biological Phenomenon," in *Revisiting Race*, 80–81.

26. Jenny Reardon, "Decoding Race and Human Difference in a Genomic Age," *Differences: A Journal of Feminist Cultural Studies* 15 (2004): 38, 53.

27. Neil Risch, Esteban Burchard, Elad Ziv, and Hua Tang, "Categorization of Humans in Biomedical Research: Genes, Race, and Disease," *Genome Biology* 3 (2002): 1–12.

28. Luca Cavalli-Sforza and Walter Bodmer, *The Genetics of Human Populations* (Minneola, NY: Dover, 1999), 698.

29. Risch et al., "Categorization of Humans," 4.

30. T.M. Baye and R.A. Wilke, "Mapping Genes That Predict Treatment Outcome in Admixed Populations," *Pharmacogenomics Journal* 10 (2010): 465, 473.

31. Risch et al., "Categorization of Humans."

32. S.O.Y. Keita and Rick A. Kittles, "The Persistence of Racial Thinking and the Myth of Racial Divergence," *American Anthropologist* 99 (1997): 534; Charles N. Rotimi and Lynn B. Jorde, "Ancestry and Disease in the Age of Genomic Medicine," *New England Journal of Medicine* 363 (2010): 1551, 1552.

33. Rick A. Kittles and Kenneth M. Weiss, "Race, Ancestry, and Genes: Implications for Defining Disease Risk," *Annual Review of Genomics and Human Genetics* 4 (2003): 33, 38.

34. Analabha Basul et al., "Ethnic India: A Genomic View, with Special Reference to Peopling and Structure," *Genome Research* 13 (2003): 2277.

35. Jenny Reardon, *Race to the Finish: Identity and Governance in an Age of Genomics* (Princeton, NJ: Princeton University Press, 2005).

36. L. Luca Cavalli-Sforza, "The Human Genome Diversity Project: Past, Present, and Future," *Nature Reviews Genetics* 6 (2005): 333.

37. Reardon, *Race to the Finish*, 2; Debra Harry, "The Human Genome Diversity Project and Its Implications for Indigenous Peoples," Indigenous Peoples Council on Biocolonialism, Jan. 1995, http://ipcb.org/publications/briefing_papers/files/hgdp.html.

38. Joseph L. Graves Jr., *The Race Myth* (New York: Dutton, 2004), 113.

39. Ibid.; Michael Bamshad et al., "Deconstructing the Relationship Between Genetics and Race," *Nature Reviews Genetics* 5 (2004): 598.

40. Michael J. Bamshad, Stephen Wooding, W. Scott Watkins, Christopher T. Ostler, Mark A. Batzer, and Lynn B. Jorde, "Human Population Structure and Inference of Group Membership," *American Journal of Human Genetics* 72 (2003): 578.

41. Mary-Claire King and Arno G. Motulsky, "Mapping Human History," *Science* 298 (2002): 2342.

42. L. Luca Cavalli-Sforza, Paolo Menozzi, and Alberto Piazza, *The History and Geography of Human Genes* (Princeton, NJ: Princeton University Press, 1994), 19.

43. Joan H. Fujimura, Ramya Rajagopalan, Pilar N. Ossorio, and Kjell A. Doksum, "Race and Ancestry: Operationalizing Populations in Human Genetic Variation Studies," in *What's the Use of Race? Modern Governance and the Biology of Difference*, ed. Ian Whitmarsh and David S. Jones (Cambridge, MA: MIT Press, 2010), 169, 172.

44. Hua Tang, Marc Coram, Pei Wang, Xiaofeng Zhu, and Neil Risch, "Reconstructing Genetic Ancestry Blocks in Admixed Individuals," *American Journal of Human Genetics* 79 (2006): 1.

45. Pritchard, "Where Does Diversity Come From?"

46. Baye and Wilke, "Mapping Genes," 466, figure 1.

47. Duana Fullwiley, "The Molecularization of Race: Institutionalizing Human Difference in Pharmacogenetics Practice," *Science as Culture* 16 (2007): 1.

48. Duana Fullwiley, "The Molecularization of Race and Institutions of Difference: Pharmacy and Public Science after the Genome," in *Revisiting Race*, 159.

49. Duana Fullwiley, "The Biologistical Construction of Race: 'Admixture' Technology and the New Genetic Medicine," *Social Studies of Science* 38, no. 5 (2008): 722.

50. Fullwiley, "Molecularization of Race and Institutions of Difference," 161–62.

51. George D. Kelsey, *Racism and the Christian Understanding of Man* (New York: Charles Scribner's, 1965), 36–43.

52. Anthony G. Greenwald and Mahzarin R. Banaji, "Implicit Social Cognition: Attitudes, Self Esteem, and Stereotypes," *Psychological Review* 102 (1995): 4; Phillip Atiba Goff et al, "Not Yet Human: Implicit Knowledge, Historical Dehumanization, and Contemporary Consequences," *Journal of Personality and Social Psychology* 94 (2008): 292.

53. Kelsey, *Racism and the Christian Understanding of Man*, 41.

54. Linda M. Hunt and Mary S. Megyesi, "The Ambiguous Meaning of the Racial/Ethnic Categories Routinely Used in Human Genetics Research," *Social Science & Medicine* 66 (2008): 349, 353.

55. Ibid., 355.

56. Ibid., 357.

57. Alexandra E. Shields, Michael Fortun, Evelynn M. Hammonds, Patricia A. King, Caryn Lerman, Rayna Rapp, and Patrick F. Sullivan, "The Use of Race Variables in Genetic Studies of Complex Traits and the Goal of Reducing Health Disparities: A Transdisciplinary Perspective," *American Psychologist* 60 (2005): 77, 82.

58. "Census, Race and Science," *Nature Genetics* 24 (2000): 97, 99.

59. Margaret A. Winker, "Race and Ethnicity in Medical Research: Requirements Meet Reality," *Journal of Law, Medicine & Ethics* 34 (2006): 520.

60. Hunt and Megyesi, "Ambiguous Meaning," 359.

61. Rick J. Carlson, "The Case of BiDil: A Policy Commentary on Race and Genetics," *Health Affairs* Web Exclusive, Oct. 11, 2005, W5-464, W5-466, http://content.healthaffairs.org/cgi/content/full/hlthaff.w5.464/DC1.

62. Hunt and Megyesi, "Ambiguous Meaning," 356.

63. Fullwiley, "Molecularization of Race and Institutions of Difference," 165.

64. Joan Fujimura and Ramya Rajagopalan, "Different Differences: The Use of 'Genetic Ancestry' Versus Race in Biomedical Human Genetic Research," *Social Studies of Science*, published online, Dec. 7, 2010, http://sss.sagepub.com/content/early/2010/11/30/0306312710379170.full.pdf+html.

65. Ibid.

66. Jonathan Marks, *What It Means to Be 98% Chimpanzee* (Berkeley: University of California Press, 2002), 66.

67. Lundy Braun and Evelynn Hammonds, "Race, Populations, and Genomics: Africa as Laboratory," *Social Science & Medicine* 67 (2008): 1580, 1581.

68. Ibid.

69. Kelsey, *Racism and the Christian Understanding of Man*, 9.

70. Ashley Montagu, *Man's Most Dangerous Myth: The Fallacy of Race* (1942; Walnut Creek, CA: AltaMira Press, 1997), 23.

71. Judith S. Neulander, "Folk Taxonomy, Prejudice and the Human Genome: Using Disease as a Jewish Ethnic Marker," *Patterns of Prejudice* 40 (2006): 381, 390.

72. "Randomised Trial of Intravenous Streptokinase, Oral Aspirin, Both or Neither Among 17,187 Cases of Suspected Acute Myocardial Infarction: ISIS-2," *Lancet* (1988): 349–60.

73. Peter Sleight, "Debate: Subgroup Analyses in Clinical Trials—Fun to Look At, but Don't Believe Them!" *Current Controlled Trials in Cardiovascular Medicine* 1 (2000): 25; Jonah Lehrer, "The Truth Wears Off," *New Yorker*, Dec. 13, 2010, 52, 57 (discussing "scientific accidents"; "when the experiments are done, we still have to choose what to believe").

74. Stephen Jay Gould, "American Polygeny and Craniometry Before Darwin: Blacks and Indians as Separate, Inferior Species," in *The "Racial" Economy of Science: Toward a Democratic Future*, ed. Sandra Harding (Bloomington: Indiana University Press, 1993); Joseph L. Graves Jr., *The Emperor's New Clothes: Biological Theories of Race at the Millennium* (New Brusnwick, NJ: Rutgers University Press, 2001).

75. Richard Thaler, "Thaler's Question," *Edge*, Nov. 23, 2010, available at http://edge.org/3rd_culture/thaler10/thaler10_index.html.

Chapter 4: Medical Stereotyping

1. David Satcher et al., "What If We Were Equal? A Comparison of the Black–White Mortality Gap in 1960 and 2000," *Health Affairs* 24 (2005): 459.

2. David Williams, "Racism and Health: Needed Contributions by Social and Biological Scientists," talk delivered at "Genetics and Its Other: Minorities, Race, and Health Inequities in Medicine," Center for the Study of Gene Structure and Function, Hunter College, and New York Academy of Sciences, December 9, 2005.

3. Robert S. Levine, James E. Foster, and Robert E. Fullilove, "Black–White Inequalities in Mortality and Life Expectancy, 1933–1999: Implications for Healthy People 2010," *Public Health Reports* 116 (2001): 474.

4. Nicholas Bakalar, "Life Expectancy Is Declining in Some Pockets of the Country," *New York Times*, Apr. 22, 2008, D5.

5. Centers for Disease Control and Prevention, *CDC Health Disparities and Inequalities Report—United States, 2011*, MMWR 60 (Supplement) (Jan. 14, 2011); National Center for Health Statistics. *Health, United States, 2009: With Special Feature on Medical Technology* (Hyattsville, MD: Department of Health and Human Services, 2010).

6. Harriet Washington, Martin Luther King Jr. Day talk, Northwestern University School of Law, Jan. 18, 2010; see also Harriet A. Washington, *Medical Apart-*

heid: The Dark History of Medical Experimentation on Black Americans from Colonial Times to the Present (New York: Doubleday, 2006).

7. Esteban Gonzáles Burchard et al., "Latino Populations: A Unique Opportunity for the Study of Race, Genetics, and Social Environment in Epidemiological Research," *American Journal of Public Health* 95 (2005): 2161, 2163; John W. Reich, Alex J. Zautra, and John Stuart Hall, "Resilience: A New Definition of Health for People and Communities," in *Handbook of Adult Resilience*, ed. John W. Reich, Alex J. Zautra, and John Stuart Hall (New York: Guilford Press, 2010), 3; Robert A. Hummer, Daniel A. Powers, Starling G. Pullum, Ginger L. Gossman, and W. Parker Frisbie, "Paradox Found (Again): Infant Mortality Among the Mexican-Origin Population in the United States," *Demography* 44 (2007): 441.

8. Jose A. Serpa et al., "Tuberculosis Disparity Between US-Born Blacks and Whites," *Emerging Infectious Diseases* 15 (2009): 899; Brian D. Smedley, Adrienne Y. Stith, and Alan R. Nelson, eds., *Unequal Treatment: Confronting Racial and Ethnic Disparities in Health Care* (Washington, DC: National Academy Press, 2003), 29, available at www.nap.edu/catalog.php?record_id=10260#toc; Kevin Sack, "Research Finds Wide Disparities in Health Care by Race and Region," *New York Times*, June 5, 2008, A18.

9. Pam Belluck, "New Hopes for Reform in Indian Health Care," *New York Times*, Dec. 2, 2009, A1.

10. Lundy Braun, "Race, Ethnicity and Health: Can Genetics Explain Disparities?" *Perspectives in Biology and Medicine* 45 (2002): 159, 161.

11. David Waldstreicher, ed., *Notes on the State of Virginia by Thomas Jefferson* (New York: Palgrave 2002), 176–77.

12. Ibid.

13. Benjamin Banneker's letter and Thomas Jefferson's reply are reprinted in Evelynn M. Hammonds and Rebecca M. Herzig, *The Nature of Difference: Sciences of Race in the United States from Jefferson to Genomics* (Cambridge, MA: MIT Press, 2008), 33–36.

14. Lerone Bennett Jr., *Before the Mayflower: A History of the Negro in America* (New York: Penguin Books, 1964), 66.

15. Samuel Kelton Roberts Jr., *Infectious Fear: Politics, Disease, and the Health Effects of Segregation* (Chapel Hill: University of North Carolina Press, 2009), 45.

16. Allan M. Brandt, "Racism and Research: The Case of the Tuskegee Syphilis Experiment," in *Tuskegee's Truths: Rethinking the Tuskegee Syphilis Study*, ed. Susan M. Reverby (Chapel Hill: University of North Carolina Press, 2000), 15; Washington, *Medical Apartheid*, 157–85.

17. Washington, *Medical Apartheid*, 157.

18. James Jones, "The Tuskegee Syphilis Experiment: 'A Moral Astigmatism,'" in *The "Racial" Economy of Science: Toward a Democratic Future*, ed. Sandra Harding (Bloomington: Indiana University Press, 1993), 275.

19. Ibid., 284.

20. Thomas W. Murrell, "Syphilis and the American Negro: A Medico-Sociologic Study," *Journal of the American Medical Association* 54 (1910): 846.

21. Ibid., 847.

22. Roberts, *Infectious Fear*, 47.

23. Ibid., 52.

24. On the black medical profession's resistance to racist medical practices, 1861–1900, see W. Michael Byrd and Linda A. Clayton, *An American Health Dilemma*, vol. 1; *A Medical History of African Americans and the Problem of Race: Beginnings to 1900* (New York: Routledge, 2000), 384–406.

25. Roberts, *Infectious Fear*, 49.

26. Ibid., 187.

27. Sander Gilman, *Jewish Frontiers: Essays on Bodies, Histories, and Identities* (New York: Palgrave McMillan, 2003), 89.

28. Ibid.; Shelley Z. Reuter, "The Genuine Jewish Type: Racial Ideology and Anti-Immigrationism in Early Medical Writing about Tay-Sachs Disease," *Canadian Journal of Sociology* 31 (2006): 291, 307; Sherry I. Brandt-Rauf et al., "Ashkenazi Jews and Breast Cancer: The Consequences of Linking Ethnic Identity to Genetic Disease," *American Journal of Public Health* 96 (2006): 1979.

29. Reuter, "Genuine Jewish Type," 310.

30. Ibid., 308.

31. Ibid., 314.

32. Natalia Molina, *Fit to Be Citizens? Public Health and Race in Los Angeles, 1879–1939* (Berkeley: University of California Press, 2006).

33. Ibid., 26.

34. *Ex parte Quong Wo*, 161 Cal. 220, 118 Pac. 714 (Cal. 1911).

35. Molina, *Fit to Be Citizens?* 63.

36. Ibid., 95–96.

37. Ibid., 179.

38. Samuel A. Cartwright, "Report on the Diseases and Physical Peculiarities of the Negro Race," *New Orleans Medical and Surgical Journal* 7 (1851): 691.

39. Ibid.

40. Jonathan M. Metzl, *The Protest Psychosis: How Schizophrenia Became a Black Disease* (Boston: Beacon Press, 2009), xiv.

41. Keith Wailoo, *Dying in the City of the Blues: Sickle Cell Anemia and the Politics of Race and Health* (Chapel Hill: University of North Carolina Press, 2001); Lisa C. Ikemoto, "Race to Health: Racialized Discourses in a Transhuman World," *DePaul Journal of Health Care Law* 9 (2005): 1101, 1115, n. 77.

42. Margaret A. Winker, "Race and Ethnicity in Medical Research: Requirements Meet Reality," *Journal of Law, Medicine & Ethics* 34 (2006): 520; Richard S. Garcia, "The Misuse of Race in Medical Diagnosis," *Pediatrics* 113 (2004): 1394.

43. Sally L. Satel, "I Am a Racially Profiling Doctor," *New York Times Magazine*, May 5, 2002.

44. Quoted in Washington, *Medical Apartheid*, 58.

45. Robert F. Kaiko, Stanley L. Wallenstein, Ada G. Rogers, and Raymond W. Houde, "Sources of Variation in Analgesic Responses in Cancer Patients with Chronic Pain Receiving Morphine," *Pain* 15 (1983): 191.

46. Donald A. Barr, *Health Disparities in the United States: Social Class, Race, Ethnicity, and Health* (Baltimore: Johns Hopkins University Press, 2008), 187.

47. Knox Todd, Nigel Samaroo, and Jerome R. Hoffman, "Ethnicity as a Risk Factor for Inadequate Emergency Department Analgesia," *Journal of the American Medical Association* 269 (1993): 1537.

48. Knox H. Todd, Christi Deaton, Anne P. D'Adamo, and Leon Goe, "Ethnicity and Analgesic Practice," *Annals of Emergency Medicine* 35 (2000): 11.

49. Joshua H. Tamayo-Sarver, Susan H. Hinze, Rita K. Cydulka, and David W. Baker, "Racial and Ethnic Disparities in Emergency Department Analgesic Prescription," *American Journal of Public Health* 93 (2003): 2067.

50. Barr, *Health Disparities in the United States*, 190.

51. Stephen Crystal et al., "Broadened Use of Atypical Antipsychotics: Safety, Effectiveness, and Policy Challenges," *Health Affairs* 28 (2009): 770.

52. Duff Wilson, "Poor Children Likelier to Get Antipsychotics," *New York Times*, Dec. 12, 2009, A1.

53. Sally L. Satel, "Erratum?" April 2006, www.sallysatelmd.com/html/a-nytimes3.html.

54. Smedley, Stith, and Nelson, eds., *Unequal Treatment*.

55. M. Gregg Bloche, "American Medicine and the Politics of Race," *Perspectives in Biology and Medicine* 48 (2005): S54, S55.

56. Richard A. Epstein, "Disparities and Discrimination in Health Care Coverage: A Critique of the Institute of Medicine Study," *Perspectives in Biology and Medicine* 48 (2005): S26, S40.

57. See Byrd and Clayton, *An American Health Dilemma*, vol. 1; W. Michael Byrd and Linda A. Clayton, *An American Health Dilemma*, vol. 2; *Race, Medicine, and Health Care in the United States, 1900–2000* (New York: Routledge, 2002).

58. Author interview of Michael Byrd and Linda Clayton, Nov. 10, 2007, Boston, MA.

59. Jay N. Cohn, "The Use of Race and Ethnicity in Medicine: Lessons from the African-American Heart Failure Trial," *Journal of Law, Medicine & Ethics* 34 (2006): 552.

60. Garcia, "Misuse of Race in Medical Diagnosis."

61. Jay S. Kaufman and Richard S. Cooper, "Use of Racial and Ethnic Identity in Medical Evaluations and the Biology of Difference," in *What's the Use of Race?* ed. Ian Whitmarsh and David S. Jones (Cambridge, MA: MIT Press, 2010): 187, 201–2.

62. Warwick Anderson, "Teaching 'Race' at Medical School: Social Scientists on the Margin," *Social Studies of Science* 38 (2008): 785.

63. Jonathan Klick and Sally Satel, *The Health Disparities Myth: Diagnosing the Treatment Gap* (Washington, DC: American Enterprise Institute Press, 2006), 23.

64. Paul S. Chan et al., "Delayed Time to Defibrillation After In-Hospital Cardiac Arrest," *New England Journal of Medicine* 358 (2008): 9.

65. Elizabeth H. Bradley et al., "Racial and Ethnic Differences in Time to Acute Reperfusion Therapy for Patients Hospitalized with Myocardial Infarction," *Journal of the American Medical Association* 292 (2004): 1563.

66. Ibid., 1572.

67. Peter B. Bach et al., "Primary Care Physicians Who Treat Blacks and Whites," *New England Journal of Medicine* 351 (2004): 575.

68. Ashwin N. Ananthakrishnan and Kia Saeian, "Racial Differences in Liver Transplantation Outcomes in the MELD Era," *American Journal of Gastroenterology* 103 (2001): 901; R.S.D. Higgins and J.A. Fishman, "Disparities in Solid Organ Transplantation for Ethnic Minorities: Facts and Solutions," *American Journal of Transplantation* 6 (2006): 2556–62; Robert D. Gibbons, "Racial Disparities in Liver Transplantation," *Liver Transplantation* 10 (2004): 842–43.

69. Kevin Sack, "Research Finds Wide Disparities in Health Care by Race and Region," *New York Times*, June 5, 2008, A18.

70. Marian E. Gornick et al., "Effects of Race and Income on Mortality and Use of Services Among Medicare Beneficiaries," *New England Journal of Medicine* 355 (1996): 791.

71. Evelynn M. Hammonds and Rebecca M. Herzig, introduction to chapter 2, in *Nature of Difference*, 15, 16.

72. Samuel A. Cartwright, "Report on the Diseases and Physical Peculiarities of the Negro Race," *New Orleans Medical and Surgical Journal* 7 (1851): 691.

73. Rebecca Skloot, *The Immortal Life of Henrietta Lacks* (New York: Crown, 2010), 256.

Chapter 5: The Allure of Race in Biomedical Research

1. Steven Epstein, *Inclusion: The Politics of Difference in Medical Research* (Chicago: University of Chicago Press, 2007).

2. Ibid., 5.

3. U.S. Department of Health & Human Services, *Report of the Secretary's Task Force on Black and Minority Health*, vol. 1: *Executive Summary* (Washington, DC: U.S. Government Printing Office, 1985).

4. U.S. Department of Health and Human Services, *NIH Policy and Guidelines on the Inclusion of Women and Minorities as Subjects in Clinical Research—Amended, October, 2001*, available at http://grants.nih.gov/grants/funding/women_min/guidelines_amended_10_2001.htm.

5. Otis W. Brawley, "Response to 'Inclusion of Women and Minorities in Clinical Trials and the NIH Revitalization Act of 1992—the Perspective of NIH Clinical Trialists,'" *Controlled Clinical Trials* 16 (1995): 293.

6. Patricia A. King, "Race, Justice, and Research," in *Beyond Consent: Seeking Justice in Research*, ed. Jeffrey P. Kahn, Anna C. Mastroianni, and Jeremy Sugarman (New York: Oxford University Press, 1998), 88, 91; Michael Omi, "Racial Identity and the State: The Dilemmas of Classification," *Law & Inequality* 15 (1997): 7; Martha Minow, *Making All the Difference: Inclusion, Exclusion, and American Law* (Ithaca, NY: Cornell University Press, 1990).

7. Epstein, *Inclusion*, 151.

8. Ibid., 150.

9. Author's interview of Mildred Cho, June 11, 2008, Stanford, CA.

10. Epstein, *Inclusion*, 79.

11. John D. Lantos, "The 'Inclusion Benefit' in Clinical Trials," *Journal of Pediatrics* 134 (1999): 130; see also Steven Epstein, *Impure Science: AIDS, Activism, and the Politics of Knowledge* (Berkeley: University of California Press, 1996).

12. Jennifer M. Orsi, Helen Margellos-Anast, and Steven Whitman, "Black–White Health Disparities in the United States and Chicago: A 15-Year Progress Analysis," *American Journal of Public Health* 100 (2010): 349.

13. U.S. Department of Health & Human Services, *National Health Care Disparities Report 2007* (Rockville, MD: Agency for Health Care Research and Quality, 2007), 1.

14. Jonathan Kozol, *Ordinary Resurrections: Children in the Years of Hope* (New York: HarperCollins, 2001), 95.

15. Elissa Ely, "House Dust Yields Clue to Asthma: Roaches," *New York Times*, Apr. 7, 2009, D5.

16. Diane R. Gold and Rosalind J. Wright, "Population Disparities in Asthma," *Annual Review of Public Health* 26 (2005): 89; Rosalind J. Wright and S.V. Subramanian, "Advancing a Multilevel Framework for Epidemiological Research on Asthma Disparities," *Chest* 132 (2007): 757, 758.

17. Jiyoun Kim, Andrew C. Merry, Jean A. Nemzek, Gerry L. Bolgos, Javed Siddiqui, and Daniel G. Remick, "Eotaxin Represents the Principal Eosinophil Chemoattractant in a Novel Murine Asthma Model Induced by House Dust Containing Cockroach Allergens," *Journal of Immunology* 167 (2001): 2808.

18. Ariel Spira-Cohen, Lung Chi Chen, Michaela Kendall, Rebecca Sheesley, and George D. Thurston, "Personal Exposures to Traffic-Related Particle Pollution Among Children with Asthma in the South Bronx, NY," *Journal of Exposure Science and Environmental Epidemiology* 20 (2009): 446–56; Manny Fernandez, "Study Links Truck Exhaust to Schoolchildren's Asthma," *New York Times*, Oct. 29, 2006, A26; Wright and Subramanian, "Advancing a Multilevel Framework," 760.

19. Author interview of Esteban Burchard, June 17, 2008, San Francisco, CA. For a critical analysis of the use of racially classified DNA in the search for genetic susceptibility for type 2 diabetes among Mexicans, see Michael J. Montoya, "Bioethnic Conscription: Genes, Race, and Mexicana/o Ethnicity in Diabetes Research," *Cultural Anthropology* 22 (2007): 94.

20. Max A. Siebold, Bin Wang, Celeste Eng, et al., "An African-Specific

Functional Polymorphism in KCNMB1 Shows Sex-Specific Association with Asthma Severity," *Human Molecular Genetics* 17 (2008): 2681.

21. Duana Fullwiley reports a more nuanced statement about race and gender in her earlier interview with Burchard: "Race is a defined set of individuals that have genetic similarity. However, race is not purely biological like gender is—X or Y chromosomes. Race has a biological component that is largely shaped by environmental factors, sociologic factors, geographic factors—mountains, large oceans separating populations—and those aren't biologic in nature, but because they are affecting a biologic process, they are of biologic importance." Duana Fullwiley, "Race and Genetics: Attempts to Define the Relationship," *Biosocieties* 2 (2007): 221, 234. See also Duana Fullwiley, "The Biologistical Construction of Race: 'Admixture' Technology and the New Genetic Medicine," *Social Studies of Science* 38, no. 5 (2008): 695.

22. Zachary A.-F. Kistka, Lisanne Palomar, Kirstin A. Lee, et al., "Racial Disparity in the Frequency of Recurrence of Preterm Births," *American Journal of Obstetrics & Gynecology* 196 (2007): 131.

23. David M. Stamilio, Gilad A. Gross, Anthony Shanks, et al., "Discussion: 'Racial Disparity in Preterm Birth' by Kistka et al.," *American Journal of Obstetrics & Gynecology* 196 (2007): e1.

24. Nicholas Bakalar, "Study Points to Genetics in Disparities in Preterm Births," *New York Times*, Feb. 27, 2007, F5.

25. For example, in a 2000 article in *Demography*, behavioral geneticists Edwin van den Oord and David Rowe, in their words, "speculated that maternal genes could play a role in addition to the traditional sociodemographic risk factors" in an effort to "advance research on birth weight differences between black and white infants." Edwin J.C.G. van den Oord and David Rowe, "Racial Differences in Birth Health Risk: A Quantitative Genetic Approach," *Demography* 37 (2000): 285. They defined races as "genetic entities" that formed because "generations of 'reproductive isolation' have led to differences in gene frequency across racial groups." In response to criticism of their "anachronistic theory of race," van den Oord and Rowe pointed to differences between African Americans and European Americans in the frequencies of alleles for skin color, cystic fibrosis, sickle cell, and "a candidate gene for psychiatric conditions." Edwin J.C.G. van den Oord and David Rowe, "A Step in Another Direction: Looking for Maternal Genetic and Environmental Effects on Racial Differences in Birth Weight," *Demography* 38 (2001): 573.

26. Keith Wailoo, *Dying in the City of the Blues: Sickle Cell Anemia and the Politics of Race and Health* (Chapel Hill: University of North Carolina Press, 2001), 78; see also Keith Wailoo and Stephen Pemberton, *The Troubled Dream of Genetic Medicine: Ethnicity and Innovation in Tay-Sachs, Cystic Fibrosis, and Sickle Cell Disease* (Baltimore, MD: Johns Hopkins University Press, 2006).

27. Troy Duster, *Backdoor to Eugenics* (New York: Routledge 1990); Dorothy Roberts, *Killing the Black Body: Race, Reproduction and the Meaning of Liberty* (New York: Vintage, 1998), 256–58.

28. Lundy Braun, "Race, Ethnicity, and Health," *Perspectives in Biology and Medicine* 45 (2002): 159, 167.

29. Jared Diamond, "Race Without Color," *Discover*, Nov. 1994, 82.

30. Jay S. Kaufman and Richard S. Cooper, "In Search of the Hypothesis," *Public Health Reports* 110 (1995): 664.

31. Richard S. Cooper and Jay S. Kaufman, "Race and Hypertension," *Hypertension* 32 (1998): 813.

32. Lizzy M. Brewster, Joseph F. Clark, and Gert A. van Montfrans, "Is Greater Tissue Activity of Creatine Kinase the Genetic Factor Increasing Hypertension Risk in Black People of Sub-Saharan African Descent?" *Journal of Hypertension* 18 (2000): 1537.

33. Mehmet Oz, "Your Questions Answered!" Jan. 1, 2006, www.oprah.com/health/Your-Questions-Answered/12.

34. Philip D. Curtin, "Hypertension Among African Americans: The Historical Evidence," *American Journal of Public Health* 82 (1992): 1681; Fatimah Linda Collier Jackson, "An Evolutionary Perspective on Salt, Hypertension, and Human Genetic Variability," *Hypertension* 17 (1991), I129–32; Jay Kaufman, "The Anatomy of a Medical Myth," *Is Race Real?* Social Science Research Council, June 7, 2006, http://raceandgenomics.ssrc.org/Kaufman.

35. Richard S. Cooper et al., "An International Comparative Study of Blood Pressure in Populations of European vs. African Descent," *BMC Medicine* 3 (2005).

36. Kaufman, "Anatomy of a Medical Myth."

37. Stephen J. Dubner, "Toward a Unified Theory of Black America," *New York Times Magazine*, Mar. 20, 2005.

38. David M. Cutler, Roland G. Fryer Jr., and Edward L. Glaeser, "Racial Differences in Life Expectancy: The Impact of Salt, Slavery, and Selection," Mar. 1, 2005, available at http://isites.harvard.edu/fs/docs/icb.topic185351.files/salt_science_submission_3-01.pdf.

39. Dubner, "Toward a Unified Theory."

40. Hasan Shanawani et al., "Non-Reporting and Inconsistent Reporting of Race and Ethnicity in Articles That Claim Association Among Genotype, Outcome and Race or Ethnicity," *Journal of Medical Ethics* 32 (2006): 724.

41. Author's telephone interview of Charles Rotimi, Nov. 19, 2008.

42. Joseph L. Graves Jr., *The Race Myth* (New York: Dutton, 2004), 142–43.

43. Richard Cooper and Jay Kaufman have written extensively on the problem of residual confounding in research on racial health disparities. See, for example, Jay S. Kaufman, Richard S. Cooper, and D.L. McGee, "Socioeconomic Status and Health in Blacks and Whites: The Problem of Residual Confounding and the Resiliency of Race," *Epidemiology* 8 (1997): 621; Cooper and Kaufman, "Race and Hypertension"; Jay S. Kaufman and Richard S. Cooper, "Seeing Causal Explanations in Social Epidemiology," *American Journal of Public Health* 150 (1999): 113; Jay S. Kaufman and Richard S. Cooper, "Commentary: Considerations for Use of Racial/

Ethnic Classification in Etiologic Research," *American Journal of Epidemiology* 154 (2001): 291; Richard S. Cooper, Jay S. Kaufman, and Ryk Ward, "Race and Genomics," *New England Journal of Medicine* 348 (2003): 1166; Jay S. Kaufman, "Epidemiologic Analysis of Racial/Ethnic Disparities: Some Fundamental Issues and a Cautionary Example," *Social Science & Medicine* 66 (2008): 1659.

44. Kurt Bauman, *Direct Measures of Poverty as Indicators of Economic Need: Evidence from the Survey of Income and Program Participation* (Washington, DC: U.S. Census Bureau, 1998).

45. David R. Williams, Michelle Sternthal, and Rosalind J. Wright, "Social Determinants: Taking the Social Context of Asthma Seriously," *Pediatrics* 123 (2010): S174, S175.

46. Michael Luo, "'Whitening' the Resume," *New York Times*, Dec. 6, 2009, Week in Review, 3; Michael Luo, "In Job Hunt, Even a College Degree Can't Close the Racial Gap," *New York Times*, Dec. 1, 2009, A1.

47. Otis W. Brawley, "Is Race Really a Negative Prognostic Factor for Cancer?" *Journal of the National Cancer Institute* 101 (2009): 970.

48. See Camara Jones, "'Race,' Racism, and the Practice of Epidemiology," *American Journal of Epidemiology* 154 (2001): 299.

49. Catherine Elton, "Why Racial Profiling Persists in Medical Research," *Time*, Aug. 22, 2009.

50. Kaufman and Cooper, "In Search of the Hypothesis."

51. Nancy Krieger, "Stormy Weather: Race, Gene Expression, and the Science of Health Disparities," *American Journal of Public Health* 95 (2005): 2155, 2156, quoting Theresa Overfield, *Biologic Variation in Health and Illness: Race, Age, and Sex* (Boca Raton, FL: CRC Press, 1995).

52. Krieger, "Stormy Weather," 2157.

53. Paul Ehrlich and Marcus Feldman, "Genes, Environments & Behaviors," *Daedalus* 136 (2007): 5–12.

54. Cooper and Kaufman, "Race and Hypertension," 815.

55. Francis S. Collins, "What We Do and Don't Know About 'Race,' 'Ethnicity,' Genetics and Health at the Dawn of the Genome Era," *Nature Genetics* 36 (2004): S13.

56. Krieger, "Stormy Weather," 2159.

57. Daniel Weintraub, "Workers' Comp System Needs Small Adjustments," *San Jose Mercury News*, Feb. 15, 2008.

Chapter 6: Embodying Race

1. Jocelyn Hirschman, Steven Whitman, and David Ansell, "The Black–White Disparity in Breast Cancer Mortality: The Example of Chicago," *Cancer Causes & Control* 18 (2007): 323.

2. Shane Tritsch, "The Deadly Difference," *Chicago*, Oct. 2007, 120.

3. Author interview of Steven Whitman, July 9, 2009, Chicago, IL.

4. U.S. Census Bureau, 2006–2008 American Community Survey, http://factfinder.census.gov/servlet/ADPTable?_bm=y&-geo_id=16000US1714000&-qr_name=ACS_2008_3YR_G00_DP3YR3&-ds_name=ACS_2008_3YR_G00_&-_lang=en&-_sse=on.

5. Tritsch, "Deadly Difference," 182.

6. Amal N. Trivedi, William Rakowski, and John Z. Ayanian, "Effect of Cost Sharing on Screening Mammography in Medical Health Plans," *New England Journal of Medicine* 358 (2008): 375.

7. "Health Insurance Co-Payments Deter Mammography Use," *Medical News*, Jan. 24, 2008.

8. David Ansell et al., "A Community Effort to Reduce the Black/White Breast Cancer Mortality Disparity in Chicago," *Cancer Causes & Control* 20 (2009): 1681.

9. Tritsch, "Deadly Difference," 183.

10. Ansell, "Community Effort."

11. William P. McWhorter and William J. Mayer, "Black/White Differences in Type of Initial Breast Cancer Treatment and Implications for Survival," *American Journal of Public Health* 77 (1987): 1515.

12. Mary Jo Lund et al., "Parity and Disparity in First Course Treatment of Invasive Breast Cancer," *Breast Cancer Research and Treatment* 109 (2008): 545.

13. Nancy N. Baxter, "Equal for Whom? Addressing Disparities in the Canadian Medical System Must Become a National Priority," *Canadian Medical Association Journal* 177 (2007): 1522.

14. See WHO Commission on Social Determinants of Health, *Closing the Gap in a Generation: Health Equity Through Action on Social Determinants of Health* (Geneva: World Health Organization, 2008).

15. Donald A. Barr, *Health Disparities in the United States: Social Class, Race, Ethnicity, and Health* (Baltimore, MD: Johns Hopkins University Press, 2008); Evelyn M. Kitagawa and Philip M. Hauser, *Differential Mortality in the United States: A Study in Socioeconomic Epidemiology* (Cambridge, MA: Harvard University Press, 1973); Nancy E. Adler and David H. Rehkopf, "U.S. Disparities in Health: Descriptions, Causes, and Mechanisms," *Annual Review of Public Health* 29 (2008): 235; Michael G. Marmot, Manolis Kogevinas, and Mary Anne Elston, "Social/Economic Status and Disease," *Annual Review of Public Health* 8 (1987): 111.

16. Michael G. Marmot, Martin J. Shipley, and Peter J. Hamilton, "Employment Grade and Coronary Heart Disease in British Civil Servants," *Journal of Epidemiology and Community Health* 32 (1978): 244; Michael G. Marmot, Martin J. Shipley, and Geoffrey Rose, "Inequalities in Death: Specific Explanations of a General Pattern?" *Lancet* 323 (1984): 1003.

17. Thomas A. LaViest, "On the Study of Race, Racism, and Health: A Shift from Description to Explanation," *International Journal of Human Services* 30 (2000): 217, 218.

18. See Clarence C. Gravlee, "How Race Becomes Biology: Embodiment of Social Inequality," *American Journal of Physical Anthropology* 139 (2009): 47.

19. Troy Duster, "Buried Alive: The Concept of Race in Science," *Chronicle of Higher Education* 48 (2001): B11; see also Pilar Ossorio and Troy Duster, "Controversies in Biomedical, Behavioral, and Forensic Sciences," *American Psychologist* 60 (2005): 115.

20. Nancy Krieger and Mary Bassett, "The Health of Black Folk: Disease, Class and Ideology," *Monthly Review* 38 (1986): 74; Madeline Drexler, "The People's Epidemiologists," *Harvard Magazine*, Mar.–Apr. 2006, 25.

21. Nancy Krieger, "Stormy Weather: Race, Gene Expression, and the Science of Health Disparities," *American Journal of Public Health* 95 (2005): 2155, 2159.

22. Nancy Krieger, "Does Racism Harm Health? Did Child Abuse Exist Before 1962? On Explicit Questions, Critical Science, and Current Controversies: An Ecosocial Perspective," *American Journal of Public Health* 93 (2003): 194.

23. Author interview of Nancy Krieger, Apr. 26, 2008, Cambridge, MA.

24. Nancy Krieger and Stephen Sidney, "Racial Discrimination and Blood Pressure: The CARDIA Study of Young Black and White Adults," *American Journal of Public Health* 86 (1996): 1370.

25. Nancy Krieger, "Embodying Inequality: A Review of Concepts, Measures, and Methods for Studying Health Consequences of Discrimination," *International Journal of Health Services* 29 (1999): 295.

26. Quoted in Drexler, "People's Epidemiologists," 33.

27. Sarah Mustillo, Nancy Krieger, Erica P. Gunderson, et al., "Self-Reported Experiences of Racial Discrimination and Black–White Differences in Preterm and Low-Birthweight Deliveries: The CARDIA Study," *American Journal of Public Health* 94 (2004): 2125.

28. James W. Collins, Richard J. David, Arden Handler, et al., "Very Low Birthweight in African American Infants: The Role of Maternal Exposure to Interpersonal Racial Discrimination," *American Journal of Public Health* 94 (2004): 2132.

29. Nancy Krieger, Jarvis T. Chen, Pamela D. Waterman, David H. Rehkopf, and S.V. Subramanian, "Painting a Truer Picture of US Socioeconomic and Racial/Ethnic Health Inequalities: The Public Health Disparities Geocoding Project," *American Journal of Public Health* 95 (2005): 312–23; Nancy Krieger, Jarvis T. Chen, Pamela D. Waterman, Mah-Jabeen Soobader, S.V. Subramanian, and Rosa Carson, "Choosing Area Based Socioeconomic Measures to Monitor Social Inequalities in Low Birth Weight and Childhood Lead Poisoning: The Public Health Disparities Geocoding Project (US)," *Journal of Epidemiology and Community Health* 57 (2003): 186–99.

30. Madeline Drexler, "How Racism Hurts—Literally," *Boston Globe*, July 15, 2007, E1; see Gravlee, "How Race Becomes Biology," 52.

31. Barr, *Health Disparities in the United States*, 157.

32. Adler and Rehkopf, "U.S. Disparities in Health," 245–46.

33. Troy Duster, "The Molecular Reinscription of Race in Medicine and Forensic Science," talk delivered at "Genetics and Its Other: Minorities Race, and Health

Inequities in Medicine," sponsored by Center for the Study of Gene Structure and Function, Hunter College, and New York Academy of Sciences, Dec. 9, 2005.

34. Robert S. Hogg, Eric F. Druyts, Scott Burris, et al., "Years of Life Lost to Prison: Racial and Gender Gradients in the United States," *Harm Reduction Journal* 5 (2008): 4.

35. Thomas W. McDade, "Measuring Immune Function: Markers of Cell-Mediated Immunity and Inflammation in Dried Blood Spots," in *Measuring Stress in Humans: A Practical Guide for the Field*, ed. Gillian H. Ice and Gary D. James (Cambridge, UK: Cambridge University Press, 2007), 181.

36. Tené T. Lewis et al., "Self-Reported Experiences of Everyday Discrimination Are Associated with Elevated C-reactive Protein Levels in Older African-American Adults," *Brain, Behavior, and Immunity* 24, no. 3 (2010): 438; Michael Greenwood, "The Unhealthy Sting of Racism," *Yale Public Health*, Fall 2010, 24.

37. Blase N. Polite and Olufunmilayo I. Olopade, "Breast Cancer and Race: A Rising Tide Does Not Lift All Boats Equally," *Perspectives in Biology and Medicine* (2005): S166.

38. Jennifer Couzin, "Probing the Roots of Race and Cancer," *Science* 315 (2007): 592.

39. Gretchen L. Hermes et al., "Social Isolation Dysregulates Endocrine and Behavioral Stress While Increasing Malignant Burden of Spontaneous Mammary Tumors," *Proceedings of the National Academy of Sciences* 106 (2009): 22393.

40. J. Bradley Williams, Diana Pang, Bertha Delgado, Masha Kocherginsky, Maria Tretiakova, et al., "A Model of Gene-Environment Interaction Reveals Altered Mammary Gland Gene Expression and Increased Tumor Growth Following Social Isolation," *Cancer Prevention Research* 2(10) (2009): 850, 860.

41. Amal Melhem-Berstrandt and Suzanne D. Conzen, "The Relationship Between Psychosocial Stressors and Breast Cancer Biology," *Current Breast Cancer Report* 2 (2010): 130, 135.

42. David R. Williams, Michelle Sternhal, and Rosalind J. Wright, "Social Determinants: Taking the Social Context of Asthma Seriously," *Pediatrics* 123 (2010): S174, S177.

43. Douglas S. Massey and Nancy A. Denton, *American Apartheid: Segregation and the Making of the Underclass* (Cambridge, MA: Harvard University Press, 1998), 41.

44. Barr, *Health Disparities in the United States*, 161; David R. Williams and Chiquita Collins, "Racial Residential Segregation: A Fundamental Cause of Racial Disparities in Health," *Public Health Reports* 116 (2001): 404.

45. Gravlee, "How Race Becomes Biology," 52.

46. NAACP Legal Defense and Educational Fund, "When Discrimination Gets Deadly, LDF Fights Back," *Defender*, Fall 2007.

47. Ying-Ying Meng, Rudolph P. Rull, Michelle Wilhelm, et al., "Living Near Heavy Traffic Increases Asthma Severity," UCLA Center for Health Policy Research (2006), www.healthpolicy.ucla.edu/pubs/files/Traffic_Asthma_PB.081606.pdf.

48. Ruchi S. Gupta, Xingyou Zhang, Lisa K. Sharp, et al., "Geographic Variability

in Childhood Asthma Prevalence in Chicago," *Journal of Allergy & Clinical Immunology* 121 (2008): 639.

49. Dawn Turner Trice, "Asthma Crisis Is a Call for All Minorities," *Chicago Tribune*, Jan. 21, 2008.

50. "Research Links Asthma Rates to Neighborhood Violence," *Eight Forty-Eight*, Chicago Public Radio, May 3, 2010, available at www.chicagopublicradio.org/content.aspx?audioid=41713.

51. "Living in a High-Crime Neighborhood May Worsen Children's Asthma," *Science Daily*, May 2, 2010. See also Edith Chen, Hannah M.C. Schreier, Robert C. Strunk, and Michael Brauer, "Chronic Traffic-Related Air Pollution and Stress Interact to Predict Biologic and Clinical Outcomes in Asthma," *Environmental Health Perspectives* 116 (2008): 970.

52. S.V. Subramanian, Jarvis T. Chen, David H. Rehkopf, Pamela D. Waterman, and Nancy Krieger, "Racial Disparities in Context: A Multilevel Analysis of Neighborhood Variations in Poverty and Excess Mortality among Black Populations in Massachusetts," *American Journal of Public Health* 95 (2005): 260; Felicia B. LeClere, Richard G. Rogers, and Kimberley D. Peters, "Ethnicity and Mortality in the United States: Individual and Community Correlates," *Social Forces* 76 (1997): 169.

53. Angus S. Deaton and Darren Lubotsky, "Mortality, Inequality, and Race in American Cities and States," *Social Science and Medicine* 56 (2003): 1139.

54. Kari Haskell, "A Mother Again, a Generation Later," *New York Times*, Nov. 23, 2008.

55. Janice D. Hamlet, "Fannie Lou Hamer: The Unquenchable Spirit of the Civil Rights Movement," *Journal of Black Studies* 26 (1996): 572.

56. Jessica Arons and Dorothy Roberts, "Sick and Tired: Working Women and Their Health," in *The Shriver Report: A Woman's Nation Changes Everything*, ed. Heather Boushey and Ann O'Leary (Washington, DC: Center for American Progress, 2009), 122; Annette Dula, Sabrina Kurtz, and Maria-Luz Samper, "Occupational and Environmental Reproductive Hazards Education and Resources for Communities of Color," *Environmental Health Perspectives Supplements* 101 (1993): 181.

57. Xu Xiong, Emily W. Harville, Donald Mattison, Karen Elkind-Hirsch, Gabriella Pridjian, and Pierre Buekens, "Exposure to Hurricane Katrina, Post-Traumatic Stress Disorder and Birth Outcomes," *American Journal of Medical Science* 336, no. 2 (2008): 111.

58. Laura M. Glynn, Pathik D. Wadhwa, Christine Dunkel-Schetter, Aleksandra Chicz-Demet, and Curt A. Sandman, "When Stress Happens Matters: Effects of Earthquake Timing on Stress Responsivity in Pregnancy," *American Journal of Obstetrics and Gynecology* 184, no. 4 (2001): 637; Annie Murphy Paul, *Origins: How the Nine Months Before Birth Shape the Rest of Our Lives* (New York: Free Press, 2010), 41–56.

59. Diane S. Lauderdale, "Birth Outcomes for Arabic-Named Women in California Before and After September 11," *Demography* 43 (2006): 185.

60. Quoted in ibid., 188.

61. Author interview of Christopher Kuzawa, Nov. 18, 2008, Evanston, IL.

62. Christopher W. Kuzawa and Elizabeth Sweet, "Epigenetics and the Embodiment of Race: Developmental Origins of US Racial Disparities in Cardiovascular Health," *American Journal of Human Biology* 21 (2008): 2.

63. Martin Hult, et al, "Hypertension, Diabetes, and Overweight: Looming Legacies of the Biafran Famine," *PLoS One* 5 (2010): el 3582.

64. Mark A. Rothstein, Yu Cai, and Gary E. Marchant, "The Ghost in Our Genes: Legal and Ethical Implications of Epigenetics," *Health Matrix* 19 (2009): 1.

65. Ibid., 21.

66. Craig A. Cooney, Apurva A. Dave, and George L. Wolff, "Maternal Methyl Supplements in Mice Affect Epigenetic Variation and DNA Methylation of Offspring," *Journal of Nutrition* 132 (2002): 23935.

67. Rothstein, Cai, and Marchant, "The Ghost in Our Genes," 11–14.

68. Marcus E. Pembrey, Lars Olav Bygren, Gunnar Kaati, et al., "Sex-Specific Male Line Transgenerational Responses in Humans," *European Journal of Human Genetics* 14 (2006): 159.

69. Kuzawa and Sweet, "Epigenetics and the Embodiment of Race."

70. Nancy Krieger, "The Science and Epidemiology of Racism and Health: Racial/Ethnic Categories, Biological Expressions of Racism, and the Embodiment of Inequality—an Ecosocial Perspective," in *What's the Use of Race? Modern Governance and the Biology of Difference*, ed. Ian Whitmarsh and David S. Jones (Cambridge, MA: MIT Press, 2010), 225, 235.

71. Nancy Krieger, David H. Rehkopf, Jarvis T. Chen, et al., "The Fall and Rise of US Inequities in Premature Mortality: 1960–2002," *PLoS Medicine* 5 (2008): e46.

72. Robert Pear, "Gap in Life Expectancy Widens for the Nation," *New York Times*, Mar. 32, 2008, A14.

73. Douglas V. Almond, Kenneth Y. Chay, and Michael Greenstone, "The Civil Rights Act of 1964, Hospital Desegregation and Black Infant Mortality in Mississippi," National Bureau of Economic Research working paper (March 2008).

74. Rucker C. Johnson, "Long-Run Impacts of School Desegregation and School Quality on Adult Health," unpublished manuscript, 2010, http://socrates.berkeley.edu/~ruckerj/manuscripts.html.

75. C. Andrew Aligne et al., "Risk Factors for Pediatric Asthma: Contributions of Poverty, Race, and Urban Residence," *American Journal of Respiratory Critical Care Medicine* 162 (2000): 873.

76. Reuters, "Genetic Link in Ailment of Bones," *New York Times*, Apr. 30, 2008.

Chapter 7: Pharmacoethnicity

1. Jonathan Marks, "Race: Past, Present, and Future," in *Revisiting Race in a Genomic Age*, ed. Barbara A. Koenig, Sandra Soo-Jin Lee, and Sarah S. Richardson (New Brunswick, NJ: Rutgers University Press, 2008), 27–28.

2. Richard C. Lewontin, *It Ain't Necessarily So: The Dream of the Human Genome and Other Illusions* (New York: New York Review Books, 2001), 162.

3. Ibid., 163.

4. "Fortune 500 2009, Top Industries: Most Profitable," *Fortune*, May 4, 2009, available at http://money.cnn.com/magazines/fortune/fortune500/2009/performers/industries/profits/; Bill Berkrot, "U.S. Prescription Drug Sales Hit $300 Billion in 2009," Reuters, Apr. 2, 2010, available at www.reuters.com/article/idUSN3122364020100401.

5. Christopher Anderson, "Controversial NIH Genome Researcher Leaves for New $70-Million Institute," *Nature* 358 (1992): 95.

6. Available at www.g2conline.info/content/c15/15071/clinton324_01.ogg.

7. Jason Lazarou, Bruce H. Pomeranz, and Paul N. Corey, "Incidence of Drug Reactions in Hospitalized Patients: A Meta-Analysis of Prospective Studies," *Journal of the American Medical Association* 278 (1998): 1200.

8. Toshiyuki Sakaeda, Tsutomu Nakamura, and Katsuhiko Okumura, "Pharmacogenetics of Drug Transporters and Its Impact on Pharmacotherapy," *Current Topics in Medical Chemistry* 4 (2004): 1383.

9. Michael M. Hopkins et al., "Putting Pharmacogenetics into Practice," *Nature Biotechnology* 24 (2006): 403; B. Michael Silber, "Pharmacogenomics, Biomarkers, and the Promise of Personalized Medicine," in *Pharmacogenomics*, ed. Werner Kalow, Urs A. Meyer, and Rachel F. Tyndale (New York: Marcel Dekker, 2001), 11.

10. Andrew Smart and Paul Martin, "The Promise of Pharmacogenetics: Assessing the Prospects for Disease and Patient Stratification," *Studies in History and Philosophy of Biological and Biomedical Sciences* 37 (2006): 583; Sandra Soo-Jin Lee, "The Ethical Implications of Stratifying by Race in Pharmacogenomics," *Nature* 81 (2007): 122.

11. "Medicine Gets Personal: Sidney Taurel Discusses Tailored Therapeutics and the Future of the Drug Industry," *Knowledge@WP Carey*, Dec. 3, 2008, http://knowledge.wpcarey.asu.edu/article.cfm?articleid=1716.

12. Ibid.; Carlene Olsen, "Better, Faster, Smaller: Personalized Medicine Could Bring R&D Savings," *Pink Sheet*, Mar. 3, 2008.

13. Andrew Pollack, "ImClone Says Its Cancer Drug Works as Initial Treatment," *New York Times*, June 2, 2008, C2.

14. Hopkins et al., "Putting Pharmacogenetics into Practice," 406; Smart and Martin, "Promise of Pharmacogenetics," 592.

15. Smart and Martin, "Promise of Pharmacogenetics," 593; Malorye Allison, "Is Personalized Medicine Finally Arriving?" *Nature Biotechnology* 26 (2008): 509.

16. Personalized Medicine Coalition, *The Case for Personalized Medicine* (New York, 2009).

17. Nicholas Wade, "Disease Cause Is Pinpointed with Genome," *New York Times*, Mar. 10, 2010, A1.

18. Brandon Keim, "The Wild Genetic Goose Chase," *GeneWatch* 17 (2004): 6.

19. Nicholas Wade, "Research Teams Identify Gene Seen as Tied to Multiple Sclerosis," *New York Times*, July 30, 2007, A9.

20. Nicholas Wade, "A Dissenting Voice as the Genome Is Sifted to Fight Disease," *New York Times*, Sept. 16, 2008, F3.

21. Nicholas Wade, "A Decade Later, Gene Map Yields Few New Cures," *New York Times*, June 13, 2010, sec. 1, 1, 22; Sharon Begley, "Back to the Genetic Future: Why Family Medical History Is Key," *Newsweek*, Nov. 22, 2010, 26; Brendan Maher, "The Case of the Missing Heritability," *Nature* 456 (2008) no. 6: 18.

22. Nicholas Wade, "Out of Bankruptcy, Genetic Company Drops Drug Efforts," *New York Times*, Jan. 22, 2010, B2.

23. Nina P. Paynter, Daniel I. Chasman, Guillaume Pare, et al., "Association Between Literature-Based Genetic Risk Score and Cardiovascular Events in Women," *Journal of the American Medical Association* 303 (2010): 631.

24. Wade, "A Decade Later, Gene Map Yields Few New Cures," 1.

25. American Society of Human Genetics, "New Research Validates Clinical Use of Family Health History as the 'Gold Standard' for Assessing Personal Disease Risk," press release, Oct. 22, 2010, http://ashg.org/pdf/PR_FamilyHealthHistory_110510.pdf.

26. Nicholas Wade, "A New Way to Look for Diseases' Genetic Roots," *New York Times*, Jan. 26, 2010, D4.

27. Wade, "Disease Cause Is Pinpointed."

28. Author's interview of Mildred Cho, June 11, 2008, Stanford, CA.

29. William E. Evans and Mary V. Relling, "Pharmacogenomics: Translating Functional Genomics into Rational Therapeutics," *Science* 286 (1999): 487, 490.

30. Ibid., 488; see also Werner Kalow, "Interethnic Differences in Drug Response," in Kalow, Meyer and Tyndale, *Pharmacogenomics*, 109.

31. William Saletan, "Unfinished Race: Race, Genes, and the Future of Medicine," *Slate*, Aug. 27, 2008, http://www.slate.com/id/2198731.

32. Richard S. Cooper, Jay S. Kaufman, and Ryk Ward, "Race and Genomics," *New England Journal of Medicine* 348 (2003): 1166, 1168.

33. Rick Weiss, "The Promise of Precision Prescriptions; 'Pharmacogenomics' Also Raises Issues of Race, Privacy," *Washington Post*, June 24, 2000, A1.

34. Pauline C. Ng, Qi Zhao, Samuel Levy, Robert L. Strausberg, and J. Craig Venter, "Individual Genomes Instead of Race for Personalized Medicine," *Clinical Pharmacology & Therapeutics* 84 (2008): 306.

35. Vence L. Bonham, Esther Warshauer-Baker, and Francis S. Collins, "Race and Ethnicity in the Genome Era," *American Psychologist* 60 (2005): 9, 13; see also Francis S. Collins, "What We Do and Don't Know About 'Race,' 'Ethnicity,' Genetics, and Health at the Dawn of the Genome Era," *Nature Genetics* 36 (2004): S13.

36. Mildred K. Cho, "Racial and Ethnic Categories in Biomedical Research: There Is No Baby in the Bathwater," *Journal of Law, Medicine & Ethics* 34 (2006): 497, 499.

37. Sarah K. Tate and David B. Goldstein, "Will Tomorrow's Medicines Work for Everyone?" *Nature Genetics* 36, supplement (2004): S34, S34.

38. James Wilson et al., "Population Genetic Structure of Variable Drug Response," *Nature Genetics* 29 (2001): 265.

39. Craig R. Lee, Joyce A. Goldstein, and John A. Pieper, "Cytochrome P450 2C9 Polymorphisms: A Comprehensive Review of the In-Vitro and Human Data," *Pharmacogenetics* 12 (2002): 251, 252.

40. Ng et al., "Individual Genomes Instead of Race," 307.

41. Ibid., 306.

42. Charles N. Rotimi and Lynn B. Jorde, "Ancestry and Disease in the Age of Genomic Medicine," *New England Journal of Medicine* 363 (2010): 1551, 1555.

43. Lori Kay Mattison et al., "Increased Prevalence of Dihydropyrimidine Dehydrogenase Deficiency in African-Americans Compared with Caucasians," *Clinical Cancer Research* 12 (2006): 5491.

44. Ibid., 5495; Muhammad Wasif Saif et al., "Dihydropyrimidine Dehydrogenase Deficiency (DPD) in GI Malignancies: Experience of 4 Years," *Pakistani Journal of Medical Science Quarterly* 23 (2007): 832.

45. DNA Direct, Drug Metabolism and Ancestry, available at www.dnadirect.com.

46. M. Teichert et al., "Genotypes Associated with Reduced Activity VKORC1 and CYP2C9 and Their Modification of Acenocoumarol Anticoagulation During the Initial Treatment Period," *Clinical Pharmacology & Therapeutics* 85 (2009): 379.

47. Jonathan Kahn, "Beyond BiDil: The Expanding Embrace of Race in Biomedical Research and Product Development," *Saint Louis University Journal of Health Law & Policy* 3 (2010): 61, 72–73.

48. Ibid., 76.

49. Kalow, "Interethnic Differences in Drug Response," 114.

50. Author interview of Jay Kaufman, Sept. 4, 2008, Chapel Hill, NC.

51. Ben Harder, "The Race to Prescribe: Drug for African Americans May Debut Amid Debate," *Science News* 167 (2005): 247.

52. Jonathan Kahn, "How a Drug Becomes Ethnic: Law, Commerce, and the Production of Racial Categories in Medicine," *Yale Journal of Health Law, Policy & Ethics* 4 (2004): 1.

53. Steven Epstein, *Inclusion: The Politics of Difference in Medical Research* (Chicago: University of Chicago Press, 2007), 73, 179.

54. President's Council of Advisors on Science and Technology (PCAST), "Priorities for Personalized Medicine," Sept. 2008, quoted in Personalized Medicine Coalition, *Case for Personalized Medicine*, 2.

55. Smart and Martin, "Promise of Pharmacogenetics," 597.

56. Kristin Choo, "Personalized Prescription," *ABA Journal*, Sept. 2006, 42, 46.

57. Quoted in Kahn, "How a Drug Becomes 'Ethnic,'" 25.

58. Quoted in Jonathan Kahn, "From Disparity to Difference: How Race-Specific

Medicines May Undermine Policies to Address Inequalities in Health Care," *Southern California Interdisciplinary Law Journal* 15 (2005): 105, 118–19.

59. Timothy Caulfield, "Genesis of Neo-Racism? Biogenetics Discoveries About Race-Specific Idiosyncracies Have a Dark Side," *Edmonton Journal*, Dec. 22, 2007.

60. Ibid.

61. Brian A. Primack et al., "Volume of Tobacco Advertising in African American Markets: Systematic Review and Meta-Analysis," *Public Health Reports* 122 (2007): 607.

62. Diana Hackbarth, Barbara Silvestri, and William Cosper, "Tobacco and Alcohol Billboards in 50 Chicago Neighborhoods: Market Segmentation to Sell Dangerous Products to the Poor," *Journal of Public Health Policy* 16 (1995): 213, 215. See also Caroline Schooler, Michael D. Basil, and David G. Altman, "Alcohol and Cigarette Advertising on Billboards: Targeting with Social Cues," *Health Education Research* 6 (1991): 487 (the authors found black neighborhoods in San Francisco had the highest rate of alcohol and cigarette billboards per thousand people in the city).

63. Nicholas Wade, "Gene Study Identifies 5 Main Human Populations, Linking Them to Geography," *New York Times*, Dec. 20, 2002, A37.

64. Gregory M. Lamb, "A Place for Race in Medicine?" *Christian Science Monitor*, Mar. 3, 2005.

Chapter 8: Color-Coded Pills

1. Legal scholar Jonathan Kahn was the first to expose the commercial reasons for BiDil's conversion to a race-specific medicine and has written several articles detailing the history of BiDil's development and FDA approval. See Jonathan Kahn, "How a Drug Becomes Ethnic: Law, Commerce, and the Production of Racial Categories in Medicine," *Yale Journal of Health Law, Policy & Ethics* 4 (2004): 1; Pamela Sankar and Jonathan Kahn, "BiDil: Race Medicine or Race Marketing?" *Health Affairs* 24 (2005): 455; Jonathan Kahn, "Race in a Bottle," *Scientific American*, July 15, 2007. See also Stephanie Saul, "U.S. to Review Drug Intended for One Race," *New York Times*, June 13, 2005, A1.

2. Steven Stiles, "Dr. Jay N. Cohn and A-HeFT: Persistence Rewarded," *Heartwire*, Nov. 5, 2004, www.theheart.org/article/359499.do.

3. Marcia Angell, *The Truth About the Drug Companies: How They Deceive Us and What to Do About It* (New York: Random House, 2005), 9.

4. Peter Carson et al., "Racial Differences in Response to Therapy for Heart Failure: Analysis of the Vasodilator-Heart Failure Trials," *Journal of Cardiac Failure* 5 (1999): 178, 182.

5. George T.H. Ellison, Jay S. Kaufman, Rosemary F. Head, Paul A. Martin, and Jonathan D. Kahn, "Flaws in the U.S. Food and Drug Administration's Rationale for Supporting the Development and Approval of BiDil as a Treatment for Heart Failure Only in Black Patients," *Journal of Law, Medicine & Ethics* 36 (2008): 449.

6. Kirsten Bibbins-Domingo and Alicia Fernandez, "BiDil for Heart Failure in Black Patients: Implications of the U.S. Food and Drug Administration Approval," *Annals of Internal Medicine* 146 (2007): 52, 53.

7. Ellison et al., "Flaws in the U.S. Food and Drug Administration's Rationale," 451.

8. Jonathan Kahn, "BiDil and Racialized Medicine," *GeneWatch* 22 (2009): 7.

9. Stephanie Saul, "Maker of Heart Drug Intended for Blacks Bases Price on Patients' Wealth," *New York Times*, July 8, 2005, C3; Andrew Pollack, "Drug Approved for Heart Failure in Black Patients," *New York Times*, July 20, 2004.

10. Robert Temple and Norman L. Stockbridge, "BiDil for Heart Failure in Black Patients: The U.S. Food and Drug Administration Perspective," *Annals of Internal Medicine* 146 (2007).

11. "Nitromed Inc," Initial Public Offerings (IPO) Deal Data, EDGAROnline, Nov. 6, 2003, http://ipoportal.edgar-online.com/ipo/displayFundamentals.asp?cikid=95331&fnid=35036&IPO=1&coname=NITROMED+INC.

12. Aaron Lorenzo, "FDA Panel Votes 9–0 to Support BiDil's Clearance," *Bioworld Today*, June 17, 2005.

13. Steven E. Nissen, "Report from the Cardiovascular and Renal Drugs Advisory Committee: US Food and Drug Administration, June 15–16, 2005, Gaithersburg, MD.," *Circulation* 112 (2005): 2043.

14. Susan M. Reverby, "Special Treatment: BiDil, Tuskegee, and the Logic of Race," *Journal of Law, Medicine & Ethics* 36 (2008): 478.

15. Department of Health and Human Services Food and Drug Administration Center for Drug Evaluation and Research, Cardiovascular and Renal Drugs Advisory Committee, June 16, 2005, meeting transcript, 203, 208, www.fda.gov/ohrms/dockets/ac/05/transcripts/2005-4145t2.pdf (hereafter "FDA transcript").

16. Ibid., 228–29.

17. Ibid., 232.

18. Ibid., 240.

19. Ibid., 99.

20. Ibid., 357.

21. Ibid., 355, 360.

22. Ibid., 354.

23. Aaron Lorenzo, "FDA Panel Votes 9–0 to Support BiDil's Clearance," *Bioworld Today*, June 17, 2005.

24. Jonathan Kahn, "Exploiting Race in Drug Development: BiDil's Interim Model of Pharmacogenomics," *Social Studies of Science* 38 (2008): 737, 746.

25. FDA transcript, 304.

26. Ibid., 305–6.

27. Reverby, "Special Treatment," 481.

28. FDA transcript, 393, 402.

29. Turna Ray, "HHS Draft Report Suggests Genetic Test for BiDil," *Pharmacogenomics Reporter*, Apr. 4, 2007.

30. Renee Bowser, "Race as a Proxy for Drug Response: The Dangers and Challenges of Ethnic Drugs," *De Paul Law Review* 53 (2004): 1111, 1126; Rene Bowser, "Racial Profiling in Health Care: An Institutional Analysis of Medical Treatment Disparities," *Michigan Journal of Race and Law* 7 (2001): 79, 113.

31. "Methods of Treating and Preventing Congestive Heart Failure with Hydralazine Compounds and Isosorbide Dinitrate or Isosorbide Mononitrate," U.S. Patent No. 658261 (issued Oct. 15, 2002).

32. NitroMed, Inc., "NitroMed Receives FDA Letter on BiDil NDA, a Treatment for Heart Failure in Black Patients," press release, Mar. 8, 2001.

33. NitroMed, Inc., "NitroMed Initiates Confirmatory BiDil Trial in African American Heart Failure Patients," press release, Mar. 17, 2001.

34. Anne L. Taylor, Susan Ziesche, Clyde Yancy, et al., "Combination of Isosorbide Dinitrate and Hydralazine in Blacks with Heart Failure," *New England Journal of Medicine* 351 (2004): 2049, 2054.

35. Reverby, "Special Treatment," 480.

36. Temple and Stockbridge, "BiDil for Heart Failure in Black Patients," 57.

37. "FDA May OK Advisory Panel Call for African American Heart Drug," *FDA Week*, June 17, 2005.

38. Robert Sade, "What's Right (and Wrong) with Racially Stratified Research and Therapies," *Journal of the National Medical Association* 99 (2007): 693.

39. Taylor et al., "Combination of Isosorbide Dinitrate and Hydralazine," 2049.

40. Sade, "What's Right (and Wrong) with Racially Stratified Research," 693.

41. Steve Stiles, "BiDil, Regulatory Milestone but a Tough Sell for Insurers and Patients, Charts Uncertain Course," *Heartwire*, Dec. 13, 2006, www.theheart.org/article/759547.do.

42. Donald A. Barr, "The Practitioner's Dilemma: Can We Use a Patient's Race to Predict Genetics, Ancestry, and the Expected Outcomes of Treatment," *Annals of Internal Medicine* 143 (2005): 809, 813.

43. Simon Crompton, "Medicine That's Only Skin Deep: Should Treatments Be Tailored to Race?" *Times* (London), Oct. 20, 2007, http://women.timesonline.co.uk/tol/life_and_style/women/body_and_soul/article2693996.ece.

44. Robin Marantz Henig, "The Genome in Black and White (and Gray)," *New York Times Magazine*, Oct. 10, 2004.

45. "FDA BiDil Approval Poses Tough Policy, Ethical Issues for FDA," *FDA Week*, June 23, 2005.

46. HCD Research, "Physicians Believe Drugs Targeted for Ethnic and Racial Groups May Provide Therapeutic Advantages," press release, June 23, 2005.

47. Doris Gellene, "Heart Pill Intended Only for Blacks Sparks Debate," *Los Angeles Times*, June 6, 2005.

48. Harvard Medical School, "Gray Area for New Heart Failure Drug: Although the FDA Approved BiDil for Blacks with Heart Failure, It May Work in Anyone," *Harvard Heart Letter* 16 (2005): 1, 2.

49. "BiDil Maker NitroMed Responds to 'Race in a Bottle,'" *Scientific American*, July 30, 2007.

50. Jannette J. Witmyer, "Prescription for Prejudice?" *Essence*, Feb. 2005, 34, quoting Clyde Yancy.

51. Keith Wailoo, *Dying in the City of the Blues: Sickle Cell Anemia and the Politics of Race and Health* (Chapel Hill: University of North Carolina Press, 2001), 22.

52. National Minority Health Month Foundation, "Organizations Unite to Support Bidil's Approval for Heart Failure, Rebuff Designation as 'Race-Only' Drug," press release, June 15, 2005, www.nmhm.org/newsroom06142005.htm.

53. Thomas Ginsberg, "Marketing Medicine for Black Community," *Philadelphia Inquirer*, July 29, 2005; Dan Devine, "NAACP Goes to the Grassroots for BiDil," *Bay State Banner*, Oct. 5, 2006.

54. Jocelyn Uhl, "Program to Focus on Cardiovascular Disease Among Minorities," *Pitt Chronicle*, July 18, 2005.

55. NAACP, Emergency Resolution No. 7, "BiDil in Treatment Plan of Black Patients with Congestive Heart Failure," Nov.–Dec. 2005.

56. "NAACP and NitroMed Announce Partnership to Narrow Disparities in Cardiovascular Healthcare," *Business Wire*, Dec. 14, 2005.

57. Zenitha Price, "NAACP to Partner in Bridging Health Care Gap," *Baltimore Afro-American* Dec. 17–23, 2005, A1.

58. Devine, "NAACP Goes to the Grassroots."

59. NAACP, "NAACP and NitroMed Announce Partnership to Narrow Disparities in Cardiovascular Healthcare," press release, Dec. 14, 2005.

60. Devine, "NAACP Goes to the Grassroots."

61. Leah Sammons, "Racial Profiling: Not Always a Bad Thing," *Chicago Defender*, Feb. 9, 2006.

62. Gary Puckrein, "BiDil: From Another Vantage Point," *Health Affairs* 25 (2006): w368.

63. Kevin Freking, "First Medication Targeted at Specific Racial Group Prompts Some Wariness," *Chicago Defender*, June 27, 2005.

64. Stephanie Saul, "F.D.A. Approves a Heart Drug for African-Americans," *New York Times*, June 24, 2005, C2.

65. Marcia Angell, "Drug Companies and Medicine: What Money Can Buy," lecture delivered at Edmond J. Safra Center for Ethics, Harvard University, Dec. 10, 2009, available at http://link.brightcove.com/services/player/bcpid59085832001?bclid=58806604001&bctid=58725921001.

66. Jay Cohn, director, Rasmussen Center for Cardiovascular Disease Prevention, commentary at the Deinard Memorial Lecture on Law & Medicine, Feb. 3, 2010.

67. NitroMed, Inc., 2005 annual report, 1.

68. Stephanie Saul, "2 Officials Quit Amid Slow Sales of Heart Drug for Blacks," *New York Times*, Mar. 22, 2006.

69. Steve Stiles, "BiDil, Regulatory Milestone but a Tough Sell for Insurers and Patients, Charts Uncertain Course," *Heartwire*, Dec. 13, 2006, www.theheart.org/article/759547.do.

70. Jacob Goldstein, "First Racially Targeted Drug Is a Flop," WSJ.com, Jan. 16, 2008.

71. Saul, "Maker of Heart Drug Intended for Blacks."

72. Sylvia Pagan Westphal, "Heart Medication for Blacks Faces Uphill Battle," *Wall Street Journal*, Oct. 16, 2006.

73. Devine, "NAACP Goes to the Grassroots."

74. Washington Legal Foundation, "WLF Petitions CMS to Revise Reimbursement Policies for BiDil," press release, Aug. 7, 2008.

75. Matt Witten, "Humpty Dumpty," *House MD*, Fox Television, aired on Sept. 27, 2005, available at www.twiztv.com/scripts/house/season2/house-203.htm.

76. Keith Knight, "BiDil Me This, Blackman," cartoon, *K Chronicles*, 2005, available at www.buzzle.com/showImage.asp?image=3273.

77. Celeste Condit, Alan Templeton, Benjamin Bates, et al., "Attitudinal Barriers to Delivery of Race-Targeted Pharmacogenomics Among Informed Lay Persons," *Genetics in Medicine* 5 (2003): 385. See also Benjamin R. Bates, Kristin Poirot, Tina M. Harris, Celeste M. Condit, and Paul J. Achter, "Evaluating Direct-to-Consumer Marketing of Race-Based Pharmacogenomics: A Focus Group Study of Public Understandings of Applied Genomic Medication," *Journal of Health Communication* 9 (2004): 541.

78. Osagie Obasogie, "Slooooooow Sales for BiDil," *Biopolitical Times*, Oct. 18, 2006, www.biopoliticaltimes.org/article.php?id=2018.

79. Ta-Nehisi Paul Coates and Sora Song, "Suspicious Minds," *Time*, July 4, 2005, 36.

80. Jeanne Whalen and Ron Winslow, "Cash-Poor Biotech Firms Cut Research, Seek Aid," *Wall Street Journal*, Oct. 29, 2008, http://online.wsj.com/article/SB122523819921178005.html; Andrew Pollack, "For Biotech Tax Break Spells Hope," *New York Times*, Dec. 10, 2008, B1.

81. Jonathan Kahn, "Exploiting Race in Drug Development: BiDil's Interim Model of Pharmacogenomics," *Social Studies of Science* 38 (2008): 737–38; Jonathan Kahn, "Beyond BiDil: The Expanding Embrace of Race in Biomedical Research and Product Development," *Saint Louis University Journal of Health Law & Policy* 3 (2010): 61.

82. Jonathan Kahn, "Patenting Race in a Genomic Age," in *Revisiting Race in a Genomic Age*, 129, 131.

83. Shubha Ghosh, "Race-Specific Patents, Commercialization, and Intellectual Property Policy," *Buffalo Law Review* 56 (2008): 409.

84. Ibid., 441.

85. Ibid., 432.

86. "Dopamine D2 Receptor Gene Variants," U.S. Patent Application No. 20090104598 (filed Apr. 23, 2009).

87. "Method of Screening for Drug Hypersensitivity Reaction," U.S. Patent Application No. 7550261 (filed June 23, 2009).

88. "BRCA1/BRCA2 Screening Panel," U.S. Patent Application No. 7384743 (filed June 10, 2008).

89. "Report: 691 New Medicines in Development for Black Population," *Memphis Business Journal*, Dec. 10, 2007.

90. Billy Tauzin, "Report: Medicines in Development for Major Diseases Affecting African Americans," Pharmaceutical Research & Manufacturers of America, 2007.

91. Sarah K. Tate and David B. Goldstein, "Will Tomorrow's Medicines Work for Everyone?" *Nature Genetics* 36 (2004): S34.

92. Sandra Soo-Jin Lee, Joanna Mountain, and Barbara A. Koenig, "The Meanings of 'Race' in the New Genomics: Implications for Health Disparities Research," *Yale Journal of Health Policy, Law, & Ethics* 1 (2001): 33, 57.

93. Gillian K. McDermott, "The Phospholipids: In Comparison and in Use," *Review of Ophthalmology* 9 (2002): 54.

94. Robert Mitchell, Elena Rochtchina, Anne Lee, et al., "Iris Color and Intraocular Pressure: The Blue Mountains Eye Study," *American Journal of Opthalmology* 135(3) (2003): 384.

95. Crestor Label, available at www.accessdata.fda.gov/drugsatfda_docs/label/2005/21366slr005lbl.pdf; "FDA Warns of Risks from AstraZeneca's Crestor," MSNBC.com, Mar. 2, 2005.

96. Jay Cohn, "The Use of Race and Ethnicity in Medicine: Letters from the African-American Heart Failure Trial," *Journal of Law, Medicine, & Ethics* 34 (2006): 552.

97. Jay S. Cohen, "The Truth About Crestor: Is Crestor Dangerous and If So, Why?" *Medication Sense*, July–Sept. 2004.

98. "The Statin Wars: Why AstraZeneca Must Retreat," *Lancet* 362 (2003): 1341.

99. Public Citizen, "Cases of Kidney Failure, Muscle Damage Should Prompt FDA to Ban Crestor," press release, Mar. 4, 2004, www.citizen.org/pressroom/release.cfm?ID=1657.

100. "The Statin Wars," 1341.

101. Naveed Satter, David Preiss, Heather M. Murray, et al., "Statins and Risk of Incident Diabetes: A Collaborative Meta-Analysis of Randomised Statin Trials," *Lancet* 375 (2010): 735.

102. Kahn, "Race in a Bottle."

103. Duff Wilson, "Plan to Widen Use of Statins Has Skeptics," *New York Times*, Mar. 31, 2010, A1.

104. Ibid., A3.

105. Jim Kaput et al., "The Case for Strategic International Alliances to Harness Nutritional Genomics for Public and Personal Health," *British Journal of Nutrition* 94 (2005): 623.

106. Anne Fausto-Sterling, "The Bare Bones of Race," *Social Studies of Science* 38 (2008): 657.

107. Department of Health & Human Services Advisory Committee on Genetics, Health and Society, "Realizing the Promise of Pharmacogenomics," May 2008, 44.

108. Dennis M. McNamara, S. William Tarn, Michael Sabolinksi, et al., "Aldosterone Synthase Promoter Polymorphism Predicts Outcome in African Americans with Heart Failure: Results from the A-HeFT Trial," *Journal of the American College of Cardiology* 48 (2006): 1277.

109. Ibid., 1281.

110. "Race-Based Medicine: A Recipe for Controversy: Is Race-Based Medicine a Boon or Boondoggle?" *Scientific American*, July 31, 2007.

111. Cohn, Commentary at the Deinard Memorial Lecture on Law & Medicine.

112. Epstein, *Inclusion*, 214.

113. Temple and Stockbridge, "BiDil for Heart Failure in Black Patients," 60.

114. David E. Winickoff and Osagie K. Obasogie, "Race-Specific Drugs: Regulatory Trends and Public Policy," *Trends in Pharmacological Sciences* 29 (2008).

Chapter 9: Race and the New Biocitizen

1. Amy Harmon, "Gene Map Becomes a Luxury Item," *New York Times*, Mar. 4, 2008, F1.

2. Ellen Nakashima, "Genome Database Will Link Genes, Traits in Public View," *Washington Post*, Oct. 18, 2008, A1.

3. Evelyn Fox Keller, *The Century of the Gene* (Cambridge, MA: Harvard University Press, 2000), 57.

4. Ibid.; Richard Lewontin, *The Triple Helix: Gene, Organism and Environment* (Cambridge, MA: Harvard University Press, 2000).

5. John Dupre, "What Genes Are and Why There Are No Genes for Race," in *Revisiting Race in a Genomic Age*, ed. Barbara A. Koenig, Sandra Soo-Jin Lee, and Sarah S. Richardson (New Brunswick, NJ: Rutgers University Press, 2008), 39, 43–45.

6. Chris Berdik, "Genetic Tests Give Consumers Hints About Disease Risk; Critics Have Misgivings," *Washington Post*, Jan. 26, 2010.

7. Jane E. Brody, "Buyer Beware of At-Home Genetic Tests," *New York Times*, Sept. 1, 2009, D6.

8. U.S. General Accountability Office, "Direct-to-Consumer Genetic Tests: Misleading Test Results Are Further Complicated by Deceptive Marketing and Other Questionable Practices," July 22, 2010.

9. Allen Salkin, "When in Doubt, Spit It Out," *New York Times*, Sept. 12, 2008, ST1; Michael Schulman, "Ptooey!" *New Yorker*, Sept. 22, 2008.

10. "Mission," Facing Our Risk, www.facingourrisk.org/about_us/mission.html.

11. Nikolas Rose, *The Politics of Life Itself: Biomedicine, Power, and Subjectivity*

in the Twenty-first Century (Princeton, NJ: Princeton University Press, 2007), 40. The term biocitizenship can be traced to the concept of biopower developed by French philosopher Michel Foucault to describe how "power/knowledge" is increasingly becoming "an agent of transformation of human life." Michel Foucault, *History of Sexuality*, vol. 1 (New York: Random House, 1978), 143; Michel Foucault, *Power/Knowledge: Selected Interviews and Other Writings 1972–1977*, ed. Colin Gordon (New York: Pantheon Books, 1980).

12. Paul Rabinow, *Essays on the Anthropology of Reason* (Princeton, NJ: Princeton University Press, 1996).

13. Rose, *Politics of Life Itself*, 4; Sarah Franklin and Celia Roberts, *Born and Made: An Ethnography of Preimplantation Genetic Diagnosis* (Princeton, NJ: Princeton University Press, 2006).

14. "What We Do," NuGene, www.nugene.org/about.htm.

15. Tamar Lewin, "College Bound, DNA Swab in Hand," *New York Times*, May 18, 2010; Larry Gordon, "UC Berkeley Offer to Test DNA of Incoming Students Sparks Debate," *Los Angeles Times*, June 1, 2010; Jennifer Epstein, "The DNA Assignment," *Inside Higher Ed*, May 18, 2010, www.insidehighered.com/news/2010/05/18/berkeley.

16. University of California, Berkeley, "On the Same Page: Bring Your Genes to Cal," http://onthesamepage.berkeley.edu/.

17. Nancy Scheper-Hughes, "The Poisoned Gift: 'Fortune Cookie' Genomics at UC Berkeley," *Anthropology Today* 26, no. 6 (Dec. 2010): 1, 2. Sociologist Troy Duster also criticized the program on scientific and ethical grounds. Troy Duster, "Welcome, Freshmen. DNA Swabs, Please," *Chronicle of Higher Education*, May 28, 2010.

18. University of California, Berkeley, "California Department of Public Health Instructs UC Berkeley to Alter DNA Testing Program," news release, Aug. 8, 2010, available at http://onthesamepage.berkeley.edu/archive/2010-genes/release.php.

19. Amy Harmon, "Taking a Peek at the Experts' Genetic Secrets," *New York Times*, Oct. 19, 2008, A1.

20. Ellen Nakashima, "Genome Database Will Link Genes, Traits in Public View," *Washington Post*, Oct. 18, 2008, A1.

21. Richard Powers, "The Book of Me," *GQ*, Oct. 24, 2008.

22. "Core Values," 23andMe, www.23andme.com/about/values.

23. Salkin, "When in Doubt."

24. Brody, "Buyer Beware of At-Home Genetic Tests."

25. 154 Cong. Rec. H2956-01 (May 1, 2008); Andrew Pollack, "Genetic-Discrimination Ban Moves Ahead in Congress," *New York Times*, Apr. 23, 2008, C1, C4.

26. Amy Harmon, "Congress Passes Bill to Bar Bias Based on Genes," *New York Times*, May 2, 2008, A1.

27. 154 Cong. Rec. E849-02 (May 1, 2008).

28. Pollack, "Genetic-Discrimination Ban Moves Ahead in Congress," C4. See also Pew Charitable Trusts, Genetics & Public Policy Center, "U.S. Public Opinion on Uses of Genetic Information and Genetic Testing," Apr. 24, 2007, www.pewtrusts.org/our_work_report_detail.aspx?id=53112&category=344 (reporting that that 92 percent of Americans were afraid that their genetic records could be used against them).

29. Pollack, "Genetic-Discrimination Ban Moves Ahead in Congress."

30. Rick Weiss, "Genetic Testing Gets Personal," *Washington Post*, Mar. 25, 2008, A1.

31. Katrina Armstrong, Ellyn Micco, and Amy Carney, et al., "Racial Differences in the Use of BRCA1/2 Testing Among Women with a Family History of Breast or Ovarian Cancer," *Journal of the American Medical Association* 293 (2010): 1729.

32. John Schwartz and Andrew Pollack, "Cancer Genes Cannot Be Patented, U.S. Judge Rules," *New York Times*, Mar. 30, 2010, B1.

33. Sherry I. Brandt-Rauf et al., "Ashkenazi Jews and Breast Cancer: The Consequences of Linking Ethnic Identity to Genetic Disease," *American Journal of Public Health* 96 (2006): 1979.

34. Ibid., 1984.

35. Esther M. John, et al, "Prevalence of Pathogenic BRCA1 Mutation Carriers in 5 US Racial/Ethnic Groups, *Journal of the American Medical Association* 298, no. 24 (2007): 2869; Jay S. Kaufman and Richard S. Cooper, "Use of Racial and Ethnic Identity in Medical Evaluations and the Biology of Difference" in *What's the Use of Race?* ed. Ian Whitmarsh and David S. Jones (Cambridge, MA: MIT Press, 2010): 187.

36. Stephanie Saul, "Building a Baby, with Few Ground Rules," *New York Times*, Dec. 13, 2009, A1.

37. Erik Parens and Lori P. Knowles, "Reprogenetics and Public Policy: Reflections and Recommendations," in *Reprogenetics: Law, Policy, and Ethical Issues*, ed. Lori P. Knowles and Gregory E. Kaebnick (Baltimore: Johns Hopkins University Press, 2007); Debora L. Spar, *The Baby Business: How Money, Science and Politics Drive the Commerce of Conception* (Cambridge, MA: Harvard Business School Press, 2006).

38. "About Us," Reprogenetics, www.reprogenetics.com/about.html.

39. David Plotz, *The Genius Factory* (New York: Random House, 2005).

40. David Brooks, "The National Pastime," *New York Times*, June 15, 2007, A23.

41. Margaret Atwood, *The Handmaid's Tale* (Toronto: McClelland & Stewart, 1985); Gena Corea, *The Mother Machine: Reproductive Technologies from Artificial Insemination to Artificial Wombs* (New York: Harper & Row, 1985).

42. Rayna Rapp, *Testing Women, Testing the Fetus: The Social Impact of Amniocentesis in America* (New York: Routledge, 1999); Dorothy Roberts, *Killing the Black Body: Race, Reproduction, and the Meaning of Liberty* (New York: Vintage, 1998); Anna Marie Smith, *Welfare Reform and Sexual Regulation* (New York: Cambridge University Press, 2007).

43. Quoted in Madeline Drexler, "Reproductive Health in the Twenty-first Century," *Radcliffe Quarterly* 90 (2005), http://radcliffe.edu/print/about/quarterly/w05_health.htm.

44. T.J. Mathews and Marian F. MacDorman, "Infant Mortality Statistics from the 2004 Period Linked Birth/Infant Death Data Set," *National Vital Statistics Reports* 55 (2007): 1; Amnesty International, *Deadly Delivery: The Maternal Health Care Crisis in the USA*, Mar. 2010.

45. Robin Schatz, "Sperm Bank Mix-Up Claim: Woman Sues Doctor, Bank; Says Wrong Deposit Used," *Newsday*, Mar. 9, 1990, 5; Ronald Sullivan, "Mother Accuses Sperm Bank of a Mixup," *New York Times*, Mar. 9, 1990, B1.

46. See Hawley Fogg-Davis, "Navigating Race in the Market for Human Gametes," *Hastings Center Report* 31 (2001): 13, 15.

47. Barbara Kantrowitz and David A. Kaplan, "Not the Right Father," *Newsweek*, Mar. 19, 1990, 50.

48. Beverly Hills Egg Donation, advertisement, Los Angeles Craigslist, SF Valley, etcetera jobs, Nov. 22, 2008.

49. F. Williams Donor Services, advertisement, Inland Empire Craigslist, etcetera jobs, Nov. 24, 2008.

50. Happy Beginnings, LLC, advertisement, Reno Craigslist, etcetera jobs, Nov. 13, 2008.

51. Asian Egg Donation, www.aed-web.com.

52. Dov Fox, "Racial Classification in Assisted Reproduction," *Yale Law Journal* 118 (2009): 1844, 1853–54; Rene Almeling, "Selling Genes, Selling Gender: Egg Agencies, Sperm Banks, and the Medical Market in Genetic Material," *American Sociological Review* 72 (2007): 319, 327.

53. "About Egg Donation," Pacific Fertility Center, www.pacificfertilitycenter.com/treat/agency_donation.php.

54. "First Trimester Screening," Reproductive Genetics Institute, www.reproductivegenetics.com/first_trimester.html.

55. Fox, "Racial Classification in Assisted Reproduction," 1852–53.

56. Almeling, "Selling Genes, Selling Gender," 333.

57. Ibid., 337.

58. Brooke Lea Foster, "The Hunt for Golden Eggs," *Washingtonian*, July 1, 2007, 101, 103.

59. Patricia J. Williams, "Colorstruck," *Nation*, Apr. 23, 2007, 9; Carmen Van Kerckhove, "Fertility Clinic Mixup Results in 'Black' Baby for 'White' Parents," Racialicious, Mar. 23, 2007, www.racialicious.com/2007/03/23/fertility-clinic-mixup-results-in-black-baby-for-white-parents.

60. Gautam Naik, "A Baby Please. Blond, Freckles, Hold the Colic," *Wall Street Journal*, Feb. 12, 2009; "Designer Baby Row over US Clinic," BBC News, Mar. 2, 2009.

61. Coco Ballantyne, "Can Babies Be Made-to-Order?" *Scientific American*, Mar. 4, 2009.

62. Quoted in Mimi Rohr, "Fertility Institutes: The Clinic That Helps Couples Choose the Sex of Their Babies," Gamma, 2006, www.editorial.fnphoto.com/stories/Texts/2332-text.html.

63. Cynthia M. Powell, "The Current State of Prenatal Genetic Testing in the United States," in *Prenatal Testing and Disability Rights*, ed. Erik Parens and Adrienne Asch (Washington, DC: Georgetown University Press, 2007), 44–53; Marsha Saxton, "Why Members of the Disability Community Oppose Prenatal Diagnosis and Selective Abortion," in *Prenatal Testing and Disability Rights*, 147.

64. Consumers Union, "The Telltale Gene," *Consumer Reports* 55 (1990): 483.

65. Lynda Beck Fenwick, *Private Choices, Public Consequences: Reproductive Technology and the New Ethics of Conception, Pregnancy, and Family* (New York: Dutton Adult, 1998), 95.

66. See Ronald Wapner et al., "First-Trimester Screening for Trisomies 21 and 18," *New England Medical Journal* 349 (2003): 1405.

67. Quoted in David T. Helm, Sara Miranda, and Naomi Angoff Chedd, "Prenatal Diagnosis of Down Syndrome: Mothers' Reflections on Supports Needed from Diagnosis to Birth," *Mental Retardation* 36 (1998): 55.

68. Brian Skotko, "Mothers of Children with Down Syndrome Reflect on Their Postnatal Support," *Pediatrics* 115 (2005): 64.

69. Amy Harmon, "Genetic Testing + Abortion = ???" *New York Times*, May 13, 2007, sec. 4, 1.

70. Elizabeth Weil, "A Wrongful Birth?" *New York Times Magazine*, Mar. 12, 2006, 52; Pilar Ossorio, "Prenatal Genetic Testing and the Courts," in *Prenatal Testing and Disability Rights*, 330.

71. *Schirmer v. Mt. Auburn Obstetrics & Gynecologic Associates, Inc.*, 108 Ohio St.3d 494, 2006-Ohio-942 (Ohio Supreme Court 2006).

72. Joshua Kleinfeld, "Tort Law and In Vitro Fertilization: The Need for Legal Recognition of 'Procreative Injury,'" *Yale Law Journal* 115 (2005): 237.

73. Andrew Pollack, "Firm Brings Gene Tests to Masses," *New York Times*, Jan. 29, 2010.

74. Ibid.

75. California Department of Public Health, "Genetic Disease Screening Program," www.cdph.ca.gov/programs/GDSP/Pages/default.aspx.

76. Wolfgang Van den Daele, "The Spectre of Coercion: Is Public Health Genetics the Route to Policies of Enforced Disease Prevention?" *Community Genetics* 9 (2006): 40; Kristin Bumiller, "The Geneticization of Autism: From New Reproductive Technologies to the Conception of Genetic Normalcy," *Signs: Journal of Women in Culture and Society* 34 (2009): 875. For an extensive review of insurance coverage of infertility treatments, see Jessica Arons, *Future Choices: Assisted Reproductive Technologies and the Law* (Washington, DC: Center for American Progress, 2007), 8.

77. Franklin and Roberts, *Born and Made*, xx, 97. Preimplantation genetic diagnosis also serves to increase fertility when it is undertaken to improve IVF success rates.

78. Roxanne Mykitiuk, "The New Genetics in the Post-Keynesian State," unpublished manuscript, Canadian Women's Health Network (2000), www.cwhn.ca/groups/biotech/availdocs/15-mykitiuk.pdf; Lealle Ruhl, "Dilemmas of the Will: Uncertainty, Reproduction, and the Rhetoric of Control," *Signs* 27 (2002): 641.

79. Silja Samerski, "Genetic Counseling and the Fiction of Choice: Taught Self-Determination as a New Technique of Social Engineering," *Signs* 34 (2009): 735.

80. Fenwick, *Private Choices, Public Consequences*, 113.

81. Laura Hershey, "Choosing Disability," *Ms.*, July–Aug. 1994, 26–32. See also Susan Wendell, *The Rejected Body: Feminist Philosophical Reflections on Disability* (New York: Routledge, 1996), 154.

82. Amy Harmon, "Couples Cull Embryos to Halt Heritage of Cancer," *New York Times*, Sept. 3, 2006, A1.

83. Kenneth Offit, Michal Sagi, and Karen Hurley, "Preimplantation Genetic Diagnosis for Cancer Syndrome," *Journal of the American Medical Association* 296 (2006): 2727; Kenneth Offit, Kelly Kohut, Bartholt Clagett, et al., "Cancer Genetic Testing and Assisted Reproduction," *Journal of Clinical Oncology* 24, no. 29 (2006): 4775.

84. Nicholas Johnson, Phil Oliff, and Erica Williams, "An Update on State Budget Cuts," Center on Budget and Policy Priorities, Washington, DC, Nov. 5, 2010, www.cbpp.org/files/3-13-08sfp.pdf. See, for example, Emily Ramshaw, "Budget Shortfall Forces Big Cuts for Disabled," *Texas Tribune*, Oct. 12, 2010.

85. Centers for Disease Control and Prevention, "Racial Disparities in Median Age at Death of Persons with Down Syndrome—United States, 1968–1997," *Morbidity and Mortality Weekly Report* 50, no. 22 (June 8, 2001): 463. See also Sonja A. Rasmussen, Lee-Yang Wong, Adolfo Correa, et al., "Survival in Infants with Down Syndrome, Metropolitan Atlanta, 1979–1998," *Journal of Pediatrics* 148 (2006): 806 (finding disparity in survival of white and black children with Down syndrome).

86. Roberts, *Killing the Black Body*; Elena R. Gutierrez, *Fertile Matters: The Politics of Mexican Origin Women's Reproduction* (Austin: University of Texas Press, 2008).

87. Barbara Katz Rothman, Professor of Sociology, City University of New York, comments at Zora Neale Hurston Lecture by Dorothy Roberts, Institute for Research in African American Studies, Columbia University, Mar. 10, 2009. See also Barbara Katz Rothman, *The Book of Life: A Personal and Ethical Guide to Race, Normality and the Implications of the Human Genome Project* (Boston: Beacon Press, 2000).

88. Amelia Gentleman, "India Nurtures Business of Surrogate Motherhood," *New York Times*, Mar. 10, 2008, A9.

89. Amrita Pande, "'It May Be Her Eggs But It's My Blood': Surrogates and Everyday Forms of Kinship in India," *Qualitative Sociology* 32 (2009): 397, 381–82. See also Kalindi Vora, "Indian Transnational Surrogacy and the Commodification of Vital Energy," *Subjectivity* 28 (2009): 266.

90. Vora, "Indian Transnational Surrogacy," 275–76.

91. Steven Pinker, "My Genome, My Self," *New York Times Magazine*, Jan. 11, 2009.

92. Ibid.

93. Richard Tutton, "Constructing Participation in Genetic Databases: Citizenship, Governance, and Ambivalence," *Science, Technology & Human Values* 32 (2007): 172.

Chapter 10: Tracing Racial Roots

1. DNAPrint Genomics, "DNAPrint Genomics AncestrybyDNA Product Helps New York City Seventh Graders Learn About Their Pasts," press release, Feb. 6, 2007, www.dnaprint.com/welcome/press/press_recent/2007/0206/DNAG-MS-223.pdf.

2. Charmaine D. Royal, John Novembre, Stephanie M. Fullerton, et al., "Inferring Genetic Ancestry: Opportunities, Challenges, and Implications," *American Journal of Human Genetics* 86 (2010): 661. See also Osagie Obasogie, *Playing the Gene Card? A Report on Race and Human Biotechnology* (Oakland, CA: Center for Genetics and Society, 2009), 25, www.geneticsandsociety.org/downloads/complete_PTGC.pdf.

3. DNAPrint Genomics, "DNAPrint Genomics."

4. "FAQs for AncestrybyDNA," AncestrybyDNA, www.ancestrybydna.com/ancestry-by-dna-faq.php.

5. Mark D. Shriver, Esteban J. Parra, Sonia Dios, et al., "Skin Pigmentation, Biogeographical Ancestry, and Admixture Mapping," *Human Genetics* 112 (2003): 387.

6. Aravinda Chakravarti, "Kinship: Race Relations," *Nature* 457 (2009): 380.

7. Quoted by Kim TallBear, "Narratives of Race and Indigeneity in the Genographic Project," *Journal of Law, Medicine & Ethics* 35 (2007): 412, 414–15.

8. Henry Louis Gates Jr., "Family Matters," *New Yorker*, Dec. 1, 2008, 34, 36.

9. Alondra Nelson, "The Factness of Diaspora," in *Revisiting Race in a Genomic Age*, ed. Barbara A. Koenig, Sandra Soo-Jin Lee, and Sarah S. Richardson (New Brunswick, NJ: Rutgers University Press, 2008), 254.

10. Tatsha Robertson, "Test May Be New 'Roots': DNA Readings to Put Us Blacks on Africa's Map," *Boston Globe*, April 24, 2000, A1.

11. "Testimonials," African Ancestry, www.africanancestry.com/testimonials/index.html.

12. Gondobay Manga Foundation, www.gondobaymangafoundation.org/#/home/.

13. Teresa Watanabe, "Called Back to Africa by DNA," *Los Angeles Times*, Feb. 18, 2009.

14. Peter Shanks, "DNA Ancestry Testing on TV," *Biopolitical Times*, Mar. 10, 2010, www.psychologytoday.com/blog/genetic-crossroads/201003/dna-ancestry-testing-tv.

15. Ron Nixon, "DNA Tests Find Branches but Few Roots," *New York Times*, Nov. 25, 2007.

16. S.O.Y. Keita and Rick A. Kittles, "The Persistence of Racial Thinking and the Myth of Racial Divergence," *American Anthropologist* 99 (1997): 534.

17. David Dudley, "Action Heroes: Georgia Dunston," *AARP Magazine*, Mar.–Apr. 2004.

18. Jenny Reardon, *Race to the Finish: Identity and Governance in an Age of Genomics* (Princeton, NJ: Princeton University Press, 2005), 163–64; Georgia Dunston, "G-RAP: A Model HBCU Genomic Research and Training Program," in *Plain Talk About the Human Genome Project*, ed. Edward Smith and Walter Sapp (Tuskegee, AL: Tuskegee University Press, 1998). Charmaine D.M. Royal and Georgia M. Dunston, "Changing the Paradigm from 'Race' to Human Genome Variation," *Nature Genetics* 36 (2004): S5; Charles N. Rotimi, "Are Medical and Nonmedical Uses of Large-Scale Genomic Markers Conflating Genetics and 'Race'?" *Nature Genetics* 36 (2004): S43.

19. "Can DNA Testing Confirm Jewish or 10-Israel Ancestry?" JewsandJoes.com, July 27, 2007, http://jewsandjoes.com/can-dna-tell-me-if-i-have-jewish-or-10-israel-hebrew-ancestry.html.

20. Gil Atzmon, Li Hao, Itsik Pe'er, et al, "Abraham's Children in the Genome Era: Major Jewish Diaspora Populations Comprise Distinct Genetic Clusters with Shared Middle Eastern Ancestry," *American Journal of Human Genetics* 86 (2010): 850; Doron M. Behar, Bayazit Yunusbayev, Mait Metspalu, et al., "The Genome-Wide Structure of the Jewish People," *Nature* 466 (2010): 238.

21. David B. Goldstein, *Jacob's Legacy: A Genetic View of Jewish History* (New Haven, CT: Yale University Press, 2008).

22. Ibid., 55.

23. Nicholas Wade, "DNA Backs a Tribe's Tradition of Early Descent From the Jews," *New York Times*, May 9, 1999.

24. Goldstein, *Jacob's Legacy*, 59.

25. Jon Entine, "Letter to the Editor," *Commentary*, Dec. 2008, 10. See also Jon Entine, *Abraham's Children: Race, Identity, and the DNA of the Chosen People* (New York: Grand Central, 2007).

26. Hillel Halkin, "Jews and Their DNA," *Commentary*, Sept. 2008, 37, 42.

27. Sarah Lyall, "British Case Raises Issue of Identity for Jews," *New York Times*, Nov. 8, 2009, A8; Sarah Lyall, "British High Court Says Jewish School's Ethnic-Based Admissions Policy Is Illegal," *New York Times*, Dec. 17, 2009, A10.

28. Lyall, "British High Court," A10.

29. Lyall, "British Case Raises Issue," 18.

30. Laurie Zoloth, "Yearning for the Long Lost Home: The Lemba and the Jewish Narrative of Genetic Return," *Developing World Bioethics* 3 (2003): 127, 131.

31. Kimberly TallBear, "Native-American-DNA.com: In Search of Native American Race and Tribe," in *Revisiting Race*, 235, 243.

32. Ibid., 236.

33. Catherine Nash, "Genetic Kinship," *Cultural Studies* 18 (2004): 1, 4.

34. TallBear, "Native-American-DNA.com," 237–38.

35. Kimberly TallBear, "DNA, Blood, and Racializing the Tribe," *Wicazo Sa Review* 18 (2003): 81, 83–84.

36. Tallbear, "Native-American-DNA.com," 237.

37. TallBear, "DNA, Blood, and Racializing the Tribe," 89.

38. Karen Kaplan, "Ancestry in a Drop of Blood," *Los Angeles Times*, Aug. 30, 2005.

39. Brendan I. Koerner, "Blood Feud," *Wired*, Sept. 2005, www.wired.com/wired/archive/13.09/seminoles.html; Anita Crane, "Who's Racist Now?" Spero News, Aug. 21, 2009, www.speroforum.com/a/20176/Whos-the-Racist-Now.

40. Kaplan, "Ancestry in a Drop of Blood."

41. Pilar Ossorio, "Race, Genetic Variation, and the Haplotype Mapping Project," *Louisiana Law Review* 66 (2005): 131, 141.

42. Royal, Novembre, Fullerton, et al., "Inferring Genetic Ancestry," 667–68.

43. Mark D. Shriver and Rick A. Kittles, "Genetic Ancestry and the Search for Personalized Genetic Histories," *Nature Reviews Genetics* 5 (2004): 611, 615.

44. Duana Fullwiley, "The Biological Construction of Race: 'Admixture' Technology and the New Genetic Medicine," *Social Studies of Science* 38 (2008): 695, 704.

45. Royal, Novembre, Fullerton, et al., "Inferring Genetic Ancestry," 668.

46. Ibid., 667.

47. Steven Pinker, "My Genome, My Self," *New York Times Magazine*, Jan. 11, 2009.

48. Dana Rosenblatt, "DNA Provides Clues to Family's African Heritage," CNN.com, July 23, 2008; Robert Ely, Jamie L. Wilson, Fatimah Jackson, and Bruce Jackson, "African-American Mitochrondrial DNA Often Match mtDNAs Found in Multiple African Ethnic Groups," *BMC Biology* 4 (2006): 1.

49. John Travis, "Scientists Decry 'Flawed' and 'Horrifying' Nationality Tests," *Science*, Sept. 29, 2009, http://news.sciencemag.org/scienceinsider/2009/09/border-agencys.html.

50. Troy Duster, "Ancestry Testing and DNA: Uses, Limits—and *Caveat Emptor*," *GeneWatch* 22 (2009): 16; full paper available at www.councilforresponsiblegenetics.org/pageDocuments/O7HIKRKXYB.pdf.

51. Nixon, "DNA Tests Find Branches but Few Roots."

52. Charles N. Rotimi, "Genetic Ancestry Tracing and the African Identity: A Double-Edged Sword?" *Developing World Bioethnics* 3 (2003): 152, 155.

53. Shriver and Kittles, "Genetic Ancestry and the Search for Personalized Genetic Histories," 611, 612.

54. Hank Greely, "Genetic Genealogy: Genetics Meets the Marketplace," in *Revisiting Race*, 215–16.

55. Deborah Bolnick, Duana Fullwiley, Troy Duster, et al., "The Science and Business of Genetic Ancestry Testing," *Science* 318 (2007): 399, 400.

56. Amy Harmon, "Seeking Ancestry in DNA Ties Uncovered by Tests," *New York Times*, Apr. 12, 2006.

57. Ibid.

58. Randy Cohen, "Hurricane Hospitality," *New York Times Magazine*, Aug. 6, 2006, 15.

59. Harmon, "Seeking Ancestry in DNA Ties."

60. *In re African-American Slave Descendants Litigation*, 304 F.Supp.2d 1027 (N.D.Ill., 2004).

61. "Slave Descendants File $1 Billion Lawsuit Against Companies with Alleged Ties to Slavery Trade," *Jet*, Apr. 26, 2004, 36.

62. African Ancestry Web site, www.AfricanAncestry.com.

63. Mark Horowitz, "Education: The Detective: Henry Louis Gates Jr., Ancestry-Based Curriculum," *Wired*, Apr. 24, 2007, www.wired.com/culture/lifestyle/multimedia/2007/04/ss_raves?slide=3.

64. Alondra Nelson, "Genetic Genealogy and the Pursuit of African Ancestry," *Social Studies of Science* 38 (2008): 759, 762, 763, 775. See also Nash, "Genetic Kinship," 27, noting: "Of course, the meaning of genetic kinship is not fixed by this work of marketing genetics in genealogy. These accounts provide cultural resources for constructing senses of personal and collective identity in myriad ways with different political implications."

65. African Ancestry advertised the Community Testing Day during February 2008 on its Web site, www.AfricanAncestry.com.

66. David Gordon Nielsen, *Black Ethos: Northern Urban Negro Life and Thought, 1890–1930* (Westport, CT: Greenwood, 1977).

67. John L. Gwaltney, *Drylongso: A Self-Portrait of Black America* (New York: The New Press, 1980), xxvii.

68. Tommie Shelby, *We Who Are Dark: The Philosophical Foundations of Black Solidarity* (Cambridge, MA: Harvard University Press, 2005).

69. James Weldon Johnson, "Lift Every Voice and Sing" (1900), quoted in ibid., xv. On "linked fate," see Michael C. Dawson, *Behind the Mule: Race and Class in African-American Politics* (Princeton, NJ: Princeton University Press, 1994): 45–68.

70. Shelby, *We Who Are Dark*, 206.

71. TallBear, "Native-American-DNA.com.," 245–46, 249–50.

Chapter 11: Genetic Surveillance

1. John Travis, "Scientists Decry 'Flawed' and 'Horrifying' Nationality Tests," *Science*, Sept. 29, 2009, http://news.sciencemag.org/scienceinsider/2009/09/border-agencys.html.

2. DNAPrint Genomics, "Forensics," www.dnaprint.com/welcome/productsandservices/forensics. See also Tony Frudakis, *Molecular Photofitting: Predicting Ancestry and Phenotype Using DNA* (Burlington, MA: Academic Press, 2007); Pamela

Sankar, "Forensic DNA Phenotyping: Reinforcing Race in Law Enforcement," in *What's the Use of Race? Modern Governance and the Biology of Difference*, ed. Ian Whitmarsh and David S. Jones (Cambridge, MA: MIT Press, 2010), 49; Richard Willing, "DNA Tests Offer Clues to Suspect's Race," *USA Today*, Aug. 17, 2005, 1A.

3. DNA Print Genomics, "Forensics."

4. Duana Fullwiley, "Can DNA 'Witness' Race? Forensic Uses of an Imperfect Ancestry Testing Technology," *GeneWatch* 21 (2008): 12; Mildred Cho and Pamela Sankar, "Forensic Genetics and Ethical, Legal and Social Implications Beyond the Clinic," *Nature Genetics* 36 (2004): S8.

5. Sheldon Krimsky and Tania Simoncelli, *Genetic Justice: DNA Data Banks, Criminal Investigations, and Civil Liberties* (New York: Columbia University Press, 2011), 47, 55.

6. Pilar Ossorio, "About Face: Forensic Genetic Testing for Race and Visible Traits," *Journal of Law, Medicine and Ethics* 34 (2006): 277.

7. Jonathan Kahn, "What's the Use of Race in Presenting Forensic DNA Evidence in Court?" in *What's The Use of Race?* 27; Christian B. Sundquist, "Science Fiction and Racial Fables: Navigating the Final Frontier of Genetic Interpretation," *Harvard Blackletter Law Journal* 25 (2009): 57; Richard C. Lewontin and Daniel L. Hartl, "Population Genetics in Forensic DNA Typing," *Science* 254 (1991), 1745.

8. *People v. Wilson*, 38 Cal.4th 1237, 1241 (2006).

9. Edmund G. Brown Jr., California Attorney General, "Information Bulletin: Expansion of State's DNA Data Bank Program on January 1, 2009: Collection of DNA Samples from all Adults Arrested for Any Felony Offense," press release, Dec. 15, 2008, http://ag.ca.gov/bfs/pdf/69IB_121508.pdf.

10. See Edmund G. Brown Jr., California Attorney General, *Crime in California, 2006* (Sacramento: California Department of Justice, 2007), 62 (in 2006, 40.3 percent of arrestees were Hispanic and 17.1 percent were black).

11. Krimsky and Simoncelli, *Genetic Justice*, xvi.

12. Ibid., 28–45; Julia Preston, "U.S. Set to Begin a Vast Expansion of DNA Sampling," *New York Times*, Feb. 5, 2007, A1.

13. Tami Abdollah, "D.A.'s Database Growing in O.C.," *Los Angeles Times*, Oct. 5, 2009.

14. Office of the Attorney General, "Brown Announces Major DNA Lab Expansion," press release, May 5, 2008, http://ag.ca.gov/newsalerts/release.php?id=1553.

15. Tania Simoncelli and Barry Steinhardt, "California's Proposition 69: A Dangerous Precedent for Criminal DNA Databases," *Journal of Law, Medicine & Ethics* 33 (2005): 280; *Haskell v. Brown*, 677 F. Supp. 2d 1187, 1190 (Cal. 2009).

16. League of Women Voters of California, "Vote No on Proposition 69," http://ca.lwv.org/action/prop0411/prop69.html (listing organizations supporting and opposing Proposition 69).

17. *Haskell v. Brown*, Complaint 2009 WL 3269641, 3, available at www.aclunc.org/news/press_releases/asset_upload_file_612_8577.pdf.

18. Ibid., 19.

19. ACLU of Northern California, "ACLU Lawsuit Challenges California's Mandatory DNA Collection at Arrest," Oct. 7, 2009, www.aclu.org/technology-and-liberty/aclu-lawsuit-challenges-california-s-mandatory-dna-collection-arrest.

20. *Haskell v. Brown*, 1203.

21. Krimsky and Simoncelli, *Genetic Justice*, 64–88; Henry T. Greely, Daniel P. Riordan, Nanibaa A. Garrison, and Joanna L. Mountain, "Family Ties: The Use of DNA Offender Databases to Catch Offenders' Kin," *Journal of Law, Medicine & Ethics* 34 (2006): 250.

22. Jeremy W. Peters, "New Rule Allows Use of Partial DNA Matches," *New York Times*, Jan. 25, 2010.

23. Krimsky and Simoncelli, *Genetic Justice*, 64, 73; Jennifer Steinhauer, "Arrest in 'Grim Sleeper' Killings Fans Debate on DNA Procedure," *New York Times*, July 9, 2010, A1.

24. Quoted in Krimsky and Simoncelli, *Genetic Justice*, 83.

25. John P. Cronan, "The Next Frontier of Law Enforcement: A Proposal For Complete DNA Databanks," *American Journal of Criminal Law* 28 (2000): 119, 128–29.

26. Thomas J. Sheeran, "DNA Tests Vindicate Ohio Man Convicted of '81 Rape," *USA Today*, May 5, 2010, www.usatoday.com/news/nation/2010-05-05-dna-frees-man_N.htm.

27. Krimsky and Simoncelli, *Genetic Justice*, 275–304.

28. Nick Madigan, "Houston's Troubled DNA Lab Faces Scrutiny," *New York Times*, Feb. 9, 2003.

29. See, for example, Miles Moffeit and Susan Greene, "Trashing the Truth," *Denver Post*, July 23, 2007; Seattle Post-Intelligencer Staff, "DNA Testing Mistakes at the State Patrol Crime Labs," *Seattle Post-Intelligencer*, July 22, 2004.

30. Glenn Puit, "Man Files Lawsuit in False Imprisonment," *Las Vegas Review-Journal*, July 30, 2002.

31. William C. Thompson, "The Potential for Error in Forensic DNA Testing," *GeneWatch* 21 (2008): 5; Simoncelli and Steinhardt, "California's Proposition 69," 286–88.

32. Barry Scheck, Peter Neufeld, and Jim Dwyer, *Actual Innocence: When Justice Goes Wrong and How to Make It Right* (New York: Signet, 2001), 168–69.

33. Thompson, "Potential for Error."

34. Maura Dolan and Jason Felch, "The Verdict Is Out on DNA Profiles," *Los Angeles Times*, July 20, 2008.

35. Thompson, "Potential for Error."

36. Troy Duster, "Explaining Differential Trust of DNA Forensic Technology: Grounded Assessment or Inexplicable Paranoia," *Journal of Law, Medicine & Ethics* 34 (2006): 293.

37. Nina Totenberg, "High Court Weighs Prosecutors' Immunity," NPR, Nov. 4, 2009, www.npr.org/templates/story/story.php?storyId=120098210.

38. Innocence Project, "Understand the Causes: Improper or Invalidated Forensic Science," www.innocenceproject.org/understand/Unreliable-Limited-Science.php.

39. John Ferak and Jack Colman, "Kofoed Gets 20 Months to 4 Years," *Omaha World Herald*, June 2, 2010.

40. Brandon L. Garrett, "Judging Innocence," *Columbia Law Review* 108 (2008), 55, 116.

41. *District Attorney's Office for the Third Judicial District v. Osborne*, 129 S.Ct. 2308 (2009).

42. Sewell Chan and Colin Moynihan, "Lil Wayne Pleads Guilty to Gun Charge," *New York Times*, Oct. 23, 2009, A28.

43. George J. Annas, "Protecting Privacy and the Public—Limits on Police Use of Bioidentifiers in Europe," *New England Journal of Medicine* 361 (2009): 196, 200.

44. Krimsky and Simoncelli, *Genetic Justice*, 320.

45. Annas, "Protecting Privacy," 200.

46. Krimsky and Simoncelli, *Genetic Justice*, 161.

47. Sheldon Krimsky and Tania Simoncelli, "Genetic Privacy: New Frontiers," *GeneWatch* 20 (2007): 7–8.

48. Innocence Project, "Understand the Causes."

49. Center on Wrongful Convictions, "About Us," www.law.northwestern.edu/wrongfulconvictions/aboutus/.

50. Center on Wrongful Convictions, "Recent Case Stories," www.law.northwestern.edu/wrongful/casestories.html.

51. Annas, "Protecting Privacy and the Public," 200.

52. *S. and Marper v. The United Kingdom*, [2008] ECHR 30562/04.

53. Jack M. Balkin, "The Constitution in the National Surveillance State," *Minnesota Law Review* 93 (2008): 1, 13 (arguing that "the rise of the National Surveillance State portends the death of amnesia").

54. Michelle Alexander, *The New Jim Crow: Mass Incarceration in the Age of Colorblindness* (New York: The New Press, 2010), 6; Marc Mauer, *Race to Incarcerate* (New York: The New Press, 1999).

55. Loïc Wacquant, "Class, Race & Hyperincarceration in Revanchist America," *Daedalus* (Summer 2010): 74, 78 (emphases in original).

56. Ben Leapman, "Three in Four Young Black Men on the DNA Database," *Sunday Telegraph*, Nov. 5, 2006; Matilda MacAttram, "How the Government's DNA Crusade Is Criminalising Innocent Black Britons," *Voice Online*, July 17, 2009, www.voice-online.co.uk/content.php?show=15859.

57. Greely, Riordan, Garrison, and Mountain, "Family Ties," 258.

58. Krimsky and Simoncelli, *Genetic Justice*, 258.

59. D.H. Kaye and Michael E. Smith, "DNA Databases for Law Enforcement," in *DNA and the Criminal Justice System*, ed. David Lazer (Cambridge, MA: MIT Press, 2004), 269–70.

60. Julia Preston, "Political Battle on Immigration Shifts to States," *New York Times*, Jan. 1, 2011, A1.

61. Alexander, *New Jim Crow*, 128–36; David Cole, *No Equal Justice: Race and Class in the American Criminal Justice System* (New York: The New Press, 1998), 16–62.

62. Charles M. Blow, "Welcome to the 'Club,'" *New York Times*, July 25, 2009, A15.

63. Alexander, *New Jim Crow*, 128.

64. David Harris, *Profiles in Injustice: Why Racial Profiling Cannot Work* (New York: The New Press, 2002).

65. Jeff Brazil and Steve Berry, "Color of Drivers Is Key to Stops on I-95 Videos," *Orlando Sentinel*, Aug. 23, 1992.

66. *State v. Soto*, 324 N.J. Super. 66, 734 A.2d 350 (N.J. Super. Ct. Law. Div. 1996).

67. Harris, *Profiles in Injustice*, 80.

68. ACLU, *Driving While Black: Racial Profiling on Our Nation's Highways* (New York: American Civil Liberties Union, 1999), 3, 27–28.

69. "Substance Use Among Black Adults," *NSDUH Report*, Feb. 18, 2010, www.oas.samhsa.gov/2k10/174/174SubUseBlackAdults.htm.

70. Human Rights Watch, *Targeting Blacks* (New York: Human Rights Watch, 2008), 26, www.hrw.org/en/reports/2008/05/04/targeting-blacks-0.

71. Harry G. Levine and Deborah Peterson Small, "Marijuana Arrest Crusade: Racial Bias and Police Policy in New York City, 1997–2007," New York Civil Liberties Union, Apr. 2008, www.nyclu.org/files/MARIJUANA-ARREST-CRUSADE_Final.pdf; Harry G. Levine and John B. Gettman, "Drug Arrests and DNA: Building Jim Crow's Database," unpublished paper, June 19, 2008, available at http://dragon.soc.qc.cuny.edu/Staff/levine/Building-Jim-Crows-Database—Drug-Arrests-and-DNA–Levine.pdf.

72. Harry G. Levine, "The Epidemic of Pot Arrests in New York City," AlterNet, Aug. 10, 2009.

73. Ibid.

74. Blow, "Welcome to the 'Club,'" A15.

75. Ray Rivera, Al Baker, and Janet Roberts, "A Few City Blocks, 4 Years, 52,000 Police Stops," *New York Times*, July 12, 2010, A1.

76. Bob Herbert, "Anger Has Its Place," *New York Times*, Aug. 1, 2009, A15; Charles M. Blow, "Smoke and Horrors," *New York Times*, Oct. 23, 2010, A19.

77. Alexander, *New Jim Crow*, 139–40.

78. National Council on Crime and Delinquency, "And Justice for Some: Differential Treatment of Youth of Color in the Justice System," Jan. 23, 2007, 3.

79. Paul Hirschfield, "The Hyper-concentration of Juvenile Justice Contact Among Urban African-American Males," unpublished paper, Jan. 10, 2007, available at www.allacademic.com/meta/p_mla_apa_research_citation/1/8/4/3/6/p184362_index.html.

80. Sheryl Stolberg, "150,000 Are in Gangs, Report by D.A. Claims," *Los Angeles Times*, May 22, 1992, A1; Dirk Johnson, "2 of 3 Young Black Men in Denver Listed by Police as Suspected Gangsters," *New York Times*, Dec. 11, 1993, 8.

81. Peter Applebome, "A False Alarm Provides Plenty to Be Alarmed About," *New York Times*, Dec. 17, 2009, A37.

82. Bob Herbert, "Arrested While Grieving," *New York Times*, May 26, 2007, A25.

83. Bob Herbert, "Cops vs. Kids," *New York Times*, Mar. 6, 2010, A19. See Paul J. Hirschfield, "Preparing for Prison? The Criminalization of School Discipline in the USA," *Theoretical Criminology* 12, no. 1 (2008): 79.

84. Osagie Obasogie, *Playing the Gene Card? A Report on Race and Human Biotechnology* (Berkeley, CA: Center for Genetics and Society, 2009), 43.

85. Simon A. Cole, "How Much Justice Can Technology Afford? The Impact of DNA Technology on Equal Criminal Justice," *Science and Public Policy* 34 (2007): 100.

86. Beverly Moran and Stephanie M. Wildman, "Race and Wealth Disparity: The Role of Law and the Legal System," in *Race and Wealth Disparities: A Multidisciplinary Discourse*, ed. Beverly Moran (Lanham, MD: University Press of America, 2008), 158.

87. Brent Staples, "Even Now, There's Risk in 'Driving While Black,'" *New York Times*, June 15, 2009, A18.

88. Sheri Lynn Johnson, "Cross-Racial Identification Errors in Criminal Cases," *Cornell Law Review* 69 (1984): 934, 949–51.

89. Jennifer L. Eberhardt, Phillip A. Goff, Valerie J. Purdie, and Paul G. Davies, "Seeing Black: Race, Crime, and Visual Processing," *Journal of Personality and Social Psychology* 87 (2004): 876; Joshua Correll, Bernadette Park, Charles M. Judd, and Bernd Wittenbrink, "The Influence of Stereotypes on Decisions to Shoot," *European Journal of Social Psychology* 37 (2007): 1102.

90. Tanya Eiserer, "Faulty Eyewitness Identifications Leading Cause of Wrongful Convictions," *Dallas Morning News*, Oct. 4, 2009.

91. Devah Pager, *Marked: Race, Crime and Finding Work in an Era of Mass Incarceration* (University of Chicago Press, 2007), 91.

92. The Innocence Project, "Facts on Post-Conviction DNA Exonerations," www.innocenceproject.org/Content/351.php.

93. Cole, "How Much Justice Can Technology Afford?," 95, 99.

94. Josh Gerstein, "Obama Talks DNA on 'America's Most Wanted': Transcript," Politico, Mar. 9, 2010, www.politico.com/blogs/joshgerstein/0310/Obama_talks_DNA_on_Americas_Most_Wanted_transcript.html.

95. Declan McCullagh, "House Votes to Expand National DNA Arrest Database," *CNET News*, May 19, 2010, http://news.cnet.com/8301-13578_3-20005458-38.html.

96. Mark Murray, "On Immigration, Racial Divide Runs Deep," MSNBC.com, May 26, 2010.

97. Cole, *No Equal Justice*, 5.

98. See, for example, Michael Seringhaus, "To Stop Crime, Share Your Genes," *New York Times*, Mar. 15, 2010, A21. The arguments for and against a universal DNA databank are discussed in Krimsky and Simoncelli, *Genetic Justice*, 143–64.

99. Annas, "Protecting Privacy and the Public," 201.

Chapter 12: Biological Race in a "Postracial" America

1. *Prison Nation: The Warehousing of America's Poor*, ed. Tara Herivel and Paul Wright (New York: Routledge, 2003), 216–77; Nina Bernstein, "Officials Obscured Truth of Migrant Deaths in Jail," *New York Times*, Jan. 10, 2010; Amanda Gardner, "Many in U.S. Prisons Lack Good Health Care," HealthDay, Jan. 16, 2009, available at http://health.usnews.com/health-news/managing-your-healthcare/articles/2009/01/16/many-in-us-prisons-lack-good-health-care.html.

2. Eduardo Bonilla-Silva, *Racism without Racists: Color-Blind Racism and the Persistence of Racial Inequality in the United States* (Lanham, MD: Rowman & Littlefield, 2003); M.K. Brown et al., *White-Washing Race: The Myth of a Color-Blind Society* (Berkeley: University of California Press, 2003). See also, for example, Stephen Thernstrom and Abigail Thernstrom, *America in Black and White: One Nation, Indivisible* (New York: Simon & Schuster, 1999).

3. Bonilla-Silva, *Racism without Racists*, 3.

4. *Regents of University of California v. Bakke*, 438 U.S. 265, 402 (Marshall, J., concurring in part and dissenting in part).

5. Ian Haney Lopez, *White by Law: The Legal Construction of Race* (New York: New York University Press, 2006), 158–59.

6. *Adarand Constructors, Inc. v. Pena*, 515 U.S. 200, 240–41 (1995) (Thomas, J., concurring) (citations omitted).

7. *Parents Involved in Community Schools v. Seattle School District No. 1*, 551 U.S.; 127 S. Ct. 2738 (2007).

8. *McCleskey v. Kemp*, 481 U.S. 279, 312 (1987).

9. Ibid., 339.

10. Sally Satel, "Race and Medicine Can Mix Without Prejudice: How the Story of BiDil Illuminates the Future of Medicine," *Medical Progress Today*, Dec. 10, 2004, www.medicalprogresstoday.com/spotlight/spotlight_indarchive.php?id=449.

11. Jon Entine, "10 Questions for Jon Entine," Gene Expression, Oct. 8, 2007, www.gnxp.com/blog/2007/10/10-questions-for-jon-entine.php. Entine elaborated his views on race and genetics in *Taboo: Why Black Athletes Dominate Sports and Why We're Afraid to Talk About It* (New York: PublicAffairs, 2001).

12. Suemedha Sood, "Science and the Next President," *Washington Independent*, Nov. 4, 2008.

13. The White House, "Memorandum for the Heads of Executive Depart-

ments and Agencies," press release, Mar. 9, 2009; the White House, "Remarks of President Barack Obama—As Prepared for Delivery Signing of Stem Cell Executive Order and Scientific Integrity Presidential Memorandum," press release, Mar. 9, 2009.

14. Neil Risch, Esteban Burchard, Elad Ziv, and Hua Tang, "Categorization of Humans in Biomedical Research: Genes, Race and Disease," *Genome Biology* 3 (2002): 1, 4.

15. Ibid., 11.

16. Michael J. Malinowski, "Dealing with the Realities of Race and Ethnicity: A Bioethics-Centered Argument in Favor of Race-Based Genetic Research, *Houston Law Review* 45 (2009): 1415, 1472.

17. Hillary Rodham Clinton and Cecile Richards, "Blocking Care for Women," *New York Times*, Sept. 19, 2008, A23.

18. Richard Hayes, Executive Director, Center for Genetics and Society, letter to author dated Mar. 16, 2006.

19. For a conservative commentary on stem cell research, see Scott Gottlieb, "Stem Cells and the Truth About Medical Innovation," *Wall Street Journal*, Mar. 14, 2009, A9.

20. Tali Woodward, "Cell Divide: Proponents of a State Plan for Stem-Cell Research Would Like You to Believe Only the Religious Right Opposes It, but Some of the Strongest Critics Are Hardly Christian Conservatives," *San Francisco Bay Guardian*, Sept. 29, 2004.

21. John M. Broder and Andrew Pollack, "Californians to Vote on Stem Cell Research Funds," *New York Times*, Sept. 20, 2004, A1.

22. Debra Greenfield, "Impatient Proponents: What's Wrong with the California Stem Cell and Cures Act?" *Hastings Center Report* 34 (2004): 32, 34.

23. Constance Holden, "The Touchy Subject of 'Race,'" *Science*, Nov. 7, 2008, 839.

24. Kenneth W. Krause, "Change We Can Believe In: 'Race' and Continuing Selection in the Human Genome," *Humanist*, Jan.–Feb. 2008, 20.

25. Amnesty International, *Deadly Delivery: The Maternal Health Care Crisis in the USA* (London: Amnesty International Secretariat, 2010), www.amnestyusa.org/dignity/pdf/DeadlyDelivery.pdf.

26. Bureau of Labor Statistics, Labor Force Statistics from the Current Population Survey, "Unemployment Rate—White," Jan. 2000–Oct. 2010; Bureau of Labor Statistics, Labor Force Statistics from the Current Population Survey, "Unemployment Rate—Black or African American," Jan. 2000–Oct. 2010.

27. "Decades of Gains Vanish for Blacks in Memphis," *New York Times*, May 30, 2010; Debbie Gruenstein Bocian, Wei Li, and Keith S. Ernst, "Foreclosures by Race and Ethnicity: The Demographics of a Crisis," Center for Responsible Learning research report, June 18, 2010, www.responsiblelending.org/mortgage-lending/research-analysis/foreclosures-by-race-and-ethnicity.html. On the racially disparate

impact of the economic recession, see generally: Applied Research Center, *Race and Recession: How Inequality Rigged the Economy and How to Change the Rules*, May 2009, www.arc.org/recession.

28. Scott Shane, *Born Entrepreneurs, Born Leaders: How Your Genes Affect Your Work Life* (New York: Oxford University Press, 2010).

29. Christopher T. Dawes and James H. Fowler, "Partisanship, Voting and the Dopamine D2 Receptor Gene," *Journal of Politics* 71 (2009): 1157.

30. Kevin M. Beaver, Matt DeLisi, Michael G. Vaughn, and J.C. Barnes, "Monoamine Oxidase A Genotype Is Associated with Gang Membership and Weapon Use," *Comprehensive Psychiatry* 51 (2010): 130.

31. "Gang-Banging May Be Genetic," Associated Press, June 17, 2009, available at http://newsone.com/nation/news-one-staff/gang-banging-may-be-genetic.

32. "Genetics Linked to Early Sexual Activity in Kids," HealthDay, Sept. 18, 2009.

33. Jane Mendle, "Associations Between Father Absence and Age of First Sexual Intercourse," *Child Development* 80 (2009): 1463.

34. Phillip Atiba Goff, Jennifer Eberhardt, Melissa J. Williams, and Matthew Christian Jackson, "Not Yet Human: Implicit Knowledge, Historical Dehumanization, and Contemporary Consequences," *Journal of Personality and Social Psychology* 94 (2008): 292.

35. Jed Horne, *Breach of Faith: Hurricane Katrina and the Near Death of a Great American City* (New York: Random House, 2008).

36. Henry A. Giroux, "Reading Hurricane Katrina: Race, Class and the Biopolitics of Disposability," *College Literature* 33, no. 3 (2006): 171–96.

37. Quoted in Tyrone A. Forman and Amanda E. Lewis, "Racial Apathy and Hurricane Katrina: The Social Anatomy of Prejudice in the Post–Civil Rights Era," *Du Bois Review* 3 (2006): 175, 195.

38. David Harvey, *A Brief History of Neoliberalism* (Oxford: Oxford University Press, 2005); Noam Chomsky, *Profit Over People: Neoliberalism and Global Order* (New York: Seven Stories Press, 1999).

39. David Boaz, "Defining an Ownership Society," Cato Institute, 2003, www.cato.org/special/ownership_society/boaz.html; Adam Werbach, "A Dangerous Legacy: Bush's 'Ownership Society' Champions a Hyper-Individualism That Threatens the Commons," *In These Times*, Nov. 15, 2004, 37, www.inthesetimes.com/site/main/article/1408.

40. Judy Keen, "Bush's Choice of Words Is Telling," *USA Today*, Mar. 11, 2005, 6A; Janet Hook, "Off to a Running Start, How Far Can GOP Go?" *Los Angeles Times*, Mar. 20, 2005, A1; James Hoopes, "Don't Draw Line at Dividends, End Wage Tax Too," *Boston Globe*, Feb. 23, 2003, F4. On government support of the rich over the past three decades, see Paul Pierson and Jacob S. Hacker, *Winner-Take-All-Politics: How Washington Made the Rich Richer—and Turned Its Back on the Middle Class* (New York: Simon & Schuster, 2010).

41. Adam Liptak, "Justices Offer Receptive Ear to Business Interests," *New York Times*, Dec. 19, 2010, A1.

42. Michiko Kakutani, "Deeper Looks at the Crisis of '08 and the Oval Office," *New York Times*, Dec. 14, 2010, C1, C7, reviewing Michael Hirsh, *Capital Offense: How Washington's Wise Men Turned America's Future Over to Wall Street* (New York: John Wiley, 2010).

43. Kevin Sack, "Hospital Cuts Dialysis Care for the Poor in Miami," *New York Times*, Jan. 8, 2010.

44. ACLU, "Los Angeles County Jail Plagued by Violence and Hazardous Conditions, ACLU Report Finds," press release, May 5, 2010; Mary Tiedeman and Daniel Ballon, *Annual Report on Conditions Inside Men's Central Jail, 2008–2009* (Los Angeles: ACLU National Prison Project and ACLU of Southern California, 2010).

45. Editorial, "The Crime of Punishment," *New York Times*, Dec. 6, 2010, A24; Nina Totenberg, "Supreme Court to Hear California Prisons Case," NPR, Nov. 30, 2010, www.npr.org/2010/11/29/131679857/supreme-court-to-hear-california-prisons-case.

46. On racial apathy, see Tyrone A. Forman, "Color-Blind Racism and Racial Indifference: The Role of Racial Apathy in Facilitating Enduring Inequalities," in *Changing Terrain of Race and Ethnicity*, ed. Maria Krysan and Amanda E. Lewis (New York: Russell Sage Foundation, 2004), 43.

47. Henry A. Giroux, *The Terror of Neoliberalism: Authoritarianism and the Eclipse of Democracy* (Boulder, CO: Paradigm Publishers, 2004); Jane L. Collins, Micaela Di Leonardo, and Brett Williams, eds., *New Landscapes of Inequality: Neoliberalism and the Erosion of Democracy in America* (Santa Fe, NM: School for Advanced Research Press, 2008); Loïc Wacquant, *Punishing the Poor: The Neoliberal Government of Social Insecurity* (Durham, NC: Duke University Press, 2009).

48. Michelle Alexander, *The New Jim Crow: Mass Incarceration in the Age of Colorblindness* (New York: The New Press, 2010); Anna Marie Smith, *Welfare and Sexual Regulation* (New York: Cambridge University Press, 2007); Gwendolyn Mink, "Violating Women: Rights Abuses in the Welfare Police State," *Annals of the American Academy of Political & Social Sciences* 577, no. 2 (2001): 79–93; Dorothy Roberts, *Shattered Bonds: The Color of Child Welfare* (New York: Basic Civitas Books, 2002).

49. Alexander, *New Jim Crow*, 13.

50. Barbara Ehrenreich, "Is It Now a Crime to Be Poor?" *New York Times*, Aug. 9, 2009, sec. 1, 9.

51. Bob Herbert, "For Two Sisters, the End of an Ordeal," *New York Times*, Jan. 1, 2010, A13.

52. Naomi Klein, *The Shock Doctrine: The Rise of Disaster Capitalism* (New York: Picador, 2007), 18.

53. Terry Aguayo, "Florida: Not-Guilty Pleas in Boot Camp Death," *New York Times*, Jan. 19, 2007, A17.

54. *Prison Nation*, 216–77.

55. Dorothy E. Roberts, "Torture and the Biopolitics of Race," in *Rethinking America: The Imperial Homeland in the 21st Century*, ed. Jeff Maskovsky and Ida Susser (Boulder, CO: Paradigm Press 2009), 167.

56. *Prison Nation*, 216–77; Fox Butterfield, "Mistreatment of Prisoners Is Called Routine in U.S.," *New York Times*, May 8, 2004, A11.

57. Anne-Marie Cusac, "The Restraint Chair," in *Prison Nation*, 216–26.

58. Editorial, "Childbirth in Chains," *New York Times*, July 21, 2009, A18; editorial, "One Protection for Prisoners," *New York Times*, Oct. 14, 2009, A26.

59. Nicholas Confessore, "U.S. and New York Agree to Sweeping Changes for Youth Prison System," *New York Times*, July 15, 2010, A25.

60. On the incarceration of mentally ill children, see Solomon Moore, "Mentally Ill Offenders Stretch the Limits of Juvenile Justice," *New York Times*, Aug. 10, 2009, A1.

61. Editorial, "Sentenced to Abuse," *New York Times*, Jan. 15, 2010, A20.

62. Quoted in Giroux, *Terror of Neoliberalism*, 94.

63. Nina Bernstein, "Few Details on Immigrants Who Died in U.S. Custody," *New York Times*, May 5, 2008, A1.

64. Ibid., A18

65. Nina Bernstein, "Ill and in Pain, Detainee Dies in U.S. Hands," *New York Times*, Aug. 13, 2008, A1.

66. Ibid., A1, 6.

67. See Karen J. Greenberg and Joshua L. Dratel, eds., *The Torture Papers: The Road to Abu Ghraib* (New York: Cambridge University Press, 2005); David Luban, "Liberalism, Torture, and the Ticking Bomb," in *The Torture Debate in America*, ed. Karen J. Greenberg (New York: Cambridge University Press, 2006), 35, 52–74.

68. Elizabeth Holtzman with Cynthia L. Cooper, *The Impeachment of George W. Bush: A Practical Guide for Concerned Citizens* (New York: Nation Books, 2006), 125.

69. Paul Gilroy, *Postcolonial Melancholia* (New York: Columbia University Press, 2006); James Thuo Gathii, "Torture, Extraterrioriality, Terrorism, and International Law," *Albany Law Review* 67 (2003): 335, 359–61.

70. Leila Nadya Sadat, "Ghost Prisoners and Black Sites: Extraordinary Rendition Under International Law," *Case Western Reserve Journal of International Law* 37 (2006): 309, 311.

71. Human Rights First Primetime Torture Project, www.humanrightsfirst.org/us_law/etn/primetime; Brian Stelter, "For '24' Terror Fight (and Series) Near End," *New York Times*, Mar. 26, 2010.

72. Jessica Ramirez, "Carnage.com," *Newsweek*, May 10, 2010, 38.

73. James Bacchus, "The Garden," *Fordham International Law Journal* 28 (2005): 308, 314–15, 320.

74. Nikolas Rose, *The Politics of Life Itself: Biomedicine, Power, and Subjectivity in the Twenty-first Century* (Princeton, NJ: Princeton University Press, 2007), 177.

Conclusion: The Crossroads

1. Michael Farzan, "The Genome Project: A Balance Sheet," letter to the editor, *New York Times*, June 18, 2010, WK7.
2. Author interview of Mildred Cho, June 11, 2008, Stanford, CA.

Index